Current Topics in Microbiology and Immunology

Volume 326

Series Editors

Richard W. Compans
Emory University School of Medicine, Department of Microbiology and
Immunology, 3001 Rollins Research Center, Atlanta, GA 30322, USA

Max D. Cooper
Department of Pathology and Laboratory Medicine, Georgia Research Alliance,
Emory University, 1462 Clifton Road, Atlanta, GA 30322, USA

Tasuku Honjo
Department of Medical Chemistry, Kyoto University, Faculty of Medicine,
Yoshida, Sakyo-ku, Kyoto 606-8501, Japan

Hilary Koprowski
Thomas Jefferson University, Department of Cancer Biology, Biotechnology
Foundation Laboratories, 1020 Locust Street, Suite M85 JAH, Philadelphia,
PA 19107-6799, USA

Fritz Melchers
Biozentrum, Department of Cell Biology, University of Basel, Klingelbergstr.
50–70, 4056 Basel Switzerland

Michael B.A. Oldstone
Department of Neuropharmacology, Division of Virology, The Scripps Research
Institute, 10550 N. Torrey Pines, La Jolla, CA 92037, USA

Sjur Olsnes
Department of Biochemistry, Institute for Cancer Research, The Norwegian
Radium Hospital, Montebello 0310 Oslo, Norway

Peter K. Vogt
The Scripps Research Institute, Dept. of Molecular & Exp. Medicine, Division of
Oncovirology, 10550 N. Torrey Pines. BCC-239, La Jolla, CA 92037, USA

Anireddy S.N. Reddy • Maxim Golovkin
Editors

Nuclear pre-mRNA Processing in Plants

 Springer

Editors:
Anireddy S.N. Reddy
Department of Biology and Program in
Molecular Plant Biology
Colorado State University
Fort Collins, CO 80523
USA
reddy@colostate.edu

Maxim Golovkin,
Department of Microbiology
Thomas Jefferson University
Philadelphia, PA 19107
USA
maxim.golovkin@mail.tju.edu

ISBN 978-3-540-76775-6   e-ISBN 978-3-540-76776-3
DOI 10.1007/978-3-540-76776-3

Current Topics in Microbiology and Immunology ISSN 0070-217x

Library of Congress Catalog Number: 2008920727

© 2008 Springer-Verlag Berlin Heidelberg

This work is subject to copyright. All rights reserved, whether the whole or part of the material is concerned, specifically the rights of translation, reprinting, reuse of illustrations, recitation, broadcasting, reproduction on microfilm or in any other way, and storage in data banks. Duplication of this publication or parts thereof is permitted only under the provisions of the German Copyright Law of September, 9, 1965, in its current version, and permission for use must always be obtained from Springer-Verlag. Violations are liable for prosecution under the German Copyright Law.

The use of general descriptive names, registered names, trademarks, etc. in this publication does not imply, even in the absence of a specific statement, that such names are exempt from the relevant protective laws and regulations and therefore free for general use.

Product liability: The publisher cannot guarantee the accuracy of any information about dosage and application contained in this book. In every individual case the user must check such information by consulting the relevant literature.

Cover design: WMXDesign GmbH, Heidelberg, Germany

Printed on acid-free paper

9 8 7 6 5 4 3 2 1

springer.com

Preface

In recent years, nuclear pre-mRNA processing, especially alternative splicing, has been taking center-stage as an important regulator of gene expression and ultimately plant growth and development. Bioinformatic analyses using the sequences of completed plant genomes and available cDNAs/ESTs have revealed that alternative splicing of pre-mRNAs contributes greatly to transcriptome and proteome complexity in eukaryotes. During the last few years, considerable progress has been made in understanding various aspects of pre-mRNA processing including alternative splicing and its importance in plant growth and development as well as in plant responses to hormones and stresses. This volume of CTMI, entitled Nuclear pre-mRNA Processing in Plants, with 16 chapters from leading scientists in this area, summarizes recent advances in nuclear pre-mRNA processing and its role in plant growth and development. It provides researchers in the field, as well as those in related areas, with an up-to-date and comprehensive, yet concise, overview of the current status and future potential of this research in understanding plant biology.

The first four chapters focus on spliceosome composition, genome-wide alternative splicing, and splice site requirements for U1 and U12 introns using computational and empirical approaches. Analysis of sequenced plant genomes has revealed that 80% of all protein-coding nuclear genes contain one or more introns. The lack of an in vitro plant splicing system has made it difficult to identify general and plant-specific components of splicing machinery in plants. Plant spliceosomes have not been isolated so far. However, computational analysis of the sequenced plant genomes as reviewed here by Y. Ru, B.-B. Wang, and V. Brendel identified many of the known snRNAs and proteins components found in two types (major and minor) of non-plant spliceosomes. Although many of the core snRNA and protein components of the spliceosome are conserved between plants and animals, plants do contain some distinct proteins. Alternative splicing of pre-mRNAs provides a means to produce structurally and functionally different proteins from the same gene. Despite several reports of alternative splicing of specific genes in plants, the extent of alternative splicing in plants was not known until recently. However, the genomic era and development of advanced bioinformatics tools have changed this dramatically. The chapter by B.J. Haas reviews the computational approaches used for genome-wide identification of alternatively spliced genes in plants and summarizes the recent findings from such analyses. These recent studies revealed that at least 30% of plant genes undergo alternative splicing, suggesting that it considerably increases the transcriptome and likely proteome complexity in plants. The following chapter by M.A. Schuler describes different types of splice sites, splice site requirements, unique features of plant introns, and their functional significance. As there is no plant in vitro splicing system, many in vivo studies using various mutants were performed to identify the importance of nucleotides in splice sites, intron length, and nucleotide composition in splice site selection. The outcome of these studies is also summarized in this chapter. Intron sequences in nuclear pre-mRNAs are excised using either the major U2 snRNA-dependent spliceosomal pathway or minor U12 snRNA-dependent spliceosomal pathway. The next chapter by C.G. Simpson and J.W.S. Brown focuses on properties and splicing signals of plant U12 introns, the minor spliceosome that splices these rare introns, as well as known cis-elements and trans-factors that regulate the splicing of these introns.

The next three chapters focus on serine/arginine-rich (SR) proteins, a family of highly conserved proteins, which are known to play key roles in constitutive and regulated splicing of pre-mRNA and other aspects of RNA metabolism in metazoans. These proteins engage both in RNA binding and protein–protein interactions and function as splicing regulators at multiple stages of spliceosome assembly. This family of proteins has expanded considerably in plants with several plant-specific SR proteins. The chapter by A. Barta, M. Kalyna, and Z. Lorković provides an overview of complex interactions among SR proteins and between SR and other spliceosomal proteins, alternative splicing of pre-mRNAs of SR proteins, and evolutionary conservation of some of these splicing events. In vivo functions of plant SR proteins in regulating splicing and several aspects of plant growth and development are also presented in this chapter. Several recent studies have used fluorescence microscopy techniques to study in vivo localization and dynamics of spliceosomal proteins, especially SR proteins. The chapter by G.S. Ali and A.S.N. Reddy reviews the experimental approaches used to study localization, dynamics, and regulation of mobility of spliceosomal proteins. Differences in the regulation of mobility of plant and animal SR proteins suggest that not all SR proteins exhibit similar dynamics and that such differences are likely to have important implications in pre-mRNA splicing. Protein phosphorylation and dephosphorylation have been shown to play an important role in spliceosome assembly, splice site choice and localization, and mobility of splicing factors. The next chapter by R. Fluhr summarizes different mechanisms by which protein phosphorylation can influence pre-mRNA splicing. The role of protein kinases, especially LAMMER-type and SRPKs, in modulating alternative splicing and cellular partitioning of SR proteins is also discussed.

Several serendipitous discoveries made using forward genetics are indicating that RNA metabolism (alternative splicing, alternative polyadenylation, mRNA transport) plays an important role in many aspects of plant growth and development and in plant responses to biotic and abiotic stresses. The next seven chapters focus on these aspects of RNA metabolism.

The plant hormone abscisic acid (ABA) regulates a number of physiological processes during plant growth and development. The chapter by J.M. Kuhn, V. Hugouvieux, and J.I. Schroeder summarizes the role of mRNA cap-binding proteins and other proteins involved in mRNA processing in ABA signaling.

Messenger RNA 3' end formation is an important step in producing mature mRNA. A.G. Hunt's chapter provides an overview of plant polyadenylation signals and alternative polyadenylation of plant pre-mRNAs and its influence on flowering. Further, the author discusses the current status of our understanding of the functioning and regulation of plant polyadenylation factor subunits. The chapter by D.A. Belostotsky summarizes the recent findings pertinent to plant mRNA turnover, which is important for the maintenance of cellular and organismal homeostasis. This chapter covers different pathways of mRNA decay and describes the high-throughput approaches that could be used to study mRNA decay globally. Flowering time in plants is controlled by internal (hormones) and external (light and temperature) cues. Analysis of many flowering mutants revealed a link between mRNA metabolism and flowering. The chapter by L.C. Terzi and G.G. Simpson discusses the involvement of alternative 3' end formation, splicing, RNA export, and miRNA biogenesis in flowering. The authors discuss novel plant-specific RNA-binding proteins involved in flowering and present the network of control pathways.

The activation of plant resistance (R) proteins is a part of plant immunity, which is tightly regulated at multiple levels. The chapter by W. Gassmann discusses the role of alternative splicing of R genes in plant defense. Possible mechanisms by which R gene splicing is regulated and how different splice forms fit into the current view of R protein-mediated defense responses are presented. The chapter by G.S. Ali and A.S.N. Reddy discusses stress-induced changes in alternative splicing with emphasis on abiotic stresses. Global and focused studies suggest that stresses significantly change splicing patterns of many pre-mRNA including those of SR proteins. The transcript variants generated by stresses are likely to play a role in stress adaptation. The chapter by V. Chinnusamy, Z. Gong, and J.-K. Zhu reviews the mRNA export pathway in plants. Genetic studies indicate a central role for mRNA export in hormonal and stress responses in plants.

A.B. Rose discusses the ways introns affect gene expression both positively and negatively in plants. The mechanisms, where known (e.g., enhancer elements or alternative promoters), are presented. A model for intron-mediated enhancement is also presented in this chapter. The final chapter by J.W.S. Brown and P.J. Shaw discusses the role of the nucleolus in pre-mRNA processing. Based on proteomic and RNomic analysis, the authors summarize emerging evidence for novel roles for this compartment in mRNA export, transcriptional silencing, and nonsense-mediated decay in plants.

Despite recent findings that alternative splicing contributes significantly to transcriptome and likely proteome complexity in plants, surprisingly little is known about the mechanisms that control alternative splicing and the functions of alternatively spliced isoforms. This will be the next frontier of research in gene regulation. The availability of high-throughput global approaches to study gene expression, alternative splicing, and mRNA decay and to identify targets of RNA binding proteins and emerging computational tools are likely to significantly further our understanding of the regulatory role of nuclear pre-mRNA processing in plant growth and development and stress responses.

A.S.N. Reddy and Maxim V. Golovkin

Contents

Spliceosomal Proteins in Plants 1
Y. Ru, B.-B. Wang, and V. Brendel

Analysis of Alternative Splicing in Plants with Bioinformatics Tools 17
B.J. Haas

Splice Site Requirements and Switches in Plants 39
M.A. Schuler

U12-Dependent Intron Splicing in Plants 61
C.G. Simpson, J.W.S. Brown

Plant SR Proteins and Their Functions 83
A. Barta, M. Kalyna, and Z.J. Lorković

Spatiotemporal Organization of Pre-mRNA Splicing Proteins in Plants 103
G.S. Ali, A.S.N. Reddy

Regulation of Splicing by Protein Phosphorylation 119
R. Fluhr

mRNA Cap Binding Proteins: Effects on Abscisic Acid Signal Transduction, mRNA Processing, and Microarray Analyses 139
J.M. Kuhn, V. Hugouvieux, and J.I. Schroeder

Messenger RNA 3′ End Formation in Plants 151
A.G. Hunt

State of Decay: An Update on Plant mRNA Turnover 179
D.A. Belostotsky

Regulation of Flowering Time by RNA Processing 201
L.C. Terzi, G.G. Simpson

Alternative Splicing in Plant Defense 219
W. Gassmann

**Nuclear RNA Export and Its Importance in Abiotic Stress
Responses of Plants** 235
V. Chinnusamy, Z. Gong, and J.-K. Zhu

Regulation of Alternative Splicing of Pre-mRNAs by Stresses 257
G.S. Ali, A.S.N. Reddy

Intron-Mediated Regulation of Gene Expression 277
A.B. Rose

The Role of the Plant Nucleolus in Pre-mRNA Processing 291
J.W.S. Brown, P.J. Shaw

Index 313

Contributors

G.S. Ali
Department of Biology and Program in Molecular Plant Biology, Fort Collins, CO 80523, USA

A. Barta
Max F. Perutz Laboratories, Medical University of Vienna, Dr. Bohrgasse 9/3, 1030 Vienna, Austria, andrea.barta@meduniwien.ac.at

D.A. Belostotsky
School of Biological Sciences, University of Missouri-Kansas City, Kansas City, MO 64110, USA, belostotskyd@umkc.edu

V. Brendel
Department of Genetics, Development and Cell Biology, Iowa State University, Ames, IA 50011-3260, USA, vbrendel@iastate.edu

J.W.S. Brown
Scottish Crop Research Institute, Invergowrie, Dundee DD2 5DA, UK, john.brown@scri.ac.uk

V. Chinnusamy
Water Technology Centre, Indian Agricultural Research Institute, New Delhi 110 012, India

R. Fluhr
Weizmann Institute of Science, Rehovot 76100, Israel, robert.fluhr@weizmann.ac.il

W. Gassmann
Division of Plant Sciences and Bond Life Sciences Center, University of Missouri-Columbia, Columbia, MO 65211-7310, USA, gassmannw@missouri.edu

Z. Gong
State Key Laboratory of Plant Physiology and Biochemistry, College of Biological Sciences, China Agricultural University, Beijing 100094, China

B.J. Haas
Broad Institute, 7 Cambridge Center, Cambridge, MA, 02142
USA, bhaas@broad.mit.edu

V. Hugouvieux
Division of Biological Sciences, Cell and Developmental Biology Section, and Center for Molecular Genetics, University of California San Diego, 9500 Gilman Drive, La Jolla, CA 92093-0116, USA

A.G. Hunt
Department of Plant and Soil Sciences, 301A Plant Science Building, University of Kentucky, Lexington, KY 40546-0312, USA, aghunt00@uky.edu

M. Kalyna
Max F. Perutz Laboratories, Medical University of Vienna,
Dr. Bohrgasse 9/3, 1030 Vienna, Austria

J.M. Kuhn
Division of Biological Sciences, Cell and Developmental Biology Section, and Center for Molecular Genetics, University of California San Diego, 9500 Gilman Drive, La Jolla, CA 92093-0116, USA

Z.J. Lorković
Max F. Perutz Laboratories, Medical University of Vienna,
Dr. Bohrgasse 9/3, 1030 Vienna, Austria

A.S.N. Reddy
Department of Biology and Program in Molecular Plant Biology, Fort Collins, CO 80523, USA, reddy@colostate.edu

A.B. Rose
Molecular and Cellular Biology, University of California, Davis, CA 95616, USA, abrose@ucdavis.edu

Y. Ru
Department of Genetics, Development and Cell Biology, Iowa State University, Ames, IA 50011-3260, USA

J.I. Schroeder
Division of Biological Sciences, Cell and Developmental Biology Section, and Center for Molecular Genetics, University of California San Diego, 9500 Gilman Drive, La Jolla, CA 92093-0116, USA, julian@biomail.ucsd.edu

M.A. Schuler
Departments of Cell and Developmental Biology, Biochemistry and Plant Biology, University of Illinois, Urbana, IL 61801, USA, maryschu@uiuc.edu

P.J. Shaw
Department of Cell Biology, John Innes Centre, Norwich NR4 7UH, UK

C.G. Simpson
Scottish Crop Research Institute, Invergowrie, Dundee DD2 5DA, UK

G.G. Simpson
Scottish Crop Research Institute, Invergowrie, UK, g.g.simpson@dundee.ac.uk

L.C. Terzi
Scottish Crop Research Institute, Invergowrie, UK

B.-B. Wang
Department of Plant Pathology, University of Minnesota, St. Paul, MN 55108-6030, USA

J.-K. Zhu
Department of Botany and Plant Sciences, Institute for Integrative Genome Biology, University of California, Riverside, CA 92521, USA, Jian-kang.zhu@ucr.edu

Spliceosomal Proteins in Plants

Y. Ru, B.-B. Wang, and V. Brendel(✉)

Contents

Introduction 1
Small Nuclear RNAs 2
Small Nuclear Ribonucleoproteins 3
   Sm Core Proteins 3
   U1 snRNP-Specific Proteins 3
   Other snRNP-Specific Proteins in the Major Spliceosome 4
   U11/U12 snRNP-Specific Proteins 5
Splicing Factors 5
   Proteins Involved in Splice Site Selection 5
   Serine/Arginine-Rich Proteins 6
Regulation of Splicing in Plants 9
Conclusion and Perspective 11
References 11

Abstract The spliceosome is a large nuclear structure consisting of dynamically interacting RNAs and proteins. This chapter briefly reviews some of the known components and their interactions. Large-scale proteomics and gene expression studies may be required to unravel the many intricate mechanisms involved in splice site recognition and selection.

Introduction

The spliceosome is a large ribonucleoprotein (RNP) complex whose highly orchestrated assembly at each intron involves small nuclear RNAs (snRNAs) and hundreds of proteins. Two types of spliceosomes have been identified in both

---

V. Brendel
Department of Genetics, Development and Cell Biology, Iowa State University,
Ames, IA 50011-3260, USA
e-mail: vbrendel@iastate.edu

vertebrates and plants and are referred to as major (or U2 type) and minor (or U12 type) spliceosomes. They differ in their snRNA composition, but differences in their protein composition are not completely understood, although some specific proteins were identified from the minor spliceosome (Will et al. 1999, 2004; Lorković et al. 2005). The minor spliceosome is involved in the splicing of a small fraction of introns with distinct donor site and branchpoint motifs (see the chapter by C. G. Simpson and J. W. S. Brown, this volume). A computational screen of the Arabidopsis genome identified 74 snRNA genes and about 400 genes encoding known or putative splicing-related proteins (Wang and Brendel 2004). Here we briefly review the snRNA genes characteristic for the two types of spliceosomes and then discuss the major classes of plant splicing-related proteins.

Small Nuclear RNAs

Like its counterparts in yeasts and metazoa, the major spliceosome in plants also contains five types of U-rich snRNAs, referred to as U1, U2, U4, U5, and U6 snRNA (reviewed in Lorković et al. 2000b; Reddy 2001). All five types were identified experimentally in Arabidopsis (Vankan et al. 1988; Vankan and Filipowicz 1988, 1989; Waibel and Filipowicz 1990; Hofmann et al. 1992). A genome-wide survey in Arabidopsis identified a total of 70 genes encoding these snRNAs, including 14 U1, 18 U2, 11 U4, 14 U5, and 13 U6 snRNA genes (Wang and Brendel 2004). Most of these genes seem to be active, as their promoter regions contain both TATA box and conserved upstream element (USE) motifs (Wang and Brendel 2004). Four types of snRNA genes (U1, U2, U4, and U5) are transcribed by RNA polymerase II (Pol II) (Vankan et al. 1988; Vankan and Filipowicz 1989; Connelly and Filipowicz 1993), while U6 snRNA genes are transcribed by polymerase III (Pol III) (Waibel and Filipowicz 1990). snRNA genes were also experimentally identified from other plant species, including bean (van Santen and Spritz 1987), pea (Hanley and Schuler 1991), potato (Vaux et al. 1992), wheat (Musci et al. 1992), and maize (Leader et al. 1993). For the minor spliceosome, four type of snRNAs named U11, U12, U4atac, and U6atac replace the U1, U2, U4, and U6 snRNA in the major spliceosome, while U5 snRNA is used in both major and minor spliceosomes (Tarn and Steitz 1997). Arabidopsis U12, U6atac, and U4atac were identified both computationally and experimentally (Shukla and Padgett 1999; Wang and Brendel 2004; Lorković et al. 2005), while U11 snRNA has only been computationally predicted (Wang and Brendel 2004). The four minor spliceosome snRNA genes also have the conserved TATA box and USE motifs in their promoter regions (Wang and Brendel 2004). Plant snRNAs likely play similar functions as their mammalian homologs, because their secondary structure and all functional domains are well conserved between plants and mammals (Lorković et al. 2000b). Recently, there is accumulating evidence that the snRNAs are involved in the catalysis of splicing, indicating that the spliceosome may be an RNA-centric enzyme similar to the ribosome (reviewed in Valadkhan 2005).

Small Nuclear Ribonucleoproteins

The major spliceosome is assembled dynamically at each intron, facilitated by interactions of the spliceosomal small nuclear ribonucleoproteins (snRNPs) with the pre-mRNA. Proteomic studies isolated five different snRNPs (U1, U2, U5, U4/U6, and U4/U6.U5 tri-snRNP) from human and yeast (Fabrizio et al. 1994; Caspary et al. 1999; Gottschalk et al. 1999; Krämer et al. 1999; Stevens and Abelson 1999; Stevens et al. 2001). Spliceosome assembly starts with U1 snRNP binding to the 5'-splice site, followed by U2 snRNP binding to the intron branch site, association of the U4/U6.U5 tri-snRNP, and release of U1 and U4 snRNPs to form the catalytic complex (reviewed in Burge et al. 1999). Many common (Sm core proteins) and specific proteins binding to the snRNAs were identified in each snRNP (reviewed in Will and Lührmann 2001). No proteomics approach has been applied to plant spliceosomes so far. Computational genome screening using human homologs as query sequences identified 91 snRNP protein-coding genes conserved in Arabidopsis, fewer than 10 of which have been experimentally studied (Wang and Brendel 2004).

Sm Core Proteins

Seven common proteins were identified in U1, U2, U4, and U5 snRNPs. All of them contain an Sm domain and thus were named as SmB, SmD1, SmD2, SmD3, SmE, SmF, and SmG (reviewed in Will and Lührmann 2001). Seven "like Sm" proteins (Lsm2–8) are the counterparts of Sm proteins in U6 snRNP. The Sm proteins bind to the snRNAs to form the structural core of the snRNPs. The Sm domain mediates protein interactions with other core proteins and snRNP-specific proteins. One additional Lsm protein exists (Lsm1) but cannot bind to snRNA (reviewed in Will and Lührmann 2001). All 15 Sm core proteins are conserved in Arabidopsis, with nine of them encoded by duplicated genes. All 24 genes are expressed based on EST data (Wang and Brendel 2004), but only Lsm5 (SAD1) has been experimentally characterized. *Lsm5 (SAD1)* is expressed universally at a low level in Arabidopsis. Mutation of Lsm5 increases plant sensitivity to drought stress and ABA (Xiong et al. 2001). Lsm5 protein may not be essential to the spliceosome but instead may affect the specificity or efficiency of the splicing machinery.

U1 snRNP-Specific Proteins

Three U1 snRNP-specific proteins (U1-70K, U1-A, and U1-C) are shared between human and yeast (Gottschalk et al. 1998). All three proteins exist in plants, and two of them (U1-70K and U1-A) are well characterized (Simpson et al. 1995; Golovkin and Reddy 1996). Arabidopsis U1-70K (atU1-70K) is an essential protein of 427 amino acid residues containing an RNA binding domain and an RS domain (Golovkin and Reddy 1996). The N-terminus of atU1-70K is conserved in eukaryotic organisms, while the C-terminal RS domain has much more variation. Four SR proteins (SR33, SR45, SRZ21, SRZ22) interact with atU1-70K, possibly through their RS domains (Golovkin and Reddy 1998, 1999). Reduced expression levels of atU1-70K in floral organs cause abnormal petal and stamen development (Golovkin and Reddy 2003). atU1-70K can be alternatively spliced by retaining the long sixth intron (910 nt), which leads to a truncated protein (Golovkin and Reddy 1996). Recently the gene structure and alternative splicing pattern of U1-70K gene were found to be conserved in rice and maize (Gupta et al. 2006; Wang and Brendel 2006a). U1-A gene was identified from potato and Arabidopsis. Similar to their human counterparts, plant U1-A proteins can bind specifically to the U1 snRNA loop II sequence (Simpson et al. 1995). U1-70K, U1-A, and the yet uncharacterized U1-C gene are all single copy in Arabidopsis (Golovkin and Reddy 1996; Wang and Brendel 2004), while potato may have multiple copies of U1-A (Ibrahim et al. 2001). Several yeast-specific proteins such as SNU71, SNU65/Prp42, SNU56, NAM8, Prp39, and Prp40 make the yeast U1 snRNP a more complex structure than its human counterpart (Gottschalk et al. 1998). Computational searches identified two homologs for each of Prp39 and Prp40 in Arabidopsis (Wang and Brendel 2004). The remaining proteins are apparently absent from the plant genome, indicating unique features in 5'-splice site recognition in plants.

Other snRNP-Specific Proteins in the Major Spliceosome

Similarity searches identified 23 U2, 14 U5, 10 U4/U6, and six U4/U6.U5 snRNP-specific protein-coding genes in the Arabidopsis genome (Wang and Brendel 2004). U2-A' and U2-B" from U2 snRNP were experimentally studied in plants (Simpson et al. 1995; Lorković et al. 2004). Potato U2-B" can interact with human U2-A' to enhance sequence-specific binding (Simpson et al. 1995). Similar to the cases in human and yeast, plant U2-B" and U1-A (from U1 snRNP) are very similar to each other, with over 50% identity and 70% similarity between their protein sequences. Only one member of the U1A/U2B" family (SNF) was found in *Drosophila melanogaster*, which can interact with both U1 and U2 snRNAs (Polycarpou-Schwarz et al. 1996). Functional redundancy was recently discovered between U1-A and U2-B" in *Caenorhabditis elegans* (Saldi et al. 2007). In Arabidopsis, two copies of U2-B" exist, with 80% identity and 90% similarity between their protein sequences. Both U2-B" genes are expressed and can be alternatively spliced, increasing the complexity of this protein family two- to fourfold (Wang and Brendel 2004). Other proteins from these snRNPs are also highly conserved. A few genes, including atSF3a120/SAP114 (Prp21) from U2 snRNP, U5-116 KD (Snu114), U5-200 KD (Brr2), U4/U6-90 KD (Prp3), U4/U6-15.5 KD (Snu13), and tri-65 KD (Snu66), are duplicated multiple times in Arabidopsis (Wang and Brendel 2004). It is not known how the functions of these redundant proteins coordinate with each other in plants.

U11/U12 snRNP-Specific Proteins

Little is known about the protein composition of the minor spliceosome. The only purified snRNPs are 18S U11/U12 snRNP and 12S U11 snRNP in humans. Seven minor spliceosome specific proteins, all Sm core proteins, and seven subunits of heteromeric SF3b from U2 snRNP were identified from U11/U12 snRNP, but none of the U1 snRNP-specific proteins were detected (Will et al. 2004). Four (25KDa, 31KDa, 35KDa, and 65KDa) minor spliceosome proteins are present in the Arabidopsis genome, while the other three (20KDa, 48KDa, and 59KDa) are absent (Wang and Brendel 2004). Sequence similarity detected between At2g46200 and U11-59KD and between At3g04160 and U11-48K suggests they are possible orthologs (Lorković et al. 2005). The U11/U12-35 KD has strong sequence similarity and analogous function to U1-70K (Lorković et al. 2005). Contrasting with the duplication events in the major spliceosome, all minor spliceosome-specific genes exist as single copy. The Arabidopsis U11/U12-specific genes are highly conserved in rice, indicating a similar mechanism of splicing of AT-AC introns throughout plants (Lorković et al. 2005).

Splicing Factors

In a previous survey (Wang and Brendel 2004), splicing factors were divided into eight subgroups: splice site selection proteins; serine/arginine-rich (SR) proteins; 17S U2-associated proteins; 35S U5-associated proteins; BΔU1 complex-specific proteins; exon junction complex (EJC) proteins; second-step splicing factors; and other known splicing factors. Here we focus on the first two subgroups because they are more thoroughly studied than other subgroups.

Proteins Involved in Splice Site Selection

Cap-Binding Proteins

The nuclear cap-binding complex (CBC) is required for effective spliceosome assembly and pre-mRNA splicing in both yeast and mammalian systems (Colot et al. 1996; Lewis et al. 1996a). More detailed experiments showed that CBC facilitates the recruitment of U1 snRNP to the 5′-splice site of the first intron in HeLa cell nuclear extracts (Lewis et al. 1996b). Two subunits of CBC in Arabidopsis [cap-binding proteins (CBP) AtCBP20 and AtCBP80] were identified and characterized (Kmieciak et al. 2002). Both *AtCBP20* and *AtCBP80* are single-copy genes in the Arabidopsis genome. AtCBP20 plays a role in ABA regulation and drought response (Papp et al. 2004). Another subunit, AtCBP80, is also called abscisic acid

(ABA) hypersensitive 1 (ABH1). Studies of several *ABH1* mutants showed that ABH1 is a modulator in early ABA signal transduction (Hugouvieux et al. 2001) and acts in the flowering pathway (Bezerra et al. 2004; Kuhn et al. 2007) (see the chapter by J. M. Kuhn et al., this volume). Specifically, *ABH1* may affect flowering time by influencing the splicing patterns of some flowering pathway genes. The intron-1-retaining transcripts of *FLOWERING LOCUS C* (*FLC*) accumulate in *abh1-7* (Kuhn et al. 2007), suggesting that *ABH1* may regulate *FLC* intron 1 removal. However, it is still unknown whether plant CBC, like its animal counterparts, facilitates the recognition of the 5'-splice site of the first intron by U1 snRNP.

U2 snRNP Auxiliary Factor

U2 snRNP Auxiliary Factor (U2AF) is a dimeric splicing factor with a small 35-kDa and a large 65-kDa subunit. In animals, the small subunit U2AF^{35} binds to the 3'-splice site (Merendino et al. 1999; Wu et al. 1999; Zorio and Blumenthal 1999). The large subunit U2AF^{65} binds to the polypyrimidine tract between the branchpoint and the 3'-splice site (Zamore et al. 1992). The bindings of U2AF help to bring U2 snRNP to the branchpoint (Ruskin et al. 1988). It is not well understood, however, how U2AF functions in plants.

Two isoforms of U2AF^{65}, named U2AF^{65}a and U2AF^{65}b, have been identified in both *Nicotiana plumbaginifolia* (Tex-Mex tobacco) and Arabidopsis (Domon et al. 1998). U2AF^{65}a localizes to the nucleus. Both proteins interact with plant introns and prefer binding to poly(U) instead of poly(G), poly(C), or poly(A) (Domon et al. 1998). The U-rich sequence in plants was shown to function as either polypyrimidine tract or splicing signal (Simpson et al. 2004). In addition, both plant isoforms can complement the lack of human U2AF^{65} in HeLa cell splicing extracts and stimulate splicing (Domon et al. 1998).

U2AF^{35} homologs are found in Arabidopsis, rice, maize, and other flowering plants (Domon et al. 1998; Wang and Brendel 2004, 2006b). Two homologs were identified in Arabidopsis, atU2AF^{35}a and atU2AF^{35}b, and both have nuclear localization. Altered levels of atU2AF^{35} cause the change of *FLC* expression level and *FCA* (*FLOWERING TIME CONTROL PROTEIN*) splicing pattern and thus lead to the change of flowering time (Wang and Brendel 2006b). Plants with altered levels of atU2AF^{35} also show other pleiotropic phenotypes. A widely conserved C-terminal motif (SERE) in seed plant U2AF^{35} homologs suggests that it may have plant-specific function (Wang and Brendel 2006b).

Serine/Arginine-Rich Proteins

The serine/arginine-rich (SR) proteins are essential in several steps of both constitutive and alternative splicing (see the chapters by A. Barta et al. and G. S. Ali and A. S. N. Reddy, this volume). All SR proteins are phosphoproteins. Typical characteristics of SR proteins include one or two N-terminal RNA recognition motifs (RRMs) and a C-terminal arginine/serine-rich (RS) domain (Kalyna and Barta 2004; Reddy 2004). During splice site recognition and spliceosome formation, the RRM recognizes and binds to particular exon and/or intron motifs, and the RS domain interacts with other proteins in the spliceosome. There are 11 and seven genes encoding SR proteins in human and *Caenorhabditis elegans*, respectively (Kalyna and Barta 2004), while there are many more plant SR genes, with at least 19 in Arabidopsis and 24 in rice (Iida and Go 2006; Isshiki et al. 2006; Reddy 2007). Some SR proteins are highly conserved, whereas others are species specific. The fact that both plant- and animal-specific SR proteins exist indicates that the mechanism of pre-mRNA splicing is different by some means in the two kingdoms.

Discovery of Plant SR Proteins

Plant SR proteins were discovered by various approaches. For example, the gene of the first plant SR protein, atSRp34/atSR1, was identified by searching Arabidopsis cDNAs for highly conserved RRMs in animal SR genes (Lazar et al. 1995). Similar approaches based on sequence similarity search were applied later to discover additional SR proteins along with other plant splicing factors (Lopato et al. 1996b, 1999a, 2002, 2006; Lorković and Barta 2002; Gao et al. 2004; Wang and Brendel 2004; Iida and Go 2006). Another approach was to isolate clones interacting with U1-70K or known SR proteins in yeast two-hybrid assays (Golovkin and Reddy 1998, 1999; Lopato et al. 2002, 2006). In the third approach, SR proteins were detected with human SF2/ASF-specific antibody and monoclonal antibody mAb104, which interacts with a shared phosphoepitope in all SR proteins (Lopato et al. 1996a, 1999b).

SR Proteins in Arabidopsis

Based on sequence similarities with human SR proteins, the 19 Arabidopsis SR proteins can be grouped into four families: SF2/ASF, SC35, 9G8, and plant-specific families (Wang and Brendel 2004). Four SR proteins, atSRp34/SR1, atSRp34a, atSRp34b, and atSRp30, have gene structures similar to human SF2/ASF. They have two N-terminal RRMs and a C-terminal RS domain. Unlike human SF2/ASF, all four of these SR proteins except atSRp30 have a PSK domain that is rich in proline, serine, and lysine (Reddy 2004). The splicing activity of atSRp34/atSR1 has been shown by its influence on alternative 5'-splice site selection in HeLa nuclear extract, which is comparable with the activity of human SF2/ASF (Lazar et al. 1995). Regulation of *atSRp34/SR1* splicing is organ- and stage-specific and temperature-dependent but is not autoregulated (Lazar and Goodman 2000). Full-length *atSRp34/SR1* mRNA is downregulated when another SR gene, *atSRp30*, is overexpressed (Lopato et al. 1999b). Overexpression of *atSRp30* also changes alternative splicing of endogenous *atRSp31*, *atSRp34/RS1*, *U1-70K*, and its own pre-mRNA.

Transgenic plants overexpressing *atSRp30* show pleiotropic phenotype changes including delayed flowering and larger flowers and rosette leaves (Lopato et al. 1999b).

The SC35 family comprises five SR proteins (atSC35, atSR33/atSCL33, atSCL30, atSCL30a, and atSCL28). All of them have typical RRM and RS domains. In yeast two-hybrid assay and coprecipitation analysis, atSR33/atSCL33 was shown to interact with itself and atSR45, a plant-specific SR protein, but not with atSRZ21 or atSRZ22 from the 9G8 family (Golovkin and Reddy 1999).

Proteins from the 9G8 family are similar to human 9G8 splicing factor. Three of them (atRSZp22/atSRZ22, atRSZp22a, and atRSZp21/atSRZ21) have a glycine hinge and a CCHC-type zinc finger motif in between the RRM and RS domain. The other two members, atRSZ33 and atRSZ32, each have two CCHC-type zinc fingers between the RRM and RS domain (Reddy 2004). atRSZp22 can restore splicing activities in both SR protein-depleted HeLa cell S100 extract and 9G8-depleted HeLa nuclear extract (Lopato et al. 1999a). In yeast two-hybrid screens and in vitro binding assays, atRSZ33 interacts with atRSZp21 and atRSZp22 from the same family, atSRp34/SR1 from the SF2/ASF family, and all five members of the SC35 family (Lopato et al. 2002).

Five Arabidopsis SR proteins do not have animal counterparts and belong to the plant-specific family. atRSp41, atRSp40/atSRp35, atRSp32/atSRp31a, and atRSp31 have two RRMs and a C-terminal RS domain. Unlike all other plant SR proteins with only one RS domain, atSR45 has two RS domains separated by an RRM (Golovkin and Reddy 1999). Both atRSp31 and atSR45 can complement SR protein-deficient HeLa cell S100 extract (Lopato et al. 1996b; Ali et al. 2007). The mutant *sr45-1* shows a pleiotropic phenotype including delayed flowering and change of flower and leaf morphologies. On the molecular level, *FLC* is upregulated in *sr45-1*, consistent with the late flowering phenotype. Compared with wild type, *sr45-1* has a different splicing pattern of pre-mRNAs of other SR genes, including *atSRp30*, *atRSp31*, *atRSp31a*, *atSRp34*, and *atSRp34b* (Ali et al. 2007).

SR Proteins in Other Plant Species

Many SR proteins from maize (*Zea mays*), wheat (*Triticum aestivum*), and rice (*Oryza sativa*) have also been identified and characterized (Gao et al. 2004; Iida and Go 2006; Isshiki et al. 2006; Lopato et al. 2006). There are at least three SF2/ASF-like SR proteins in maize (zmSRp30, zmSRp31, and zmSRp32) (Gao et al. 2004). They show strong structural similarity to Arabidopsis atSRp34/SR1 and atSRp30. Like *atSRp34/SR1*, all three maize SR genes are alternatively spliced. Overexpression of *zmSRp32* in transiently transformed maize BMS cells enhances the selection of a weak 5′-splice site. In wheat, two SR proteins, taRSZ38 and taRSZ38a, have high similarity to Arabidopsis atRSZ33, a 9G8 family protein (Lopato et al. 2006). Additional wheat proteins involved in pre-mRNA splicing were identified with the yeast two-hybrid system by using taRSZ38 and several subsequent positive clones as baits. Among the resulting proteins are taU1-70K, taU2AF[65], taU2AF[35], and a number of SR proteins (taSRp30, taSRp30a, taRSZ22, and taRSZ22a) (Lopato et al. 2006). To date, 24 rice SR proteins have been found; four of them are homologs of the Arabidopsis plant-specific family (Iida and Go 2006; Isshiki et al. 2006). Several rice SR genes undergo alternative splicing events, some of which are conserved between Arabidopsis and rice (see below) (Iida and Go 2006; Kalyna et al. 2006). Transient assays in rice Oc cell protoplasts indicate that osRSp29 and osRSZp23 enhance splicing and favor different 5'-splice sites of the same intronic region (Isshiki et al. 2006). Domain-swapping experiments show that the first RRM is essential for osRSp29's efficient splicing activity (Isshiki et al. 2006). Overexpression of *osRSZ36* in transgenic rice changes its own RNA's splicing pattern, suggesting a feedback regulation. Overexpression of *osSRp33b* alters the splicing of *osSRp33a* and *osSRp32* pre-mRNAs (Isshiki et al. 2006).

Conserved Alternative Splicing of Plant SR Genes

Alternative splicing is observed for 15 of the 19 Arabidopsis SR genes, from which about 95 transcripts are produced (Palusa et al. 2007). This greatly increases the flexibility and complexity of the spliceosome. Seventeen of the 24 rice SR genes are subjected to alternative splicing as well (Iida and Go 2006). Alternative splicing events also occur in some SR genes from other plant species (see above). The observation that some events are conserved among SR homologs from different plant species (Iida and Go 2006; Kalyna et al. 2006) highlights the importance of alternative splicing in plant SR gene transcripts.

Studies reveal that long introns (>400 nt) are present in most (12 of 19) Arabidopsis SR genes, and these long introns are alternatively spliced in nine of such genes (Kalyna and Barta 2004). This is remarkable considering that the average length of Arabidopsis intron is 173 nt (Reddy 2007) and about 22% of the genes in Arabidopsis are alternatively spliced (Wang and Brendel 2006a). More detailed analyses show that the alternatively spliced long introns are also found in rice, maize, gymnosperms, moss (*Physcomitrella patens*), and green alga (*Chlamydomonas reinhardtii*) transcripts encoding SR proteins (Iida and Go 2006; Kalyna et al. 2006). These alternatively spliced long introns are located in the RRM-encoded regions, resulting in mRNAs encoding SR proteins with partial RRMs. Such conserved alternative splicing events are found in the plant-specific and SC35 families and the 9G8 members with two zinc knuckles (Iida and Go 2006; Kalyna et al. 2006) and may be critical to splicing regulation.

Regulation of Splicing in Plants

Splicing is a dynamic process affecting the expression of all intron-containing genes. Recent studies revealed that 20%–30% expressed genes are alternatively spliced in Arabidopsis and rice, demonstrating the critical roles of splicing in gene expression (Campbell et al. 2006; Wang and Brendel 2006a). Splicing must therefore be tightly regulated to ensure accuracy and efficiency. Many proteins are involved in splicing regulation. Some proteins can assist and promote splicing, while others may have a negative effect. We defined proteins assisting splicing as splicing factors, such as SR proteins and U2AF, which were discussed in the previous section. Other proteins that can either modify splicing factors or have negative effect on splicing were defined as splicing regulators (Wang and Brendel 2004). Nevertheless, this classification is tentative, and sometimes it is hard to define clearly whether a protein is a splicing factor or a splicing regulator.

The main difference in splicing-related genes between mammals and plants is the expansion of splicing regulators in plants (Wang and Brendel 2004). One type of splicing regulator is hnRNP protein, which can bind to pre-mRNA and block the binding site for splicing factors (Wang and Brendel 2004). PTB/hnRNP I, for instance, can compete with the U2AF large subunit for the polypyrimidine tract of introns (Lin and Patton 1995). Three homologs of PTB/hnRNP I were identified in Arabidopsis (Wang and Brendel 2004). Moreover, a family of 21 glycine-rich RNA binding proteins were identified in Arabidopsis as homologs of human hnRNP A1 and hnRNP A2/B1(Wang and Brendel 2004). Five of these genes were also experimentally identified, including AtGRP7, AtGRP8, and three UBA2 homologs (Heintzen et al. 1994; Lambermon et al. 2002). AtGRP7 can influence alternative splicing of its own transcripts as well as AtGRP8 transcripts (Staiger et al. 2003). UBA2 proteins can interact with UBP1 and UBA1 proteins to recognize U-rich sequences (Lambermon et al. 2002). Homologs for other hnRNP proteins such as hnRNP F and CUG-BP were also identified in Arabidopsis (Lorković and Barta 2002; Wang and Brendel 2004). In addition, 15 hnRNP-like proteins were identified in Arabidopsis, possibly representing the plant-specific hnRNPs (Lambermon et al. 2000; Lorković et al. 2000a; Landsberger et al. 2002; Wang and Brendel 2004). The UBP1 proteins can strongly enhance splicing of some introns in protoplasts (Lambermon et al. 2000), whereas UBA1, RBP45, and RBP47 proteins have no such function reported (Lorković et al. 2000a; Landsberger et al. 2002).

SR protein kinase is another type of splicing regulator (see the chapter by R. Fluhr, this volume). By phosphorylating SR proteins, SR protein kinase can modulate splicing in both constitutive and alternative splicing (reviewed in Stojdl and Bell 1999). Sequences of SR protein kinases are highly conserved in plants, suggesting the critical function of these proteins. Eight homologs corresponding to three mammalian SR protein kinases, including Lammer/CLK kinases, SRPK1, and SPRK2, were identified in Arabidopsis (Wang and Brendel 2004). Lammer/CLK kinases were proven to phosphorylate SR proteins *in vitro* in plants (Golovkin and Reddy 1999; Savaldi-Goldstein et al. 2000). Overexpression of one tobacco Lammer/CLK kinase homolog, *PK12*, causes alternative splicing pattern changes in several genes, including U1-70K and two SR proteins, atSRp30 and atSR1/atSRp34 (Savaldi-Goldstein et al. 2003). Connections between splicing and other SR protein kinase are yet to be established.

In addition to proteins, many other factors may affect splicing, either directly or indirectly. For instance, *cis*-elements in pre-mRNAs are critical for the spliceosome to define and remove introns correctly. Hormones, such as ABA and IAA, can change the alternative splicing pattern in three SR genes in Arabidopsis (Palusa et al. 2007). Ethylene can induce the expression of SR protein kinase *PK12* in tobacco (Savaldi-Goldstein et al. 2000). Environmental factors, such as heat, cold, and salt, have dramatic effects on the alternative splicing pattern of several SR genes (Palusa et al. 2007). Light can also regulate the expression of some splicing-related genes (Heintzen et al. 1994). Detailed discussion of these regulators is beyond the scope of this chapter. It is clear that plants developed specific regulatory mechanisms to control the largely conserved splicing machinery when adapting to their environments.

Conclusion and Perspective

The complex interactions of pre-mRNA, snRNAs, and spliceosomal proteins in pre-mRNA processing are critical to the accurate expression of a eukaryotic genome. It has become clear that while the core molecular entities involved in these processes are conserved throughout the animal and plant kingdoms, there are also distinct proteins that participate in splicing in plants. Much less clear is the extent to which these proteins are conserved among different phylogenetic groups of plants, and very little is known about how distinct proteins influence processing of particular pre-mRNAs and determine alternative splicing choices, for example. A recent study identified 84 putative exonic splicing enhancer motifs, of which 35 were experimentally shown to promote splicing (Pertea et al. 2007). In the years ahead it will be fascinating to correlate such motifs with specific (SR) proteins that recognize them and to seek understanding of how these proteins interact with the core constituents of the spliceosome. Proteomics studies of isolated plant spliceosomes may be key to progress toward these goals.

Acknowledgements Y. R. and V. B. were supported in part by National Science Foundation Grant DBI-0606909. B.-B. W. was supported by National Science Foundation Grants DBI-0321460 and DBI-0606966.

References

Ali GS, Palusa SG, Golovkin M, Prasad J, Manley JL, Reddy ASN (2007) Regulation of plant developmental processes by a novel splicing factor. PLoS ONE 2:e471

Bezerra IC, Michaels SD, Schomburg FM, Amasino RM (2004) Lesions in the mRNA cap-binding gene ABA HYPERSENSITIVE 1 suppress FRIGIDA-mediated delayed flowering in Arabidopsis. Plant J 40:112–119

Burge CB, Tuschl T, Sharp PA (1999) Splicing of precursors to mRNAs by the spliceosome. In: Gesteland RF, Cech TR, Atkins JF (eds) The RNA World: the nature of modern RNA suggests a prebiotic RNA. Cold Spring Harbor Laboratory Press, Cold Spring Harbor, New York, pp 525–560

Campbell MA, Haas BJ, Hamilton JP, Mount SM, Buell CR (2006) Comprehensive analysis of alternative splicing in rice and comparative analyses with Arabidopsis. BMC Genomics 7:327

Caspary F, Shevchenko A, Wilm M, Seraphin B (1999) Partial purification of the yeast U2 snRNP reveals a novel yeast pre-mRNA splicing factor required for pre-spliceosome assembly. EMBO J 18:3463–3474

Colot HV, Stutz F, Rosbash M (1996) The yeast splicing factor Mud13p is a commitment complex component and corresponds to CBP20, the small subunit of the nuclear cap-binding complex. Genes Dev 10:1699–1708

Connelly S, Filipowicz W (1993) Activity of chimeric U small nuclear RNA (snRNA)/mRNA genes in transfected protoplasts of *Nicotiana plumbaginifolia*: U snRNA 3′-end formation and transcription initiation can occur independently in plants. Mol Cell Biol 13:6403–6415

Domon C, Lorković ZJ, Valcarcel J, Filipowicz W (1998) Multiple forms of the U2 small nuclear ribonucleoprotein auxiliary factor U2AF subunits expressed in higher plants. J Biol Chem 273:34603–34610

Fabrizio P, Esser S, Kastner B, Lührmann R (1994) Isolation of *S. cerevisiae* snRNPs: comparison of U1 and U4/U6.U5 to their human counterparts. Science 264:261–265

Gao H, Gordon-Kamm WJ, Lyznik LA (2004) ASF/SF2-like maize pre-mRNA splicing factors affect splice site utilization and their transcripts are alternatively spliced. Gene 339:25–37

Golovkin M, Reddy ASN (1996) Structure and expression of a plant U1 snRNP 70K gene: alternative splicing of U1 snRNP 70K pre-mRNAs produces two different transcripts. Plant Cell 8:1421–1435

Golovkin M, Reddy ASN (1998) The plant U1 small nuclear ribonucleoprotein particle 70K protein interacts with two novel serine/arginine-rich proteins. Plant Cell 10:1637–1648

Golovkin M, Reddy ASN (1999) An SC35-like protein and a novel serine/arginine-rich protein interact with Arabidopsis U1-70K protein. J Biol Chem 274:36428–36438

Golovkin M, Reddy ASN (2003) Expression of U1 small nuclear ribonucleoprotein 70K antisense transcript using APETALA3 promoter suppresses the development of sepals and petals. Plant Physiol 132:1884–1891

Gottschalk A, Neubauer G, Banroques J, Mann M, Lührmann R, Fabrizio P (1999) Identification by mass spectrometry and functional analysis of novel proteins of the yeast [U4/U6.U5] tri-snRNP. EMBO J 18:4535–4548

Gottschalk A, Tang J, Puig O, Salgado J, Neubauer G, Colot HV, Mann M, Seraphin B, Rosbash M, Lührmann R, Fabrizio P (1998) A comprehensive biochemical and genetic analysis of the yeast U1 snRNP reveals five novel proteins. RNA 4:374–393

Gupta S, Ciungu A, Jameson N, Lal SK (2006) Alternative splicing expression of U1 snRNP 70K gene is evolutionary conserved between different plant species. DNA Seq 17:254–261

Hanley BA, Schuler MA (1991) cDNA cloning of U1, U2, U4 and U5 snRNA families expressed in pea nuclei. Nucleic Acids Res 19:1861–1869

Heintzen C, Melzer S, Fischer R, Kappeler S, Apel K, Staiger D (1994) A light- and temperature-entrained circadian clock controls expression of transcripts encoding nuclear proteins with homology to RNA-binding proteins in meristematic tissue. Plant J 5:799–813

Hofmann CJ, Marshallsay C, Waibel F, Filipowicz W (1992) Characterization of the genes encoding U4 small nuclear RNAs in *Arabidopsis thaliana*. Mol Biol Rep 17:21–28

Hugouvieux V, Kwak JM, Schroeder JI (2001) An mRNA cap binding protein, ABH1, modulates early abscisic acid signal transduction in Arabidopsis. Cell 106:477–487

Ibrahim AF, Watters JA, Brown JW (2001) Differential expression of potato U1A spliceosomal protein genes: a rapid method for expression profiling of multigene families. Plant Mol Biol 45:449–460

Iida K, Go M (2006) Survey of conserved alternative splicing events of mRNAs encoding SR proteins in land plants. Mol Biol Evol 23:1085–1094

Isshiki M, Tsumoto A, Shimamoto K (2006) The serine/arginine-rich protein family in rice plays important roles in constitutive and alternative splicing of pre-mRNA. Plant Cell 18:146–158

Kalyna M, Barta A (2004) A plethora of plant serine/arginine-rich proteins: redundancy or evolution of novel gene functions? Biochem Soc Trans 32:561–564

Kalyna M, Lopato S, Voronin V, Barta A (2006) Evolutionary conservation and regulation of particular alternative splicing events in plant SR proteins. Nucleic Acids Res 34:4395–4405

Kmieciak M, Simpson CG, Lewandowska D, Brown JW, Jarmolowski A (2002) Cloning and characterization of two subunits of *Arabidopsis thaliana* nuclear cap-binding complex. Gene 283:171–183

Krämer A, Gruter P, Groning K, Kastner B (1999) Combined biochemical and electron microscopic analyses reveal the architecture of the mammalian U2 snRNP. J Cell Biol 145:1355–1368

Kuhn JM, Breton G, Schroeder JI (2007) mRNA metabolism of flowering-time regulators in wild-type Arabidopsis revealed by a nuclear cap binding protein mutant, abh1. Plant J 50:1049–1062

Lambermon MH, Fu Y, Wieczorek Kirk DA, Dupasquier M, Filipowicz W, Lorković ZJ (2002) UBA1 and UBA2, two proteins that interact with UBP1, a multifunctional effector of pre-mRNA maturation in plants. Mol Cell Biol 22:4346–4357

Lambermon MH, Simpson GG, Wieczorek Kirk DA, Hemmings-Mieszczak M, Klahre U, Filipowicz W (2000) UBP1, a novel hnRNP-like protein that functions at multiple steps of higher plant nuclear pre-mRNA maturation. EMBO J 19:1638–1649

Landsberger M, Lorković ZJ, Oelmuller R (2002) Molecular characterization of nucleus-localized RNA-binding proteins from higher plants. Plant Mol Biol 48:413–421

Lazar G, Goodman HM (2000) The Arabidopsis splicing factor SR1 is regulated by alternative splicing. Plant Mol Biol 42:571–581

Lazar G, Schaal T, Maniatis T, Goodman HM (1995) Identification of a plant serine-arginine-rich protein similar to the mammalian splicing factor SF2/ASF. Proc Natl Acad Sci USA 92:7672–7676

Leader D, Connelly S, Filipowicz W, Waugh R, Brown JW (1993) Differential expression of U5snRNA gene variants in maize (*Zea mays*) protoplasts. Plant Mol Biol 21:133–143

Lewis JD, Gorlich D, Mattaj IW (1996a) A yeast cap binding protein complex (yCBC) acts at an early step in pre-mRNA splicing. Nucleic Acids Res 24:3332–3336

Lewis JD, Izaurralde E, Jarmolowski A, McGuigan C, Mattaj IW (1996b) A nuclear cap-binding complex facilitates association of U1 snRNP with the cap-proximal 5′ splice site. Genes Dev 10:1683–1698

Lin CH, Patton JG (1995) Regulation of alternative 3′ splice site selection by constitutive splicing factors. RNA 1:234–245

Lopato S, Borisjuk L, Milligan AS, Shirley N, Bazanova N, Langridge P (2006) Systematic identification of factors involved in post-transcriptional processes in wheat grain. Plant Mol Biol 62:637–653

Lopato S, Forstner C, Kalyna M, Hilscher J, Langhammer U, Indrapichate K, Lorković ZJ, Barta A (2002) Network of interactions of a novel plant-specific Arg/Ser-rich protein, atRSZ33, with atSC35-like splicing factors. J Biol Chem 277:39989–39998

Lopato S, Gattoni R, Fabini G, Stevenin J, Barta A (1999a) A novel family of plant splicing factors with a Zn knuckle motif: examination of RNA binding and splicing activities. Plant Mol Biol 39:761–773

Lopato S, Kalyna M, Dorner S, Kobayashi R, Krainer AR, Barta A (1999b) atSRp30, one of two SF2/ASF-like proteins from *Arabidopsis thaliana*, regulates splicing of specific plant genes. Genes Dev 13:987–1001

Lopato S, Mayeda A, Krainer AR, Barta A (1996a) Pre-mRNA splicing in plants: characterization of Ser/Arg splicing factors. Proc Natl Acad Sci USA 93:3074–3079

Lopato S, Waigmann E, Barta A (1996b) Characterization of a novel arginine/serine-rich splicing factor in Arabidopsis. Plant Cell 8:2255–2264

Lorković ZJ, Barta A (2002) Genome analysis: RNA recognition motif (RRM) and K homology (KH) domain RNA-binding proteins from the flowering plant *Arabidopsis thaliana*. Nucleic Acids Res 30:623–635

Lorković ZJ, Hilscher J, Barta A (2004) Use of fluorescent protein tags to study nuclear organization of the spliceosomal machinery in transiently transformed living plant cells. Mol Biol Cell 15:3233–3243

Lorković ZJ, Lehner R, Forstner C, Barta A (2005) Evolutionary conservation of minor U12-type spliceosome between plants and humans. RNA 11:1095–1107

Lorković ZJ, Wieczorek Kirk DA, Klahre U, Hemmings-Mieszczak M, Filipowicz W (2000a) RBP45 and RBP47, two oligouridylate-specific hnRNP-like proteins interacting with poly(A)+ RNA in nuclei of plant cells. RNA 6:1610–1624

Lorković ZJ, Wieczorek Kirk DA, Lambermon MH, Filipowicz W (2000b) Pre-mRNA splicing in higher plants. Trends Plant Sci 5:160–167

Merendino L, Guth S, Bilbao D, Martinez C, Valcarcel J (1999) Inhibition of msl-2 splicing by Sex-lethal reveals interaction between U2AF35 and the 3′ splice site AG. Nature 402:838–841

Musci MA, Egeland DB, Schuler MA (1992) Molecular comparison of monocot and dicot U1 and U2 snRNAs. Plant J 2:589–599

Palusa SG, Ali GS, Reddy ASN (2007) Alternative splicing of pre-mRNAs of Arabidopsis serine/arginine-rich proteins: regulation by hormones and stresses. Plant J 49:1091–1107

Papp I, Mur LA, Dalmadi A, Dulai S, Koncz C (2004) A mutation in the Cap Binding Protein 20 gene confers drought tolerance to Arabidopsis. Plant Mol Biol 55:679–686

Pertea M, Mount SM, Salzberg SL (2007) A computational survey of candidate exonic splicing enhancer motifs in the model plant *Arabidopsis thaliana*. BMC Bioinformatics 8:159

Polycarpou-Schwarz M, Gunderson SI, Kandels-Lewis S, Seraphin B, Mattaj IW (1996) *Drosophila* SNF/D25 combines the functions of the two snRNP proteins U1A and U2B' that are encoded separately in human, potato, and yeast. RNA 2:11–23

Reddy ASN (2001) Nuclear pre-mRNA splicing in plants. Critical Rev Plant Sci 20:523–571

Reddy ASN (2004) Plant serine/arginine-rich proteins and their role in pre-mRNA splicing. Trends Plant Sci 9:541–547

Reddy ASN (2007) Alternative splicing of pre-messenger RNAs in plants in the genomic era. Annu Rev Plant Biol 58:267–294

Ruskin B, Zamore PD, Green MR (1988) A factor, U2AF, is required for U2 snRNP binding and splicing complex assembly. Cell 52:207–219

Saldi T, Wilusz C, MacMorris M, Blumenthal T (2007) Functional redundancy of worm spliceosomal proteins U1A and U2B. Proc Natl Acad Sci USA 104:9753–9757

Savaldi-Goldstein S, Aviv D, Davydov O, Fluhr R (2003) Alternative splicing modulation by a LAMMER kinase impinges on developmental and transcriptome expression. Plant Cell 15:926–938

Savaldi-Goldstein S, Sessa G, Fluhr R (2000) The ethylene-inducible PK12 kinase mediates the phosphorylation of SR splicing factors. Plant J 21:91–96

Shukla GC, Padgett RA (1999) Conservation of functional features of U6atac and U12 snRNAs between vertebrates and higher plants. RNA 5:525–538

Simpson CG, Jennings SN, Clark GP, Thow G, Brown JW (2004) Dual functionality of a plant U-rich intronic sequence element. Plant J 37:82–91

Simpson GG, Clark GP, Rothnie HM, Boelens W, van Venrooij W, Brown JW (1995) Molecular characterization of the spliceosomal proteins U1A and U2B from higher plants. EMBO J 14:4540–4550

Staiger D, Zecca L, Wieczorek Kirk DA, Apel K, Eckstein L (2003) The circadian clock regulated RNA-binding protein AtGRP7 autoregulates its expression by influencing alternative splicing of its own pre-mRNA. Plant J 33:361–371

Stevens SW, Abelson J (1999) Purification of the yeast U4/U6.U5 small nuclear ribonucleoprotein particle and identification of its proteins. Proc Natl Acad Sci USA 96:7226–7231

Stevens SW, Barta I, Ge HY, Moore RE, Young MK, Lee TD, Abelson J (2001) Biochemical and genetic analyses of the U5, U6, and U4/U6 x U5 small nuclear ribonucleoproteins from *Saccharomyces cerevisiae*. RNA 7:1543–1553

Stojdl DF, Bell JC (1999) SR protein kinases: the splice of life. Biochem Cell Biol 77:293–298

Tarn WY, Steitz JA (1997) Pre-mRNA splicing: the discovery of a new spliceosome doubles the challenge. Trends Biochem Sci 22:132–137

Valadkhan S (2005) snRNAs as the catalysts of pre-mRNA splicing. Curr Opin Chem Biol 9:603–608 van Santen VL, Spritz RA (1987) Nucleotide sequence of a bean (*Phaseolus vulgaris*) U1 small nuclear RNA gene: implications for plant pre-mRNA splicing. Proc Natl Acad Sci USA 84:9094–9098

Vankan P, Edoh D, Filipowicz W (1988) Structure and expression of the U5 snRNA gene of *Arabidopsis thaliana*. Conserved upstream sequence elements in plant U-RNA genes. Nucleic Acids Res 16:10425–10440

Vankan P, Filipowicz W (1988) Structure of U2 snRNA genes of *Arabidopsis thaliana* and their expression in electroporated plant protoplasts. EMBO J 7:791–799

Vankan P, Filipowicz W (1989) A U-snRNA gene-specific upstream element and a −30 'TATA box' are required for transcription of the U2 snRNA gene of *Arabidopsis thaliana*. EMBO J 8:3875–3882

Vaux P, Guerineau F, Waugh R, Brown JW (1992) Characterization and expression of U1snRNA genes from potato. Plant Mol Biol 19:959–971

Waibel F, Filipowicz W (1990) U6 snRNA genes of Arabidopsis are transcribed by RNA polymerase III but contain the same two upstream promoter elements as RNA polymerase II-transcribed U-snRNA genes. Nucleic Acids Res 18:3451–3458

Wang BB, Brendel V (2004) The ASRG database: identification and survey of *Arabidopsis thaliana* genes involved in pre-mRNA splicing. Genome Biol 5:R102

Wang BB, Brendel V (2006a) Genomewide comparative analysis of alternative splicing in plants. Proc Natl Acad Sci USA 103:7175–7180

Wang BB, Brendel V (2006b) Molecular characterization and phylogeny of U2AF35 homologs in plants. Plant Physiol 140:624–636

Will CL, Lührmann R (2001) Spliceosomal UsnRNP biogenesis, structure and function. Curr Opin Cell Biol 13:290–301

Will CL, Schneider C, Hossbach M, Urlaub H, Rauhut R, Elbashir S, Tuschl T, Lührmann R (2004) The human 18S U11/U12 snRNP contains a set of novel proteins not found in the U2-dependent spliceosome. RNA 10:929–941

Will CL, Schneider C, Reed R, Lührmann R (1999) Identification of both shared and distinct proteins in the major and minor spliceosomes. Science 284:2003–2005

Wu S, Romfo CM, Nilsen TW, Green MR (1999) Functional recognition of the 3′ splice site AG by the splicing factor U2AF35. Nature 402:832–835

Xiong L, Gong Z, Rock CD, Subramanian S, Guo Y, Xu W, Galbraith D, Zhu JK (2001) Modulation of abscisic acid signal transduction and biosynthesis by an Sm-like protein in Arabidopsis. Dev Cell 1:771–781

Zamore PD, Patton JG, Green MR (1992) Cloning and domain structure of the mammalian splicing factor U2AF. Nature 355:609–614

Zorio DA, Blumenthal T (1999) Both subunits of U2AF recognize the 3′ splice site in *Caenorhabditis elegans*. Nature 402:835–838

Analysis of Alternative Splicing in Plants with Bioinformatics Tools

B.J. Haas

Contents

Introduction	18
Bioinformatics Strategies for Analyzing Alternative Splicing	19
Analysis of Expressed Transcript Sequences	19
Computational Representation of Alternatively Spliced Genes	21
Reconstructing Transcript Isoforms from Splicing Graphs	23
Analysis of Alternative Splicing Mediated by the PASA Software	26
Extended Splicing Variation Classifications	27
Analysis of Alternative Splicing in Plants	28
Quantifying Alternatively Spliced Plant Genes	28
Localizing Splicing Variations Within Gene Structures	29
Comparisons Between Alternative Splicing in Plants and Animals	30
Web Resources for Alternative Splicing in Plants	33
References	34

Abstract Alternative splicing is a molecular mechanism utilized by a broad range of eukaryotes to extend the repertoire of functions encoded by single genes and to posttranscriptionally regulate gene expression. Recent analyses of expressed transcript sequences aligned to the complete genomes of Arabidopsis and rice indicate that alternative splicing in plants is prevalent and exhibits several features similar to other higher eukaryotes including mouse and human. This chapter reviews the computational strategies employed to study alternative splicing with bioinformatics tools and the recent findings from analyses performed on plants by applying such methods.

B.J. Haas
Broad Institute, 7 Cambridge Center, Cambridge, MA, 02142 USA
e-mail: bhaas@broad.mit.edu

Introduction

Alternative splicing (AS) of mRNA transcripts provides a mechanism by which a single gene can express transcripts that encode proteins with altered functions, physiological properties, or function to postranscriptionally regulate gene expression (reviewed in: Graveley 2001; Lareau et al. 2004; Stamm et al. 2005; Stetefeld and Ruegg 2005). The mechanism of AS is not completely understood but does appear to involve regulatory motifs in the transcript sequence coupled with regulatory proteins that together influence the pattern of splicing (reviewed in: Black 2003; Maniatis and Tasic 2002; Reddy 2007). The important biological role of AS was known decades ago, such as in the molecular mechanisms responsible for *Drosophila* sexual development (Baker 1989). Initial analyses of AS were limited to individual genes, and now in the genome era high-throughput sequencing of expressed transcripts and complete genome sequences has enabled comprehensive analyses of AS, identifying many thousands of genes that are alternatively spliced across a range of eukaryotic species (notable examples include: Campbell et al. 2006; Iida et al. 2004; Kan et al. 2001; Modrek et al. 2001; Nagasaki et al. 2005; Okazaki et al. 2002; Wang and Brendel 2006; Zavolan et al. 2003). Given the surprisingly small number of protein-coding genes that have been identified in humans, totaling between 20,000 and 30,000 genes (International Human Genome Sequencing Consortium 2004; Lander et al. 2001; Venter et al. 2001), AS is implicated as a major contributor to proteome diversity and organism complexity, with evidence that more than half of all human genes are alternatively spliced (reviewed in Modrek and Lee 2002). A more recent study using exon junction microarrays reports that at least 74% of all human multiexon genes undergo AS (Johnson et al. 2003). While the majority of genes are expressed as a small assortment of transcript isoforms, there are striking examples of genes with the potential to generate thousands. Perhaps the most extraordinary example is the *DScam* gene of *Drosophila melanogaster*, capable of expressing over thirty thousand distinct protein products (reviewed in Zipursky et al. 2006). The biological consequence of AS, including the association of defects in AS with human diseases (reviewed in: Blencowe 2006; Licatalosi and Darnell 2006; Novoyatleva et al. 2006), has attracted much attention to this phenomenon.

Analyses of AS in plants are just now catching up with the more rigorous studies performed on metazoans including insects and vertebrates. Such studies in plants have been made possible by the complete sequencing of the genomes for the model dicot species *Arabidopsis thaliana* (Arabidopsis Genome Initiative 2000) and the model monocot species *Oryza sativa* (rice) (Goff et al. 2002; International Rice Genome Sequencing Project 2005; Yu et al. 2002). Results of recent studies indicate that AS is prevalent in plants, impacting at least 20%–30% of all expressed protein-coding genes (Campbell et al. 2006; Wang and Brendel 2006). Several experimental studies have shown that AS is critically important to the biology of plants, involved in processes including plant development and disease resistance (reviewed in Reddy 2007). Comparisons between the findings in plants and animals indicate that there are considerable differences in the rates and types of AS variations that exist (Brett et al. 2002; Nagasaki et al. 2005). This chapter describes the bioinformatics approaches taken to identify and characterize alternatively spliced genes with a focus on biological sequence analysis in the genome era, highlighting the recent findings relevant to plant biology.

Bioinformatics Strategies for Analyzing Alternative Splicing

Analysis of Expressed Transcript Sequences

The identification of alternatively spliced genes most often involves examining expressed transcript sequences in the form of expressed sequence tags (ESTs) or, more recently, full-length cDNAs (FL-cDNAs). Under the assumption that the transcript sequences represent fully processed messenger RNA (mRNA) molecules, differences found between such sequences expressed from the same gene are interpreted as deriving from alternative transcription and mRNA processing, including expression from alternate promoters, AS, and alternative transcriptional termination and polyadenylation. The variations resulting from these processes can be classified into the following observed events: exon skipping, alternate donor or acceptor splice sites, alternate terminal exons, retained intron, or initiating or terminating within an intron (Fig. 1).

Variations in mRNA processing including those derived from AS are evident from pairwise comparisons of individual fully processed transcript sequences derived from the same gene, or from examining alignments of the mRNA sequences to their cognate DNA sequence. The most valuable studies of AS are performed in the context of the complete genome sequence, and there are several arguments for this: The complete genome allows the separation of paralogs that might otherwise be confused for transcripts derived from the same gene, alignments to high-quality genome sequences separate EST sequencing errors, low-quality sequence regions, and contaminating vector or other artifacts from found splicing variations, and subtle splicing variations are retained and verified by the presence of consensus splice sites that might otherwise be considered dubious. Finally, insertions or deletions observed as variations between aligned mRNA sequences in the absence of the genome are identified as specific introns, exons, and splice site choice variations in the context of alignments to the genome, elucidating variations at specific gene structures.

Sequence alignment provides a foundation for these analyses. For pairwise comparisons between individual transcript sequences, a tool such as BLAT (Kent 2002) or BLAST (Altschul et al. 1997) is sufficient. Brett et al. (2002) measured the prevalence of AS among several eukaryotes including animals and plants by comparing EST sequences to mRNA sequences with BLASTN. Ner-Gaon et al. (2004) developed a method termed EST pairs gapped alignment (EPGA) to identify candidate alternatively spliced transcripts, requiring a BLAT alignment between two transcripts to contain two aligned regions of at least 20 bp straddling an unaligned region of at least 4 bp. Again, such comparisons identify variations among sequence pairs including insertions and deletions that may reflect retained introns, exon skipping events, or alternate donors or acceptors, but the exact event responsible for the found

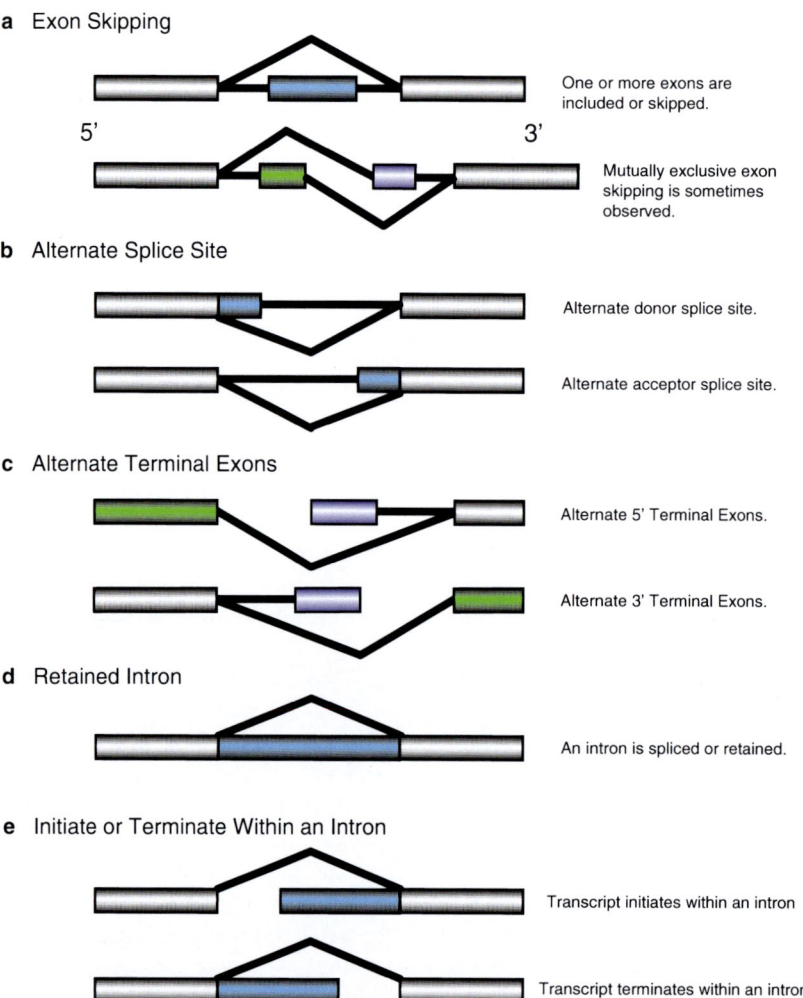

Fig. 1 Variations resulting from alternative mRNA processing. Differences between transcript isoforms resulting from alternative mRNA processing including alternative splicing, alternative transcriptional initiation, and transcriptional termination or polyadenylation can be distilled into the following categories: **a** Exon skipping: one or more exons are either included in the processed transcript or spliced as part of an intron. In some cases, exons are observed to be mutually exclusive. **b** Alternate splice sites: an alternate acceptor or donor splice site is chosen. **c** Alternate terminal exons: alternate 5′ terminal exons may derive from alternate transcription initiation coupled with alternative splicing. Likewise, alternate 3′ terminal exons may derive from alternate transcriptional termination or polyadenylation coupled with alternative splicing. **d** Retained intron: an intron is either spliced or retained in the processed transcript. **e** Initiation or termination within an intron: transcript isoforms likely derive from alternate transcriptional initiation, or alternate transcriptional termination or polyadenylation, respectively

variation is not evident in the absence of the underlying genome sequence. In the context of the genome sequence, each transcript can be aligned to reveal the intron and exon components of the gene structure. BLAST is not capable of revealing the precise details of these structures, and instead, a more specialized alignment tool is used to generate a "spliced alignment." Numerous spliced alignment tools are available, including EST_GENOME (Mott 1997), AAT (Huang et al. 1997), sim4 (Florea et al. 1998), GeneSeqer (Usuka et al. 2000), BLAT (Kent 2002), and GMAP (Wu and Watanabe 2005) (note that BLAT can be used as a general purpose nucleotide sequence alignment tool or in a spliced alignment mRNA mode). Each tool implements an algorithm that takes into account the nucleotide matches between the genome and the transcript sequence, allows for insertions within the genome sequence that correspond to intron regions spliced from the processed mRNA, and recognizes the consensus splice sites found at intron/exon boundaries. The result is that the exons of the transcript are effectively unspliced back onto the genome sequence from which it derived, revealing the intron and exon structures that reside in the genome sequence. After all transcript sequences are aligned to their respective genomic locus, variations resulting from alternate mRNA processing become evident as discrepancies between the individual spliced alignment-inferred gene structures. Figure 2 illustrates the use of the AAT spliced alignment software to align a set of Arabidopsis transcript sequences to their cognate genomic locus, revealing the choice of alternate acceptor splice sites by differentially spliced transcripts.

Computational Representation of Alternatively Spliced Genes

Although the sequence of a gene can be represented by a linear string of characters corresponding to the genome sequence, the set of processed mRNAs expressed from the gene cannot be adequately represented in a linear form because of the variations that result from alternative mRNA processing. Instead, a bioinformatics approach is to represent a gene structure as a popular data structure called a directed acyclic graph (DAG), often referred to as a splicing graph. This graph data structure forms the core of many large-scale computational analyses of AS and isoform reconstruction from transcript sequences and can be formulated in several ways (Heber et al. 2002; Kan et al. 2001; Lee 2003; Lee and Wang 2005; Leipzig et al. 2004; Malde et al. 2005; Xing et al. 2004, 2006). In the absence of the genome sequence, mRNA sequences deriving from the same gene can be clustered and assembled into a splicing graph such that common regions of the expressed mRNAs are represented as vertices and variable regions represented as edges. Given a genome sequence and spliced alignments of the transcripts to their locus of origin, a splicing graph can be constructed by representing distinct exons as vertices and distinct introns as edges that connect pairs of exons as evident from the spliced alignments. An example of a splicing graph for a rice gene structure based on spliced alignments of transcript sequences to the genome is provided in Fig. 3. The splicing graph is directional, initiating at transcriptional starts and terminating at sites of 3′ polyadenylation. Any valid path from a

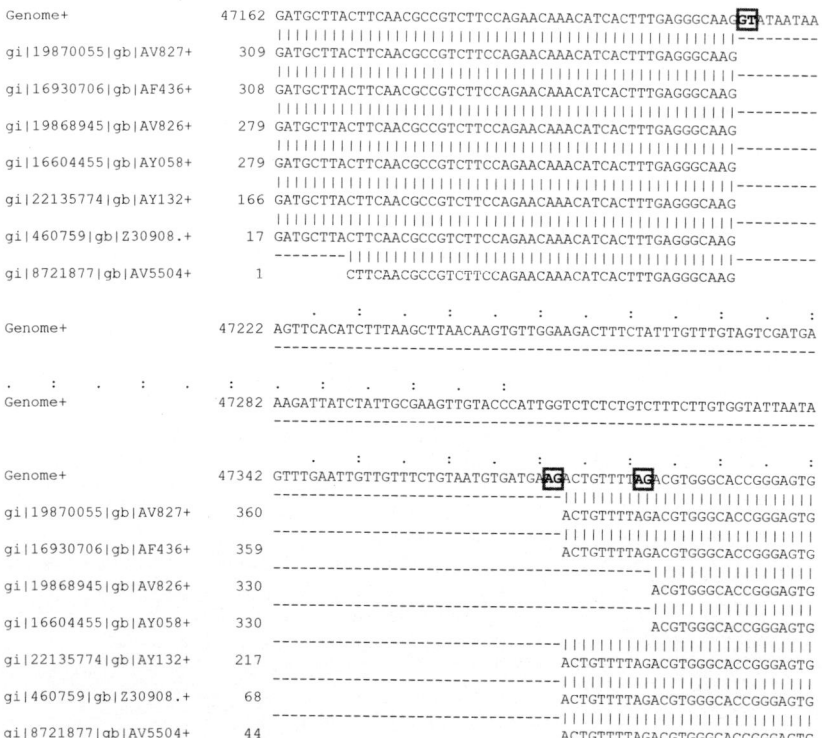

Fig. 2 Spliced genome alignment of Arabidopsis transcript sequences reveals alternative splicing. The AAT spliced alignment software was used to align seven Arabidopsis transcript sequences to their cognate genome location within a bacterial artificial chromosome (BAC) sequence. Five and two of the seven transcript sequences reveal the use of alternate acceptor splice sites separated by 10 nucleotides. The displayed alignment corresponds to only the short region of the gene At1g04870, encoding an arginine N-methyltransferase family protein, restricted to the alternative splicing variation in the context of a complete intron sequence and portions of the flanking exons. Consensus GT donor splice site and alternative AG acceptor splice sites are *boxed*

transcriptional initiation point through the graph to a 3′ polyadenylation site would yield the sequence and structure of an individual isoform, and all such isoforms are represented as distinct paths within the splicing graph (Leipzig et al. 2004).

The population of transcript sequences that correspond to a single gene can be quite diverse both in their abundance and coverage of the gene. EST sequences typically correspond to partial transcript sequences and derive from the 5′ or 3′ end of a cloned cDNA. FL-cDNAs ideally cover the complete fully processed transcript sequence from the transcriptional start site to the polyadenylation site. Individual isoforms may be unequally represented by these data both in abundance and coverage, but each of the isoforms can be readily decomposed into a splicing graph. To better understand the role of AS, we are interested in the extent to which individual genes can produce alternative isoforms, and how individual isoforms differ in their

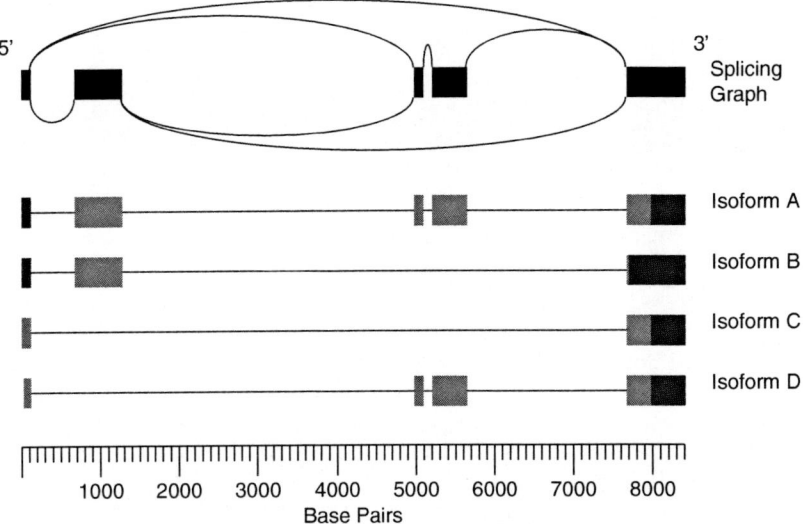

Fig. 3 Splicing graph representation of gene structure. The splicing graph provides a useful data structure for representing complex gene structures. The splicing graph corresponding to the rice gene encoding expressed protein (LOC_Os02g51140) is illustrated at top, with exons drawn as *black boxes* and introns drawn as *arcs* connecting exons, as evident from over a dozen individual transcript alignments. The four distinct transcript isoforms constructed by PASA from the individual alignments are provided, with putative protein-coding regions derived as the longest open reading frames shown in *light gray*

structure and function. This requires a mechanism to reconstruct the isoforms represented by the splicing graph, for which several algorithms have been devised, as described below.

Reconstructing Transcript Isoforms from Splicing Graphs

One approach is to report every isoform that exists as a path within the splicing graph by following every connected exon and intron (Leipzig et al. 2004). Such an approach can be very useful for identifying those genes that have the potential to contribute greatly to transcriptomic and proteomic diversity, but at the same time can create an unwieldy number of gene products, only a small number of which may represent genuine products of mRNA processing. Alternative approaches attempt to identify those fewer numbers of candidate isoforms that are best represented by the experimental evidence (expressed transcript sequences).

The Transcript Assembly Program (TAP) creates a splice graph from spliced transcript alignments and traverses the graph to find the predominant gene structure (Kan et al. 2001). TAP scores each pair of introns connected directly or transitively based on transcript alignment coverage and then reports the path through these introns and their connected exons with the highest cumulative coverage as the predominant gene structure. Directly connecting introns are those that are connected by a single exon as found within at least one transcript alignment. Transitively connecting introns are found between neighboring transcript alignments that overlap only partially within the adjoining exon. In addition to reporting the predominant gene structure, TAP can report alternative isoform structures that include introns missing from the predominant gene structure.

A similar approach is taken by Xing et al. (2004) in which a splicing graph is constructed from introns and exons inferred from transcript-genome alignments, each component is weighted based on the number of transcripts supporting it, and a "heaviest bundling" algorithm is applied to report the predominant isoform as the set of connected introns and exons with the greatest cumulative evidence support. In this scheme, transcript evidence missing from the predominant isoform is assigned an inflated weight to encourage the selection of corresponding components from the splice graph during a subsequent application of the heaviest bundling algorithm.

The above strategies are considered maximum likelihood approaches, attempting to construct the isoform that best reflects the transcript alignment evidence. Although these strategies have been demonstrated to be effective and accurate, they do not take into account couplings between distal gene structure components and so can generate combinations of introns and exons that are not observed in known gene structures.

Capturing the combinations of introns and exons as found within complete alignments and ensuring their representation within isoforms reconstructed from clustered transcript alignments was the impetus for the development of the Program to Assemble Spliced Alignments (PASA) (Haas et al. 2003). The PASA software constructs a DAG from a cluster of overlapping transcript alignments such that each complete transcript alignment is represented as an individual node within the graph. The alignment nodes are ordered by their 5'-most coordinate, and edges connect each alignment to a downstream compatible alignment; a compatible alignment corresponds to an overlapping alignment that has identical introns within the overlapping region and is not wholly contained within the former alignment. Each alignment node also maintains a list of all those alignments fully contained within it. The predominant gene structure is found as the path through the graph of alignment nodes that traverses the maximal number of compatible alignments, including those alignments that are contained within other alignments. In essence, this is the gene structure with the greatest support based upon the abundance of transcript evidence. This operation is similar to the heaviest bundling algorithm in that we are finding the heaviest bundle of complete and compatible alignment structures instead of the heaviest bundle of connected individual gene structure components. The structures for AS isoforms are reported by finding the maximal set of compatible alignments that include those alignments missing from the predominant gene structure. PASA has been applied extensively in the annotation of both the Arabidopsis (Haas et al. 2003; Haas et al. 2005;

Wortman et al. 2003) and rice (Ouyang et al. 2007) genomes, among other eukaryotes including other plants, protists, fungi, and animals, as an essential component of the eukaryotic genome annotation system at The Institute for Genomic Research. A more complete description of PASA is provided in Section 2.4, and specific insights gained into AS in plants via the application of PASA are described throughout this chapter.

The importance of retaining correlations between splice variations as found in FL-cDNAs has spurred similar innovations with similar goals. Eyras et al. (2004) describe the ClusterMerge algorithm implemented as part of the Ensembl automated gene structure annotation system. In contrast to PASA, which reports gene structures for each isoform consistent with the greatest quantity of compatible transcript evidence, ClusterMerge finds the minimum set of isoforms that fully explain the observed set of transcript sequences. The differences between the PASA and ClusterMerge isoform reconstruction algorithms are subtle. A more recent version of PASA reports isoform structures that are both maximal in complete transcript coverage while minimizing the number of isoforms reported, further blurring their distinction. Both PASA and ClusterMerge build a DAG data structure that is different from the conventional splicing graph in that each complete transcript alignment, corresponding to an individual EST or mRNA sequence, exists as an individual node in the graph. Comparisons are performed between nodes to identify compatible transcript alignments and those transcripts that are included completely within other transcript alignments, or those that allow extensions of the compatible structure. The prime difference between PASA and ClusterMerge involves implementation details. PASA employs dynamic programming to arrive at maximal isoform structures, whereas ClusterMerge involves a rigorous graph traversal with rules governing its behavior at each node.

A more recent approach by Xing et al. (2006) provides a probabilistic framework for reporting isoforms constructed from the splicing graph with an expectation-maximization algorithm that accounts for couplings between splicing variations as evident within the individual transcript alignments. Initially, all possible isoforms are generated from the splicing graph by traversing all connected vertices and edges. Each isoform of this superset of possible isoforms is then compared to each of the inputted transcript alignments, tallying the number of transcript alignments that are consistent with each putative isoform variant. The expectation-maximization algorithm is applied to identify the subset of isoforms that best explain the known transcript structures under maximum likelihood estimation.

Regardless of the approach, once the isoforms are reconstructed from the diverse set of expressed transcript sequences, these reconstructions provide a more complete view of the observable transcript diversity and more easily serve as the substrate for comprehensive analyses of alternative mRNA processing. In addition to identifying the individual genes found to be alternatively spliced, comparisons between pairs of isoforms reveal the specific variations that contribute to the observed genetic complexity. Below, PASA is highlighted as a bioinformatics tool that serves these basic needs, from isoform reconstruction to resolving the types of splicing variations exhibited by individual isoforms.

Analysis of Alternative Splicing Mediated by the PASA Software

The PASA software includes a set of tools to leverage ESTs and FL-cDNA sequences for eukaryotic gene structure annotation and for studying AS (Campbell et al. 2006; Haas et al. 2003). The inputs to PASA include the genome sequences and transcript sequences plus an indication of the transcripts that are considered to be full-length. PASA executes a series of steps to reconstruct transcript isoforms from transcript alignments and then to identify all splicing variations evidenced by differences in gene structures between individual isoforms. The complete PASA pipeline is described as follows:

- Transcript sequence cleaning: Transcript sequence termini resulting from polyadenylation are removed, contaminating cloning vector sequences are trimmed, and low-quality sequences are discarded.
- Spliced transcript to genome alignment: The transcripts are aligned to the genome with spliced alignment software such as GMAP or BLAT. The single best scoring alignment to the genome is retained.
- Alignment validation: Only those spliced alignments that align by at least 90% of their length and at least 95% sequence identity are retained (both parameters are user-specified values, defaults provided). All alignments inferring introns are required to have consensus dinucleotides at splice junctions, including GT-AG, GC-AG, or AT-AC dinucleotide pairs. To ensure that the alignments at splice junctions are accurate, identical matches are required for a range of exonic bases (three by default) abutting each splice junction.
- Isoform reconstruction: Alignments are assembled into maximal alignment assemblies such that each alignment is found within a structure containing the maximal number of compatible alignments. Two alignments are considered compatible if they overlap, are transcribed on the same DNA strand, and have identical intron positions in their region of overlap. Each maximal alignment assembly (PASA assembly) is considered either a partial or complete transcript isoform; those PASA assemblies containing at least one FL-cDNA alignment are expected to encode the complete transcript isoforms.
- Isoform clustering: Each pair of PASA assemblies (isoforms) found transcribed on the same DNA strand and overlapping by at least 40% (configurable) of either length are clustered. Single-linkage clustering of all such paired assemblies yields isoform clusters.
- Alternative splicing analysis: Each pair of assemblies within an isoform cluster is compared, and differences between their gene structures reveal splicing variations. Each assembly is labeled according to the splicing variation observed (i.e., alternate donor site), position, and the evidence for the variation.

The splicing variations cataloged by PASA include the types illustrated in Fig. 1, specifically exon skipping, alternate donor, alternate acceptor, alternate terminal exons, retained intron, and initiate within intron or terminate within intron. Variations are only realized in the context of the comparison of two different isoforms. Each isoform is labeled according to the variation exhibited in the context of that comparison. This labeling is typically symmetric, as in the case of alternate donors or acceptors or alternate terminal exons, where each isoform in the pair being compared is assigned the same label. Other variations observed are asymmetric, such as the case of exon skipping where one isoform retains the exon and the other skips it. Likewise, in the case of the retained intron, the alternate isoform splices the intron. Each isoform of the pair is labeled accordingly.

Although PASA was originally developed as a genome annotation tool, the above functions of the software provide a standard framework for analyzing AS for any eukaryote provided a complete genome sequence and database of expressed transcript sequences. In addition to identifying individual splicing variations within the isoforms, the variations are localized to the protein-coding regions of the genes, and the impact of the variation on the protein's primary structure is examined.

Extended Splicing Variation Classifications

The small number of splice variation types catalogued as described above, and as typical of most AS analyses, does not adequately capture the extensive diversity of splicing variations that are observed between isoforms. For example, the skipped exon is captured as a single type when what are sometimes observed are two or more exons skipped as a single event, or neighboring exons that are skipped in a mutually exclusive fashion. The simple "skipped exon" classification fails to capture these cases as special instances of this type.

A scheme introduced by Nagasaki et al. (2006) captures these rich details and accounts for each complex type independently. Nagasaki et al. (2006) describe variations between processed mRNAs as alternative splicing and transcriptional initiation (ASTI) units. Each ASTI is defined as "a minimal span of variably expressed genomic region flanked by common exonic or extragenic region(s)." A single ASTI is computed by comparing gene structures within a cluster of transcript isoforms. The gene structure of each isoform is decoded into a series of ones and zeros (bits) for each nucleotide of the gene. Exonic nucleotides are represented by ones, and intronic and flanking regions are represented by zeros. Each isoform is decoded into such a bit vector, and these bit vectors are compared in a pairwise fashion. At nucleotide positions where isoforms agree, the corresponding positions of the bit vector will contain identical values (both zero or both one). Where differences are found because of AS or transcriptional initiation, conflicting values exist. The pairs of bit vectors can be compressed into regions that agree or disagree, drastically shrinking the sizes of the vectors to contain only the nonredundant values. Comparisons between positions in the compressed vectors then yield the genic regions that agree or disagree between isoforms. A single ASTI unit, as defined above, can be extracted from the compressed bit vectors as a paired combination of bits with values that disagree internally and are flanked on both ends by agreeing values. The bit patterns corresponding to the individual ASTI units can be conveniently

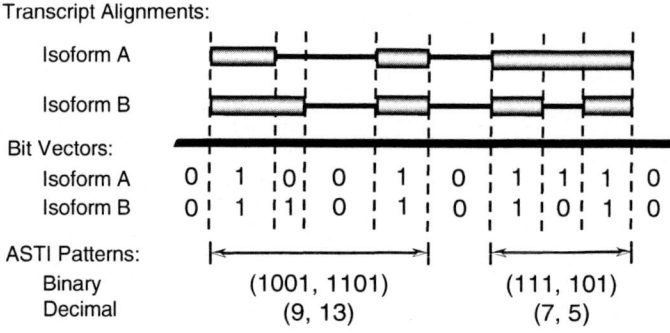

Fig. 4 Alternative splicing variations captured as ASTI units from bit vectors. Transcript alignments are converted into bit vectors where exonic regions contribute ones and introns and flanking regions contribute zeros. Comparisons between isoform bit vectors yield conflicting positions flanked by agreeing positions captured together as ASTI units. Each ASTI unit is described by a pair of binary numbers or their decimal equivalents. Two ASTI units are shown: an alternate donor splice site at left and a retained intron at right

converted to decimal numbers to provide more palatable numeric accessions for each ASTI unit. An illustration of this method is provided as Fig. 4.

The classification system provided by the ASTI algorithm is the only method described thus far that provides a robust mechanism for cataloging both the simpler AS variations such as alternate donors and alternate acceptors as well as the more complex variations involving multiple correlated features such as several skipped exons or an alternate donor site tied to a skipped exon. Over a hundred different ASTI units have been identified, the vast majority of which cannot be named in an obvious and meaningful way and are instead referenced by unique numeric identifiers corresponding to their pattern (Nagasaki et al. 2006).

Analysis of Alternative Splicing in Plants

Quantifying Alternatively Spliced Plant Genes

Alternative splicing can be quantified and presented in many ways. Unfortunately, there are currently no standard methods for computing or reporting these data, and hidden biases can introduce confusion and make it difficult to accurately quantify and compare levels of AS. With ever-increasing numbers of available EST and FL-cDNA sequences, the evidence for AS in plants continues to increase (Table 1).

Global analyses of AS in plants have been limited to a small number of studies, and the general observations are mostly consistent (Brett et al. 2002; Campbell et al. 2006; Haas et al. 2002; Nagasaki et al. 2005; Ner-Gaon et al. 2004; Wang and Brendel 2006). The most recent large-scale studies of expressed transcripts in

Table 1 Studies identifying alternatively spliced genes in Arabidopsis and rice

Arabidopsis		
Campbell et al. (2006)	5,313 genes	690,119 ESTs and 61,117 FL-cDNAs
Wang and Brendel (2006)	4,707 genes	323,340 ESTs and 62,009 FL-cDNAs
Nagasaki et al. (2005)	2,195 genes	58,376 FL-cDNAs
Iida et al. (2004)	1,764 genes	265,307 ESTs and 13,427 FL-cDNAs
Haas et al. (2003)	1,188	177,973 ESTs and 27,414 FL-cDNAs
Rice		
Campbell et al. (2006)	8,772 genes	1,156,705 ESTs and 33,799 FL-cDNAs
Wang and Brendel (2006)	6,568 genes	298,857 ESTs and 32,136 FL-cDNAs
Nagasaki et al. (2005)	1,514 genes	32,127 FL-cDNAs

plants indicate that at least 20%–30% of all expressed plant genes are alternatively spliced (Campbell et al. 2006; Wang and Brendel 2006). The rates of AS and the distribution of variation types (i.e., alternate donor or acceptor) are very similar between rice and Arabidopsis, suggesting that these are general properties of plants across monocots and dicots. Of the types of splicing variations found, retained introns are most prevalent, found in approximately half of the alternatively spliced genes (Campbell et al. 2006; Nagasaki et al. 2005; Ner-Gaon et al. 2004; Wang and Brendel 2006). A peculiar finding is that alternate acceptor sites are found at twice the frequency of alternate donor sites in both plants (Campbell et al. 2006; Nagasaki et al. 2005; Wang and Brendel 2006). The distribution of alternatively spliced genes according to variation type is provided by Fig. 5.

Localizing Splicing Variations Within Gene Structures

The location of splicing variations within gene structures is very similar between Arabidopsis and rice, suggesting additional features that are conserved across plants (Campbell et al. 2006; Nagasaki et al. 2005). Using gene structures supported by FL-cDNAs as a reference, Campbell et al. mapped identified splicing variations to the 5′ 'UTR, CDS, or 3′ UTR regions of each gene (Campbell et al. 2006). Excluding the differences due exclusively to alternative initiation or termination of transcription, splicing variations were predominantly localized to the CDS, directly impacting the protein sequence and hence structure. For example, 62%–81% of all alternate acceptor or donor sites and 87%–90% of all retained introns were localized to the CDS. Flanking the CDS, there is a strong bias for alternate acceptors and donors to be found in the 5′ UTR in contrast to the 3′ UTR. Although retained introns in Arabidopsis are found almost threefold more prevalent in the 5′ UTR as compared to the 3′ UTR, the retained introns in rice are nearly equivalently distributed between the two, suggesting a difference in the regulation of intron retention between rice and Arabidopsis, or monocots and dicots. Future studies should elucidate the significance of this finding.

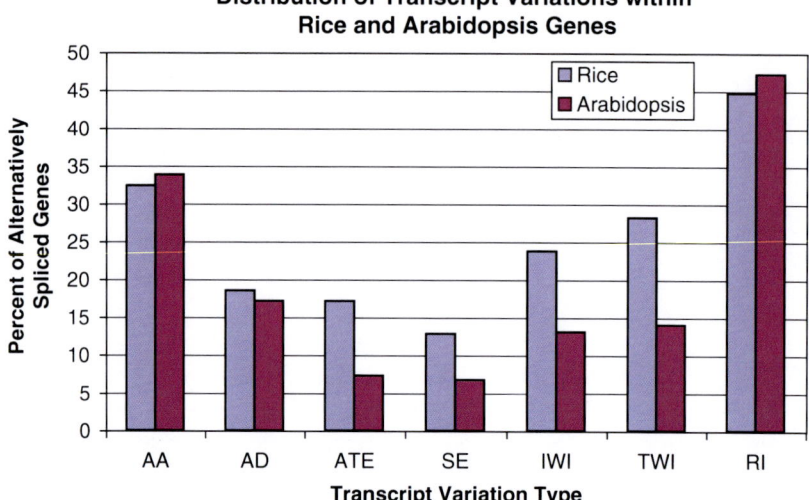

Fig. 5 Distribution of transcript variations within rice and Arabidopsis genes. The distribution of the abundance of rice and Arabidopsis genes corresponding to the various types of splicing variations are shown. *AA*, alternate acceptor splice site; *AD*, alternate donor splice site; *ATE*, alternate terminal exon; *SE*, skipped exon; *IWI*, initiate within intron; *TWI*, terminate within intron; *RI*, retained intron. Data points taken from (Campbell et al. 2006)

Comparisons Between Alternative Splicing in Plants and Animals

After the human genome was sequenced and initially examined, most everyone was surprised to find that we harbor only around 20,000 to 30,000 protein-coding genes (Lander et al. 2001; Venter et al. 2001), a number similar to other arguably less complex organisms including the plant *Arabidopsis thaliana*, with ~25,000 genes (Arabidopsis Genome Initiative 2000). Given that an increased number of genes did not seem to account for increased organism complexity, AS was thought to be a primary mechanism for generating increased transcriptomic and proteomic diversity in humans as compared to other organisms (Modrek and Lee 2002). However, claims of AS rates correlated with organism complexity have been controversial, given that the results of measurements of AS across organisms are influenced by levels of transcript coverage, which vary widely across organisms surveyed (Brett et al. 2002; Kim et al. 2004). For example, on January 1st, 2002, shortly after the draft human genome sequence was completed, there were 3.7 million human ESTs available. At that same time, there were only ~237,000 Drosophila ESTs and ~60,000 Arabidopsis ESTs. By restricting an analysis of AS to similar-sized data sets across a broad set of organisms, Brett et al. (2002) showed that the level of AS in humans was actually quite similar to other animals; however, plants (represented by Arabidopsis) exhibited a decreased level of AS.

The sets of expressed transcripts available for both rice and Arabidopsis have increased rather drastically from that time, now with over a million rice EST sequences available and many thousands of FL-cDNAs (see Table 1). The latest estimates of the abundance of AS in plants continue to indicate that alternative splicing as a whole is less prevalent than in animals and that AS is more prevalent in vertebrates than invertebrates (Kim et al. 2007a). Figure 6 illustrates the observed abundance of alternatively spliced genes (transcript clusters) as a function of the number of transcript sequences analyzed with PASA, comparing rice and Arabidopsis to human and mouse. It is clear that at this time we are approaching saturation of transcript coverage and that the prevalence of AS in Arabidopsis and rice is significantly less than that observed in vertebrates. Time will tell whether the lesser prevalence of AS is observed across other plants, since many other plant genomes are being sequenced and relevant EST resources are available or expected to accumulate.

The most striking difference between AS observed between plants and animals is the types of variations encountered (Kim et al. 2007a; Nagasaki et al. 2005). Nagasaki et al. (2005) used the bit-encoding scheme described above for identifying ASTI units across six eukaryotes, including human, mouse, fly (Drosophila), worm (*C. elegans*), and both plants (Arabidopsis and rice). From this study, it was apparent that AS generates a wide array of variations, and the types and abundances of

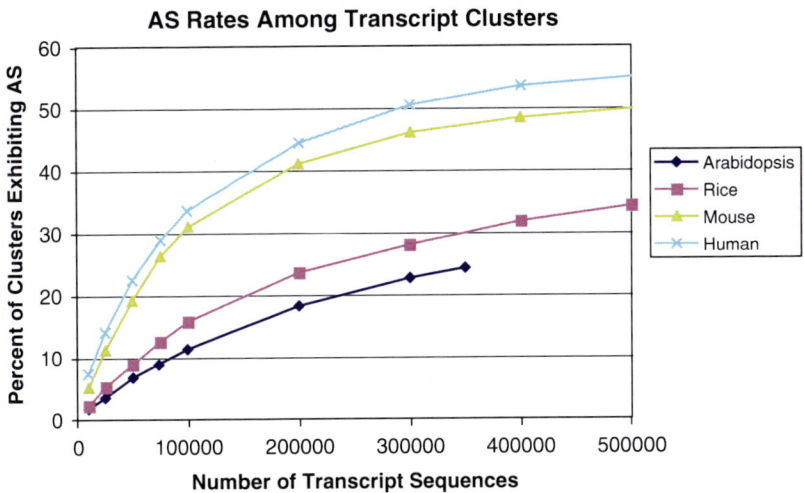

Fig. 6 Rates of AS observed as a function of transcript sequence abundance. The rate of AS observed is impacted by the number of transcript sequences available but does appear to level off toward saturation at higher transcript abundance. Human and mouse exhibit increased rates of AS compared to the plants Arabidopsis and rice, as measured with PASA and random sets of spliced nearly perfect transcript alignments (transcripts lacking evidence of intron splicing were excluded). Calculations for Arabidopsis were limited by sequence availability to a maximum of ~350,000 spliced transcript alignments

these variations appear to be similar across evolutionary lineages. Whereas in human and mouse the predominant splicing variation encountered is the skipped exon (aka. cassette exon), accounting for 25%–29% of all variations classified, this variation is much rarer in plants at 3%–9%. In contrast, the predominant splicing variation found in rice and Arabidopsis is the retained intron, a finding consistently reported in related studies (Campbell et al. 2006; Kim et al. 2007a; Nagasaki et al. 2005; Ner-Gaon et al. 2004; Wang and Brendel 2006). The complexity of AS might be assessed by the percentage of ASTI units that defy the simpler classification scheme. Of all six eukaryotes examined, rice and Arabidopsis exhibit the smallest percentage of ASTI units classified as "others," suggesting that AS contributes to genetic complexity to a lesser degree in plants.

The skipped exons, the predominant variation in human and mice, are often divisible by three, encoding an integral number of codons, and hence can readily contribute to protein diversity by their inclusion or exclusion within isoforms (Magen and Ast 2005; Nagasaki et al. 2005). Evolutionarily conserved alternatively spliced exons between human and mouse are further enriched for reading frame preservation (Sorek et al. 2004). The finding that the skipped exons are more prevalent in vertebrates supports the notion of AS contributing to protein diversity and hence increased organism complexity (Graveley 2001; Maniatis and Tasic 2002). Retained introns, as most prevalent in plants, tend to alter the reading frame and introduce a premature termination codon and hence, if translated, would generate truncated protein, intuitively inconsistent with the notion of promoting protein diversity and more suited to a role in posttranscriptional gene regulation (Campbell et al. 2006).

One possible explanation for the prevalence of retained introns in plant ESTs is that they derive from non-fully processed mRNAs and present as artifacts from cDNA library construction. Ner-Gaon et al. (2004) demonstrate otherwise by showing that retained introns are found in mRNAs recruited to ribosomes, and hence have been exported from the nucleus as processed transcripts. Also, current evidence indicates that truncated proteins expressed as isoforms containing retained introns play important biological functions, and so these retained introns appear to be functionally regulated and are not mistakes or artifacts as one might presume (Reddy 2007). Current evidence seems to indicate that the retained intron is a type of splicing variation exploited by plants to increase protein diversity, in addition to their anticipated role to posttranscriptionally regulate gene expression. Albeit there are key differences between AS in plants and animals, several features are quite similar, including the locations of splicing variations within gene structures, and their tendency to impact the coding regions (Nagasaki et al. 2005). By examining the distances between alternate acceptor and alternate donor sites, it was shown that alternate splice sites are most frequently separated by short distances, with a maximal occurrence of only three nucleotides between alternate acceptors and only four nucleotides between alternate donors (Campbell et al. 2006). An abundance of alternate splice sites separated by small distances is also observed in mammals (Chern et al. 2006; Sugnet et al. 2004; Zavolan et al. 2003). The three nucleotide difference between alternate acceptor splice sites has been termed the NAGNAG motif, a tandem splice junction where N can be any of the four nucleotides (Hiller et al. 2004). Similar to mammals, this site in plants is most often CAGCAG (Akerman and Mandel-Gutfreund 2006; Campbell et al. 2006; Hiller et al. 2004). Isoforms varying at this motif in a coding region will differ by only a single amino acid. Although a few of these have been shown to be functionally relevant, many of these variants may derive from noise generated by the splicing machinery (Hiller et al. 2006).

An analysis of alternatively spliced human transcripts indicates that as many as a third of all alternative isoforms may be derived from regulated unproductive splicing (RUST) and be targeted by the nonsense-mediated decay (NMD) pathway (Lewis et al. 2003) (see the chapter by D. A. Belostotsky, this volume). Such transcripts are identified as those containing premature termination codons, defined as stop codons that are found more than 50 nucleotides upstream from the final splice junction (Nagy and Maquat 1998). By applying this criterion to Arabidopsis and rice transcript isoforms, Wang and Brendel (2006) show that at least a third of plant transcript variants are likewise targeted to NMD. To determine the effects of splicing variations on protein primary structure, Campbell et al. (2006) compared protein sequences encoded by full-length isoforms exhibiting only a single type of splicing variation. In the vast majority of cases examined, the choice of an alternate acceptor or alternate donor splice site leads to a frameshift and truncates the protein by a quarter of its length. It remains to be shown that these transcripts are indeed NMD targets.

Web Resources for Alternative Splicing in Plants

The wide array of discoveries made possible by high-throughput analyses of AS has spurred the development of public databases to relay these findings to the scientific community and to assist researchers in follow-up studies. Most of these databases are geared toward animal species, with a strong emphasis on human and mouse. This is understandable given our interest in human genetic diversity and the role of AS in our biology, development, evolution, and disease. A small sample of these databases includes ECgene (Lee et al. 2007), ASAP II (Kim et al. 2007b), ASD (Stamm et al. 2006), and Hollywood (Holste et al. 2006) (a list of dozens of related databases can be found at http://hollywood.mit.edu/db/index.html).

Similar resources targeted to plants are less numerous. One of the earliest web-based resources, provided by The Institute for Genomic Research (TIGR), http://www.tigr.org/tdb/e2k1/ath1/altsplicing/splicing variations.shtml, includes a sampling of alternatively spliced Arabidopsis genes discovered during an early application of PASA as part of the Arabidopsis genome reannotation effort (Haas et al. 2003; Haas et al. 2005; Wortman et al. 2003). A site of particular interest is ASIP (Alternative Splicing in Plants, http://www.plantgdb.org/ASIP/index.php), a component of PlantGDB (Plant Genome Database) (Dong et al. 2005). ASIP houses the results from the genomewide study on AS in rice and Arabidopsis as described by Wang and Brendel (Wang and Brendel 2006), available in a queryable and visual form. Another site of interest is ASTRA (Alternative Splicing and TRanscription Archives, http://alterna.cbrc.jp/), which describes the ASTI patterns mined by Nagasaki et al. (2005) from eukaryotes including animals and plants (Nagasaki et al. 2005), as described above.

References

Akerman M, Mandel-Gutfreund Y (2006) Alternative splicing regulation at tandem 3′ splice sites. Nucleic Acids Res 34:23–31

Altschul SF, Madden TL, Schaffer AA, Zhang J, Zhang Z, Miller W, Lipman DJ (1997) Gapped BLAST and PSI-BLAST: a new generation of protein database search programs. Nucleic Acids Res 25:3389–3402

Arabidopsis Genome Initiative (2000) Analysis of the genome sequence of the flowering plant *Arabidopsis thaliana*. Nature 408:796–815

Baker BS (1989) Sex in flies: the splice of life. Nature 340:521–524

Black DL (2003) Mechanisms of alternative pre-messenger RNA splicing. Annu Rev Biochem 72:291–336

Blencowe BJ (2006) Alternative splicing: new insights from global analyses. Cell 126:37–47

Brett D, Pospisil H, Valcarcel J, Reich J, Bork P (2002) Alternative splicing and genome complexity. Nat Genet 30:29–30

Campbell MA, Haas BJ, Hamilton JP, Mount SM, Buell CR (2006) Comprehensive analysis of alternative splicing in rice and comparative analyses with Arabidopsis. BMC Genomics 7:327

Chern TM, van Nimwegen E, Kai C, Kawai J, Carninci P, Hayashizaki Y, Zavolan M (2006) A simple physical model predicts small exon length variations. PLoS Genet 2:e45

Dong Q, Lawrence CJ, Schlueter SD, Wilkerson MD, Kurtz S, Lushbough C, Brendel V (2005) Comparative plant genomics resources at PlantGDB. Plant Physiol 139:610–618

Eyras E, Caccamo M, Curwen V, Clamp M (2004) ESTGenes: alternative splicing from ESTs in Ensembl. Genome Res 14:976–987

Florea L, Hartzell G, Zhang Z, Rubin GM, Miller W (1998) A computer program for aligning a cDNA sequence with a genomic DNA sequence. Genome Res 8:967–974

Goff SA, Ricke D, Lan TH, Presting G, Wang R, Dunn M, Glazebrook J, Sessions A, Oeller P, Varma H, Hadley D, Hutchison D, Martin C, Katagiri F, Lange BM, Moughamer T, Xia Y, Budworth P, Zhong J, Miguel T, Paszkowski U, Zhang S, Colbert M, Sun WL, Chen L, Cooper B, Park S, Wood TC, Mao L, Quail P, Wing R, Dean R, Yu Y, Zharkikh A, Shen R, Sahasrabudhe S, Thomas A, Cannings R, Gutin A, Pruss D, Reid J, Tavtigian S, Mitchell J, Eldredge G, Scholl T, Miller RM, Bhatnagar S, Adey N, Rubano T, Tusneem N, Robinson R, Feldhaus J, Macalma T, Oliphant A, Briggs S (2002) A draft sequence of the rice genome (*Oryza sativa* L. ssp. japonica). Science 296:92–100

Graveley BR (2001) Alternative splicing: increasing diversity in the proteomic world. Trends Genet 17:100–107

Haas BJ, Volfovsky N, Town CD, Troukhan M, Alexandrov N, Feldmann KA, Flavell RB, White O, Salzberg SL (2002) Full-length messenger RNA sequences greatly improve genome annotation. Genome Biol 3:RESEARCH0029

Haas BJ, Delcher AL, Mount SM, Wortman JR, Smith RK Jr, Hannick LI, Maiti R, Ronning CM, Rusch DB, Town CD, Salzberg SL, White O (2003) Improving the Arabidopsis genome annotation using maximal transcript alignment assemblies. Nucleic Acids Res 31:5654–5666

Haas BJ, Wortman JR, Ronning CM, Hannick LI, Smith RK Jr, Maiti R, Chan AP, Yu C, Farzad M, Wu D, White O, Town CD (2005) Complete reannotation of the Arabidopsis genome: methods, tools, protocols and the final release. BMC Biol 3:7

Heber S, Alekseyev M, Sze SH, Tang H, Pevzner PA (2002) Splicing graphs and EST assembly problem. Bioinformatics 18 [Suppl 1]:S181–S188

Hiller M, Huse K, Szafranski K, Jahn N, Hampe J, Schreiber S, Backofen R, Platzer M (2004) Widespread occurrence of alternative splicing at NAGNAG acceptors contributes to proteome plasticity. Nat Genet 36:1255–1257

Hiller M, Szafranski K, Backofen R, Platzer M (2006) Alternative splicing at NAGNAG acceptors: simply noise or noise and more? PLoS Genet 2: e207; author reply e208

Holste D, Huo G, Tung V, Burge CB (2006) HOLLYWOOD: a comparative relational database of alternative splicing. Nucleic Acids Res 34:D56–62

Huang X, Adams MD, Zhou H, Kerlavage AR (1997) A tool for analyzing and annotating genomic sequences. Genomics 46:37–45

Iida K, Seki M, Sakurai T, Satou M, Akiyama K, Toyoda T, Konagaya A, Shinozaki K (2004) Genome-wide analysis of alternative pre-mRNA splicing in *Arabidopsis thaliana* based on full-length cDNA sequences. Nucleic Acids Res 32:5096–5103

International Human Genome Sequencing Consortium (2004) Finishing the euchromatic sequence of the human genome. Nature 431:931–945

International Rice Genome Sequencing Project (2005) The map-based sequence of the rice genome. Nature 436:793–800

Johnson JM, Castle J, Garrett-Engele P, Kan Z, Loerch PM, Armour CD, Santos R, Schadt EE, Stoughton R, Shoemaker DD (2003) Genome-wide survey of human alternative pre-mRNA splicing with exon junction microarrays. Science 302:2141–2144

Kan Z, Rouchka EC, Gish WR, States DJ (2001) Gene structure prediction and alternative splicing analysis using genomically aligned ESTs. Genome Res 11:889–900

Kent WJ (2002) BLAT—the BLAST-like alignment tool. Genome Res 12:656–664

Kim E, Magen A, Ast G (2007a) Different levels of alternative splicing among eukaryotes. Nucleic Acids Res 35:125–131

Kim H, Klein R, Majewski J, Ott J (2004) Estimating rates of alternative splicing in mammals and invertebrates. Nat Genet 36:915–916; author reply 916–917

Kim N, Alekseyenko AV, Roy M, Lee C (2007b) The ASAP II database: analysis and comparative genomics of alternative splicing in 15 animal species. Nucleic Acids Res 35:D93–D98

Lander ES, Linton LM, Birren B, Nusbaum C, Zody MC, Baldwin J, Devon K, Dewar K, Doyle M, FitzHugh W, Funke R, Gage D, Harris K, Heaford A, Howland J, Kann L, Lehoczky J, LeVine R, McEwan P, McKernan K, Meldrim J, Mesirov JP, Miranda C, Morris W, Naylor J, Raymond C, Rosetti M, Santos R, Sheridan A, Sougnez C, Stange-Thomann N, Stojanovic N, Subramanian A, Wyman D, Rogers J, Sulston J, Ainscough R, Beck S, Bentley D, Burton J, Clee C, Carter N, Coulson A, Deadman R, Deloukas P, Dunham A, Dunham I, Durbin R, French L, Grafham D, Gregory S, Hubbard T, Humphray S, Hunt A, Jones M, Lloyd C, McMurray A, Matthews L, Mercer S, Milne S, Mullikin JC, Mungall A, Plumb R, Ross M, Shownkeen R, Sims S, Waterston RH, Wilson RK, Hillier LW, McPherson JD, Marra MA, Mardis ER, Fulton LA, Chinwalla AT, Pepin KH, Gish WR, Chissoe SL, Wendl MC, Delehaunty KD, Miner TL, Delehaunty A, Kramer JB, Cook LL, Fulton RS, Johnson DL, Minx PJ, Clifton SW, Hawkins T, Branscomb E, Predki P, Richardson P, Wenning S, Slezak T, Doggett N, Cheng JF, Olsen A, Lucas S, Elkin C, Uberbacher E, Frazier M, et al (2001) Initial sequencing and analysis of the human genome. Nature 409:860–921

Lareau LF, Green RE, Bhatnagar RS, Brenner SE (2004) The evolving roles of alternative splicing. Curr Opin Struct Biol 14:273–282

Lee C (2003) Generating consensus sequences from partial order multiple sequence alignment graphs. Bioinformatics 19:999–1008

Lee C, Wang Q (2005) Bioinformatics analysis of alternative splicing. Brief Bioinform 6:23–33

Lee Y, Lee Y, Kim B, Shin Y, Nam S, Kim P, Kim N, Chung WH, Kim J, Lee S (2007) ECgene: an alternative splicing database update. Nucleic Acids Res 35:D99–103

Leipzig J, Pevzner P, Heber S (2004) The Alternative Splicing Gallery (ASG): bridging the gap between genome and transcriptome. Nucleic Acids Res 32:3977–3983

Lewis BP, Green RE, Brenner SE (2003) Evidence for the widespread coupling of alternative splicing and nonsense-mediated mRNA decay in humans. Proc Natl Acad Sci USA 100:189–192

Licatalosi DD, Darnell RB (2006) Splicing regulation in neurologic disease. Neuron 52:93–101

Magen A, Ast G (2005) The importance of being divisible by three in alternative splicing. Nucleic Acids Res 33:5574–5582

Malde K, Coward E, Jonassen I (2005) A graph based algorithm for generating EST consensus sequences. Bioinformatics 21:1371–1375

Maniatis T, Tasic B (2002) Alternative pre-mRNA splicing and proteome expansion in metazoans. Nature 418:236–243

Modrek B, Lee C (2002) A genomic view of alternative splicing. Nat Genet 30:13–19

Modrek B, Resch A, Grasso C, Lee C (2001) Genome-wide detection of alternative splicing in expressed sequences of human genes. Nucleic Acids Res 29:2850–2859

Mott R (1997) EST_GENOME: a program to align spliced DNA sequences to unspliced genomic DNA. Comput Appl Biosci 13:477–478

Nagasaki H, Arita M, Nishizawa T, Suwa M, Gotoh O (2005) Species-specific variation of alternative splicing and transcriptional initiation in six eukaryotes. Gene 364:53–62

Nagasaki H, Arita M, Nishizawa T, Suwa M, Gotoh O (2006) Automated classification of alternative splicing and transcriptional initiation and construction of visual database of classified patterns. Bioinformatics 22:1211–1216

Nagy E, Maquat LE (1998) A rule for termination-codon position within intron-containing genes: when nonsense affects RNA abundance. Trends Biochem Sci 23:198–199

Ner-Gaon H, Halachmi R, Savaldi-Goldstein S, Rubin E, Ophir R, Fluhr R (2004) Intron retention is a major phenomenon in alternative splicing in Arabidopsis. Plant J 39:877–885

Novoyatleva T, Tang Y, Rafalska I, Stamm S (2006) Pre-mRNA missplicing as a cause of human disease. Prog Mol Subcell Biol 44:27–46

Okazaki Y, Furuno M, Kasukawa T, Adachi J, Bono H, Kondo S, Nikaido I, Osato N, Saito R, Suzuki H, Yamanaka I, Kiyosawa H, Yagi K, Tomaru Y, Hasegawa Y, Nogami A, Schonbach C, Gojobori T, Baldarelli R, Hill DP, Bult C, Hume DA, Quackenbush J, Schriml LM, Kanapin A, Matsuda H, Batalov S, Beisel KW, Blake JA, Bradt D, Brusic V, Chothia C, Corbani LE, Cousins S, Dalla E, Dragani TA, Fletcher CF, Forrest A, Frazer KS, Gaasterland T, Gariboldi M, Gissi C, Godzik A, Gough J, Grimmond S, Gustincich S, Hirokawa N, Jackson IJ, Jarvis ED, Kanai A, Kawaji H, Kawasawa Y, Kedzierski RM, King BL, Konagaya A, Kurochkin IV, Lee Y, Lenhard B, Lyons PA, Maglott DR, Maltais L, Marchionni L, McKenzie L, Miki H, Nagashima T, Numata K, Okido T, Pavan WJ, Pertea G, Pesole G, Petrovsky N, Pillai R, Pontius JU, Qi D, Ramachandran S, Ravasi T, Reed JC, Reed DJ, Reid J, Ring BZ, Ringwald M, Sandelin A, Schneider C, Semple CA, Setou M, Shimada K, Sultana R, Takenaka Y, Taylor MS, Teasdale RD, Tomita M, Verardo R, Wagner L, Wahlestedt C, Wang Y, Watanabe Y, Wells C, Wilming LG, Wynshaw-Boris A, Yanagisawa M, et al (2002) Analysis of the mouse transcriptome based on functional annotation of 60,770 full-length cDNAs. Nature 420:563–573

Ouyang S, Zhu W, Hamilton J, Lin H, Campbell M, Childs K, Thibaud-Nissen F, Malek RL, Lee Y, Zheng L, Orvis J, Haas B, Wortman J, Buell CR (2007) The TIGR Rice Genome Annotation Resource: improvements and new features. Nucleic Acids Res 35:D883–887

Reddy AS (2007) Alternative splicing of pre-messenger RNAs in plants in the genomic era. Annu Rev Plant Biol 58:267–294

Sorek R, Shamir R, Ast G (2004) How prevalent is functional alternative splicing in the human genome? Trends Genet 20:68–71

Stamm S, Ben-Ari S, Rafalska I, Tang Y, Zhang Z, Toiber D, Thanaraj TA, Soreq H (2005) Function of alternative splicing. Gene 344:1–20

Stamm S, Riethoven JJ, Le Texier V, Gopalakrishnan C, Kumanduri V, Tang Y, Barbosa-Morais NL, Thanaraj TA (2006) ASD: a bioinformatics resource on alternative splicing. Nucleic Acids Res 34:D46–D55

Stetefeld J, Ruegg MA (2005) Structural and functional diversity generated by alternative mRNA splicing. Trends Biochem Sci 30:515–521

Sugnet CW, Kent WJ, Ares M Jr, Haussler D (2004) Transcriptome and genome conservation of alternative splicing events in humans and mice. Pac Symp Biocomput 2004:66–77

Usuka J, Zhu W, Brendel V (2000) Optimal spliced alignment of homologous cDNA to a genomic DNA template. Bioinformatics 16:203–211

Venter JC, Adams MD, Myers EW, Li PW, Mural RJ, Sutton GG, Smith HO, Yandell M, Evans CA, Holt RA, Gocayne JD, Amanatides P, Ballew RM, Huson DH, Wortman JR, Zhang Q, Kodira CD, Zheng XH, Chen L, Skupski M, Subramanian G, Thomas PD, Zhang J, Gabor Miklos GL, Nelson C, Broder S, Clark AG, Nadeau J, McKusick VA, Zinder N, Levine AJ, Roberts RJ, Simon M, Slayman C, Hunkapiller M, Bolanos R, Delcher A, Dew I, Fasulo D, Flanigan M, Florea L, Halpern A, Hannenhalli S, Kravitz S, Levy S, Mobarry C, Reinert K, Remington K, Abu-Threideh J, Beasley E, Biddick K, Bonazzi V, Brandon R, Cargill M, Chandramouliswaran I, Charlab R, Chaturvedi K, Deng Z, Di Francesco V, Dunn P, Eilbeck K, Evangelista C, Gabrielian AE, Gan W, Ge W, Gong F, Gu Z, Guan P, Heiman TJ, Higgins ME, Ji RR, Ke Z, Ketchum KA, Lai Z, Lei Y, Li Z, Li J, Liang Y, Lin X, Lu F, Merkulov GV, Milshina N, Moore HM, Naik AK, Narayan VA, Neelam B, Nusskern D, Rusch DB, Salzberg S, Shao W, Shue B, Sun J, Wang Z, Wang A, Wang X, Wang J, Wei M, Wides R, Xiao C, Yan C, et al (2001) The sequence of the human genome. Science 291:1304–1351

Wang BB, Brendel V (2006) Genomewide comparative analysis of alternative splicing in plants. Proc Natl Acad Sci USA 103:7175–7180

Wortman JR, Haas BJ, Hannick LI, Smith RK Jr, Maiti R, Ronning CM, Chan AP, Yu C, Ayele M, Whitelaw CA, White OR, Town CD (2003) Annotation of the Arabidopsis genome. Plant Physiol 132:461–468

Wu TD, Watanabe CK (2005) GMAP: a genomic mapping and alignment program for mRNA and EST sequences. Bioinformatics 21:1859–1875

Xing Y, Resch A, Lee C (2004) The multiassembly problem: reconstructing multiple transcript isoforms from EST fragment mixtures. Genome Res 14:426–441

Xing Y, Yu T, Wu YN, Roy M, Kim J, Lee C (2006) An expectation-maximization algorithm for probabilistic reconstructions of full-length isoforms from splice graphs. Nucleic Acids Res 34:3150–3160

Yu J, Hu S, Wang J, Wong GK, Li S, Liu B, Deng Y, Dai L, Zhou Y, Zhang X, Cao M, Liu J, Sun J, Tang J, Chen Y, Huang X, Lin W, Ye C, Tong W, Cong L, Geng J, Han Y, Li L, Li W, Hu G, Huang X, Li W, Li J, Liu Z, Li L, Liu J, Qi Q, Liu J, Li L, Li T, Wang X, Lu H, Wu T, Zhu M, Ni P, Han H, Dong W, Ren X, Feng X, Cui P, Li X, Wang H, Xu X, Zhai W, Xu Z, Zhang J, He S, Zhang J, Xu J, Zhang K, Zheng X, Dong J, Zeng W, Tao L, Ye J, Tan J, Ren X, Chen X, He J, Liu D, Tian W, Tian C, Xia H, Bao Q, Li G, Gao H, Cao T, Wang J, Zhao W, Li P, Chen W, Wang X, Zhang Y, Hu J, Wang J, Liu S, Yang J, Zhang G, Xiong Y, Li Z, Mao L, Zhou C, Zhu Z, Chen R, Hao B, Zheng W, Chen S, Guo W, Li G, Liu S, Tao M, Wang J, Zhu L, Yuan L, Yang H (2002) A draft sequence of the rice genome (*Oryza sativa* L. ssp. indica). Science 296:79–92

Zavolan M, Kondo S, Schonbach C, Adachi J, Hume DA, Hayashizaki Y, Gaasterland T (2003) Impact of alternative initiation, splicing, and termination on the diversity of the mRNA transcripts encoded by the mouse transcriptome. Genome Res 13:1290–1300

Zipursky SL, Wojtowicz WM, Hattori D (2006) Got diversity? Wiring the fly brain with Dscam. Trends Biochem Sci 31:581–588

Splice Site Requirements and Switches in Plants

M.A. Schuler

Contents

Splice Site Types and Consensus Sequences 40
AU Content and Length Variations 43
Mutational Analyses of Splice Site Requirements 44
Natural and Chemically Induced Splice Site Mutations 48
Exonic Elements 53
Multi-intron Transcripts 54
Conclusions 54
References 55

Abstract Intron sequences in nuclear pre-mRNAs are excised with either the major U2 snRNA-dependent spliceosomal pathway or the minor U12 snRNA-dependent spliceosomal pathway that exist in most eukaryotic organisms. While the predominant dinucleotides bordering each of these types of introns and the catalytic mechanism used in their excision are conserved in plants and animals, several features aiding in the recognition of plant introns are distinct from those in animals and yeast. Along with their short length, high AU content and high variation in their 5′ and 3′ splice sites and branchpoint consensus sequences are the most prominent characteristics of plant introns. Detailed surveys of site-directed mutant introns tested in vivo and chemically induced and naturally mutant introns analyzed *in planta* emphasize the effects of changing individual nucleotides in these splice site consensus sequences and highlight a number of noncanonical dinucleotides that are functional in plant systems.

M.A. Schuler
Departments of Cell and Developmental Biology, Biochemistry and Plant Biology,
University of Illinois, Urbana, IL 61801, USA
e-mail: maryschu@uiuc.edu

Splice Site Types and Consensus Sequences

With genomic sequences available for representative vertebrates, insects, and plants, it is becoming clear that the structural elements bounding pre-mRNA introns maintain different levels of sequence conservation in different organisms. Among the three structural elements that are used in the process of excising introns from primary transcripts, the first element located at the 5′ splice site is a series of nine semiconserved nucleotides spanning the 5′ exon/intron border that was originally represented as **AG**/GUAAGU in mammals and plants and **AG**/GUAUGU in yeast (bold designates exonic nucleotides; slashes designate splice sites; underlines designate highly conserved nucleotides) (Rymond and Rosbash 1992; Moore et al. 1993; Kramer 1996; Simpson and Filipowicz 1996; Schuler 1998). The second element is a less conserved branchpoint sequence (YNYURAC in mammals, UACUAAC in yeast, CURAY in plants) that contain an absolutely conserved branchpoint (double underline) located 19–50 nt upstream from the 3′ intron/exon boundary; this adenosine is utilized for the formation of the lariat intermediate (Brown et al. 1996; Liu and Filipowicz 1996; Simpson et al. 1996). The final element that precedes the 3′ splice site was originally represented as YAG/**G** (in mammals), CAG/**G** (in yeast), and GCAG/**G** (in plants).

The subsequent discovery of other noncanonical splice sites, such as /AU...AC/, in some vertebrate and *Drosophila* transcripts (Jackson 1991; Hall and Padgett 1994) and two classes of splicesomal components in mammalian cells (Hall and Padgett 1996; Tarn and Steitz 1996a, b; Patel and Steitz 2003) necessitated closer analysis of the splice sites, and branchpoints of plant introns as well as those in other eukaryotes. In these more recent comparisons, the major U2-dependent class of introns was subdivided into a predominant group of /GU...AG/ introns, a smaller group of /GC...AG/ introns, and a few significantly rarer /AU...AC/ introns. The minor U12-dependent class of introns was subdivided into groups of /AU...AC/ and /GU...AG/ introns, with some extremely rare variations on these sequences occurring in vertebrates (Levine and Durbin 2001; Zhu and Brendel 2003; Sheth et al. 2006; W. Zhu, personal communication) (see the chapter by C. G. Simpson and J. W. S. Brown, this volume). Detailed analysis of the splice sites for these subtypes summarized in Table 1 from work by Sheth et al. (2006) has shown that /GU...AG/ and /GC...AG/ introns represent 99.20% and 0.80%, respectively, of the U2-type introns in *Arabidopsis thaliana* (mouse-ear cress), with no U2-type /AU...AC/ introns identified. In the remaining 191 U12-type introns in *Arabidopsis* (0.17% of the total intron pool), /GU...AG/, /AU...AC/, and other splice sites account for 84.8%, 0.14%, and 0.01% of these introns, respectively. More recent compilations by W. Zhu (http://rice.tigr.org/) have shown that /GU...AG/ and /GC...AG/ introns represent 97.80% and 1.98% of the total intron pool in *Oryza sativa* (cultivated rice). The remaining 208 U12-type introns in rice (0.17% of the total intron pool) have 68% /GU...AG/, 31% /AU...AG/, and 1% other U12-type junctions. Thus, as noted in compilations of human and mouse introns

Table 1 Splice site subtypes

	H. sapiens	D. melanogaster	C. elegans	A. thaliana	O. sativa
Number of genes	23,505	11,756	19,269	22,354	~25,000
U2-type GT-AG	183,678	40,637	93,699	111,351	116,828
U2-type GC-AG	1,602	185	351	903	2,365
U2-type AT-AC	15	3	0	0	3
Total U2-type	185,295	40,825	94,072	112,254	119,196
U12-type GT-AG	469	11	0*	162	142
U12-type AT-AC	169	6	0	27	64
Other U12-type[a]	33	1	0	2	2
Total U12-type	671	18	0	191	208
Total number	185,966	40,843	94,072	112,445	119,404
% U2-type GC-AG	0.865	0.453	0.373	0.804	1.98
% U2-type AT-AC	0.008	0.007	0	0	0.0025
% U12-type	0.361	0.044	0	0.170	0.174

Data on *H. sapiens*, *D. melanogaster*, *C. elegans*, and *A. thaliana* are from Sheth et al. (2006)
Data on *O. sativa* are from W. Zhu (JCVI)
* There are 22 U12-type-like GT-AG introns in *C. elegans*, which were not included in the counts of U12-type introns recorded in Sheth et al. (2006)
[a]Other U12-type introns sorted in the RT-RN and GC-AG categories

(Burge et al. 1998; Sheth et al. 2006), /GU...AG/ introns predominate in both the U2-type and U12-type groups.

Global compilations of the *Arabidopsis* sequences spanning the 5′ cleavage site in these intron subtypes (Sheth et al. 2006) have indicated that the U2-type /GU...AG/ introns have a general 5′ splice site consensus of (**A/C**)**AG**/<u>GU</u>A(A/U)(G/C/U)(U/A)$_5$ with the underlined nucleotides present in more than 75% of the introns analyzed and the bold nucleotides present in exons. Overlapping with the plant 5′ splice site consensus initially mentioned in this review, this new sequence extends the consensus at its 3′ end with adenosines and uridines that reflect the AU-rich nature of *Arabidopsis* and other plant introns (Goodall and Filipowicz 1989; Wang and Brendel 2006). It also indicates how few other positions in U2-type introns have prevalent nucleotides in global comparisons. Contrasting with U2-type /GU...AG/ introns, plant U2-type /GC...AG/ introns show a general 5′ splice site consensus of (**C/A**)**AG**/<u>GCAAG</u>U(U/A/C)$_2$(U/C/A)$_2$ with this underlining scheme indicating that, in the absence of a canonical uridine at +2 in the 5′ splice site, particular nucleotides become prevalent at other positions (−2, +3, +4, +5). As is the case in mammalian U12-type introns (Hall and Padgett 1994; Burge et al. 1998), U12-type introns in *Arabidopsis* have much stronger consensus sequences than U2-type introns surrounding their 5′ splice site and branchpoint (Zhu and Brendel 2003; Sheth et al. 2006). Following the same underlining scheme as above, U12-type /GU...AG/ introns have a general consensus of (**U/A/G**)(**C/U**)(**U/A/C**)/<u>GUAUCCUU</u>(U/C)(U/A) with nearly absolute conservation at positions +1 to +7 downstream from the 5′ splice site. U12-type /AU...AC/ introns have a consensus of (**A/C**)(**A/U/C**)

(A/G/C)/AUAUCCUU(U/A)(U/C) with absolute conservation at positions +1 to +8. Intron compilations in rice show that their U2-type /GU...AG/ introns have the same preferences at all positions except +5, where the preferred nucleotide order is (G/U/A) instead of (G/C/U) in *Arabidopsis* (Kitamura-Abe et al. 2004). Rice U2-type /GC...AG/ introns have the same preferences as *Arabidopsis* at all positions.

In *Arabidopsis*, the general consensus sequences spanning the 3′ cleavage sites in both U2-type /GU...AG/ and /GC...AG/ introns is virtually identical, corresponding to $(U/A)_9U(G/U/A)\underline{CAG}/(G/A)(U/A/G)_2$ (Sheth et al. 2006). The 3′ cleavage sites in U12-type /GU...AG/ introns have a pronounced absence of guanosines and (U/A/C) variations from −9 to −14 (relative to the 3′ cleavage site) corresponding to a consensus of $(A/U/G)(U/A)(U/A/C)(U/A)_2(U/C)\underline{AG}/A$ **(U/C)(A/G/C)** (Sheth et al. 2006) or, more simply, (A/G/U)ACAC/ as represented in earlier comparisons (Zhu and Brendel 2003). U12-type /AU...AC/ introns have a consensus sequence of $(U/A)(A/U/G)(A/C/U)(C/U/G)(U/C/A)(U/A/G)_2(U/A)(U/G)(U/A/G)(U/A)(C/U)\underline{AC}/(G/A/U)(U/C)(A/G/U)$, with guanosines being more prevalent from −5 to −13 than in the U12-type /GU...AG/ introns.

Embedded within the high AU content of plant introns, branchpoint consensus sequences have been difficult to define globally for U2-type introns. Site-directed mutagenesis of potential branchpoints in a small number of U2-type plant introns (with CUN<u>A</u>N as a weak consensus and the branchpoint adenosine double underlined) has indicated that flexibility exists in the position and range of sequences flanking the branchpoint (Simpson and Filipowicz 1996; Simpson et al. 1996). Insertion of a consensus branchpoint sequence in a poorly spliced intron is capable of activating 3′ splice sites positioned 19–30 nucleotides downstream from it, and, often, mutagenesis of one branchpoint sequence allows for use of another (Simpson et al. 1996). Mutations in the branchpoint sequence preceding a potato invertase mini-exon (9 nt), which has a more extended distance (50 nt) between its branchpoint and 3′ splice site, have demonstrated that branchpoint sequences matching the CU(A/G)<u>A</u>(C/U) consensus are functional to varying degrees, with CUA<u>A</u>U being the one that is most efficiently recognized (Simpson et al. 2002). Because of the higher degree of conservation in the branchpoint sequences of U12-type introns and their more predictable proximity to 3′ splice sites, the branchpoint consensus in plant U12-type introns has been identified as UCCUU(A/G)<u>A</u>(C/U) (Zhu and Brendel 2003), in close agreement with vertebrate and *Drosophila* U12-type branchpoint sequences situated 10–20 nucleotides upstream from their 3′ splice sites (Hall and Padgett 1994). However, in contrast to the relatively defined distances between vertebrate U12-type branchpoints and their downstream 3′ splice sites, the branchpoints found in *Arabidopsis* U12-type introns range from 11 to 39 nucleotides (Zhu and Brendel 2003). As mentioned below, expression of *Arabidopsis* U12-type intron mutants in tobacco protoplasts has indicated that an additional adenosine is required at the position preceding the normal branchpoint adenosine and that guanosine can serve as an alternate branchpoint nucleotide (Lewandowska et al. 2004).

AU Content and Length Variations

Even while these intron categories subdivide the large number of introns annotated in *Arabidopsis* and *Oryza* and provide us with splice site and branchpoint sequences to view as consensus motifs, it is important to note that these compilations do not differentiate between constitutive splice sites, which utilize ubiquitous splicing factors, and alternative splice sites, which may need development- and/or tissue-specific factors for their recognition. In both cases, visual inspection often indicates that individual introns have substantially fewer consensus nucleotides than these general consensus sequences imply.

Recognizing that some plant 5′ splice sites maintain only a few contiguous consensus nucleotides capable of base pairing with U1 snRNA and that many lack polypyrimidine tracts upstream from their 3′ splice sites, Goodall and Filipowicz (1989) first distinguished plant introns from their counterparts in vertebrates and yeast by noting that plant introns are relatively short (average 80–139 nt) with high AU compositions (typically 11%–19% richer in adenosines and uridines than their adjacent exon sequences). Using 171 dicot introns from multiple species, their original comparisons of GC-rich exon sequences with AU-rich intron sequences served to highlight the transitions in nucleotide composition that may be more relevant to the recognition of plant introns than any specific sequence motifs at their boundaries. Later comparisons of the AU transitions in 146 monocot and 280 dicot introns indicated that the proportional increases in AU composition are maintained in the introns of these two plant groups even though the GC content of monocot exons is higher than in dicot exons (Goodall and Filipowicz 1991). As a result, the absolute level of AU-richness in the collective group of dicot introns is significantly higher (74% vs. 55% in the exons) than in the collective group of monocot introns (56% vs. 44% in the exons).

Subsequent compilations in *Arabidopsis* have indicated that the vast majority of all introns (99.2%) have a median length of 101 nucleotides and the minority (0.8%) have lengths over 1 kilobase (Wang and Brendel 2006). Calculated as having an average intron length of 173 nucleotides and 67% AU content, these introns are distinguished from adjacent exons that average 172 nucleotides in length and 58% AU content (Wang and Brendel 2006). Similar compilations in *Oryza* have indicated that the majority of introns (89.6%) have a median length of 160 nucleotides and a significant minor fraction (10.4%) have lengths over 1 kilobase. Calculated as an average intron length of 433 nucleotides and 63% AU content, these are distinguished from rice exons that average 193 nucleotides in length and 51% AU content (Wang and Brendel 2006). These calculations do not take into account the fact that, as originally noted by Yu et al. (2002) in their annotations of the rice genome, there is a substantial average decrease in GC-richness (from 75% to 50%) across the first kilobase of rice coding sequences. Subsequent analyses of *Zea mays* (another monocot) and *Nicotiana tabacum* (another dicot) have indicated that these GC fluctuations are typical of monocot coding regions but not dicot coding regions (Wong et al. 2002) and represent a feature of monocot intron recognition not typically discussed. Comparisons of *Arabidopsis* introns differentiating the characteristics of internal exons from first and last exons have indicated that their average length (112 nt) is substantially shorter than the first exon, which includes the 5′ UTR (average length 274 nt), or the last exon, which includes the 3′ UTR (average length 402 nt) (Kitamura-Abe et al. 2004), suggesting that recognition of first and last introns may involve parameters different from those used for introns in GC-rich coding regions. As in vertebrate genes, introns occur in *Arabidopsis* genes at significantly higher frequency in the 5′ UTR sequences (17%) than in 3′ UTR sequences (3%) (Alexandrov et al. 2006; Hong et al. 2006).

Comparisons of these same types of parameters in the U12-type introns of *Arabidopsis* have shown no significant difference in their lengths or AU-richness compared to U2-type introns (Zhu and Brendel 2003).

Mutational Analyses of Splice Site Requirements

Much of the mutational analysis of plant splice sites and AU compositions was done on U2-type introns before the discovery of U12-type introns. These early studies were aimed at determining the sequences defining the 5′ and 3′ boundaries of constitutively spliced introns (meaning one set of splice sites with no alternatives) in plant nuclei and the flexibility of these sites with either of two systems: an *Agrobacterium* geminivirus vector system for tobacco leaf discs or an electroporation system for tobacco and maize protoplasts (Goodall and Filipowicz 1989, 1991; McCullough et al. 1991, 1993; Lou et al. 1993a, b; McCullough and Schuler 1993, 1997; Baynton et al. 1996; Gniadkowski et al. 1996; Simpson and Filipowicz 1996; Egoavil et al. 1997; Merritt et al. 1997). On the basis of extensive splice site selection analyses, it has become apparent that plant 5′ and 3′ splice sites are recognized by their position relative to the AU transition points at the 5′ and 3′ boundaries between GC-rich exon and AU-rich intron sequences as well as their degree of complementarity to U1 snRNA (5′ splice site) and U-richness followed by GC<u>AG</u>/ (3′ splice site) (Simpson and Filipowicz 1996; Schuler 1998). Significant support for this mode of intron recognition came from the fact that various AU-rich replacements between several competing 5′ splice sites activated the 5′ splice site at the AU transition point independent of the orientation of the AU-rich replacement (McCullough et al. 1993; McCullough and Schuler 1997). Additional support came from the fact that mutations in AU-rich intronic elements upstream of the 3′ splice site in maize *Adh1* intron 3 activated upstream cryptic 3′ splice sites that sat at the new AU transition points created in the mutant introns (Lou et al. 1993a, b; Merritt et al. 1997). Other studies on a series of inefficiently spliced GC-rich synthetic introns indicated that introduction of U-rich repeats, but not A-rich repeats, at any of four places rescued splicing; the presence of small numbers of adenosine or cytosine substitutions in these U-rich repeats did not substantially affect splicing (Gniadowski et al. 1996). Similarly, the addition of U-rich motifs to a poorly spliced version of maize *Adh1* intron 1 enhanced its splicing efficiency more than A-rich motifs (Luehrsen and Walbot 1994). Conversely, reductions in the internal uridine content of the maize *Bz2* intron reduced its splicing efficiency (Ko et al. 1998). More detailed reviews of the testing of splice site requirements in these types of single intron constructs are included in Simpson and Filipowicz (1996) and Schuler (1998).

The importance of AU-rich segments in intron recognition have been further highlighted by the discovery that the central AU-rich coding sequence of the green fluorescent protein (GFP) transcript is cryptically spliced in transgenic *Arabidopsis* with several adjacent /GU...AG/ dinucleotides (Haseloff et al. 1997). The two cryptic 5′ splice sites used have five or six contiguous nucleotides complementary to the 5′ end of U1 snRNA, making them stronger than the natural splice sites in any natural constitutively spliced introns. The one cryptic 3′ splice site paired with these alternate 5′ splice sites contains the most prevalent nucleotides preceding 3′ splice junctions in *Arabidopsis* (GCAG/), but it is not preceded by the U-rich tracts found in many plant introns. Functional expression of GFP transcripts in plants has necessitated mutation of these cryptic sites and incrementing the GC content of the region between these cryptic splice sites. Somewhat similarly, expression of intron-containing GUS transcripts in tobacco protoplasts and transgenic plants has indicated that low expression of this reporter is caused by activation of a cryptic 5′ splice site four nucleotides upstream from the intron insertion site (Ibrahim et al. 2001). Mutation of this cryptic 5′ splice site, which contains three contiguous nucleotides complementary to U1 snRNA (**GU**/**GU**AC**GU** with double underlines indicating nucleotides complementary to U1 snRNA) compared to the normal 5′ splice site (**A**C/**GU**UU**GU**), has indicated that, at this AU transition point, closely spaced sites are capable of competing with one another. Mutation of /GU at the cryptic 5′ splice site to /CU was able to enhance GUS expression severalfold in protoplasts and transgenic plants (Ibrahim et al. 2001).

Further support for the recognition of AU transitions in plant pre-mRNAs is also provided by the demonstration that the soybean β-conglycinin intron 4 can proceed through both steps in the splicing process using a wide array of noncanonical dinucleotides at the position of its normal 5′ splice site and a canonical AG/ dinucleotide at its normal 3′ splice site (a series that is designated as /NN...AG/) (Schuler 1998; C. E. Baynton and M. A. Schuler, unpublished). Among these alternate 5′ dinucleotides that are used at moderate efficiency (/GC...AG/, /CU...AG/ and /GA...AG/) and one used at reduced efficiency (/GG...AG/), only the /CU and /GG dinucleotides caused alternate 5″ splice sites within the intron to be activated. Conversion of the /GU dinucleotide to /UU allowed splicing at the canonical AG/ 3′ splice site to proceed at moderate efficiency by shifting the 5′ cleavage site to a cryptic /GU occurring one nucleotide forward of the normal 5′ splice site; as a result, canonical U2-type /GU...AG/ splice sites were used in this transcript rather than the /UU...AG/ at the intron boundaries. Contrasting with these, introduction of an /AU dinucleotide activated alternate 3′ splice sites (/AU...AC/, /AU...AA/) at −9 and −7 upstream of the normal 3′ splice site in pairings that are the same as those found in the rare U2-type /AU...AC/ and /AU...AA/ introns.

It also activated the same noncanonical 5′ splice site (/UU at +27) used in other 5′ splice site variants and paired it with the canonical AG/ at the normal 3′ position.

Splicing of this β-conglycinin intron also proceeds using a canonical /GU dinucleotide at its normal 5′ splice site and a noncanonical dinucleotide at its normal 3′ splice site (a series that is designated as /GU...NN/) but, in most cases, alternate 3′ splice sites were used at efficiencies significantly lower than the wild-type AG/ dinucleotide. The exceptions to this were the /GU...UG/ and /GU...GG/ introns that spliced at their normal 5′ and 3′ positions as well as at a cryptic canonical dinucleotide 4 nucleotides downstream from the normal 3′ position (/GU...AG/) and a cryptic noncanonical dinucleotide 2 nucleotides downstream from the normal 3′ splice site (/GU...GG/). With the canonical 5′ splice site, all other noncanonical 3′ splice sites tested (CG/, AU/, AC/, AA/) were nonfunctional and splicing occurred only at cryptic downstream sites that were both canonical (AG/ at +4, AG/ at +1) and noncanonical (GG/ at +2, UG/ at +1). In this series, conversion of the /GU at the normal 5′ splice site to /AU with noncanonical dinucleotides at the normal 3′ splice site utilized AC/ and AA/ dinucleotides when present at -1 (the normal 3′ position) and the cryptic AC/ and AA/ at -9 and -7 (as in the original /AU 5′ splice site mutant). And, finally, the cryptic /UU 5′ splice site at +27 was not spliced to any other noncanonical 3′ dinucleotides in the absence of a canonical 3′ AG/. The fact that the many alternate 3′ splice sites exist within 4 to 9 nucleotides of the normal 3′ splice site indicates that branchpoint sequences embedded within the intron and AU transitions across the 3′ intron/exon boundary limit the positions of 3′ splice site cleavage.

The ability of this β-conglycinin test intron to reveal functional sets of dinucleotides used by the major U2-dependent spliceosome in plants can be attributed to several features. These include its extremely strong 5′ splice site that has 7 of 9 nucleotides complementary to U1 snRNA and its adjacent exon nucleotides (3 nucleotides upstream of the 5′ splice site and 2 nucleotides downstream of the 3′ splice site) that are complementary to the conserved loop in U5 snRNA. With these nucleotides contributing to exon tethering interactions with U5 snRNA in yeast (Newman 1997) and to a limited extent in mammals (Segault et al. 1999), the presence of strong AU-transitions (from 71% AU in the intron to 55%–58% in the adjacent exons), three potential branchpoints and a strong U-rich tract ($U_5CU_3GU_2C$) upstream from the noncanonical AC/ at -9 and AA/ at -7 identify this intron as one containing many optimal recognition motifs. In addition, the very high splicing efficiency of this intron in the tobacco leaf disc system (McCullough and Schuler 1997) ensures that usage of even moderately functional dinucleotides can be detected.

The flexibilities of nucleotides at particular branchpoint positions in U2-dependent introns have most recently been evaluated in tobacco protoplasts using potato invertase transcripts containing two introns flanking a 9-nucleotide mini-exon (Simpson et al. 2000, 2002, 2004). In these constructs, which effectively measure the competition between splice sites in adjacent introns, branchpoints that match the CU(A/G)A(C/U) consensus allow for high-efficiency splicing (Simpson et al. 2002). Deviations from this sequence at the four positions surrounding the branchpoint adenosine reduce usage of the downstream 3′ splice site, sometimes quite significantly. More specifically, mutation of the branchpoint adenosine to a guanosine essentially abolished splicing and recognition of the mini-exon (0.4% of wild type), while mutation of the branchpoint adenosine to a pyrimidine still allowed for significant splicing (7%–19% of wild type), presumably due to activation of alternate branchpoint nucleotides as in mammalian systems (Ruskin et al. 1985; Hornig et al. 1986; Query et al. 1994). The fact that alternate branchpoint nucleotides can be used is consistent with earlier analyses suggesting that the efficiency of splicing synthetic plant introns having optimal splice sites is only moderately affected by mutations in the branchpoint adenosine or elimination of all adenosines within 67 nucleotides upstream of the 3′ splice site (Goodall and Filipowicz 1989, 1991). However, the fact that a variety of naturally occurring monocot and dicot introns (e.g., wheat amylase, pea legumin) splice at reduced efficiency or not at all in tobacco protoplasts when their branchpoint adenosine is replaced with another nucleotide (Simpson et al. 1996) indicates that an adenosine is optimal.

Only a limited number of studies have actually identified the branchpoint used in the splicing process. Primer extension mappings in the *Arabidopsis* rubisco activase *rca* 5′ splice site mutant containing a /GU to /AU change have indicated that lariat intermediates form at a UUGAU sequence 32 nucleotides upstream from the AG/ at −1 in mutant seedlings as well as tobacco protoplasts (Liu and Filipowicz 1996). Mappings in a strong synthetic intron containing this same 5′ splice mutation identified lariat intermediates as forming at a CTAAC positioned 31 nucleotides from the AG/ at −1. Interestingly, this /AU 5′ splice site mutant spliced to the canonical AG/ at −1 and at lower efficiency to the noncanonical AU/ at −4 (Liu and Filipowicz 1996). Thus, in contrast to the splicing of mammalian introns in HeLa cell systems, recognition of the AG/ at the normal 3′ position is not necessarily disrupted by a G-to-A mutation at +1 when other intron consensus elements are optimal.

A variety of studies have demonstrated the importance of U-rich sequences upstream of the AG/ 3′ splice site. In the potato mini-exon construct, modification of its long U_{11} tract has indicated that it acts like the polypyrimidine tracts present in mammalian introns. In this unusual set of adjacent introns, selection of the alternative exon depends on the presence of this uridine tract within 14 nucleotides of the branchpoint (Simpson et al. 2002). Hybrid constructs have indicated that this U_{11} tract has different functionalities depending on the presence or absence of an upstream branchpoint. In its presence, U-to-C changes (maintaining pyrimidine context) allow for recognition of the normal AG/ at the downstream 3′ splice site (Simpson et al. 2004) and U-to-A changes (maintaining AU-richness) prevent recognition of the downstream 3′ splice site and activate skipping of the mini-exon (Simpson et al. 2004). In the absence of a branchpoint, U-to-A changes allowed for exon skipping at high frequency because, embedded in AU-rich sequences, the downstream AG/ at the 3′ splice site is not recognized. In contrast, U-to-C changes enhanced recognition of the downstream intron and decreased exon skipping by maintaining pyrimidine-richness and reducing AU-richness of the upstream 3′ splice site.

In single intron constructs, uridine-richness also appears to be important in the region immediately preceding the 3' splice site with some gradation in the importance of uridines versus other nucleotides. In the region just prior to the normal 3' splice site in maize *Adh1* intron 3, U-to-A and U-to-C replacements reduce its functionality and shift splicing to upstream cryptic sites within this unusually long intron (Lou et al. 1993a). In regions further upstream from the normal 3' splice site (−66 to −50), U-to-A replacements maintaining AU-richness still allow for efficient recognition of the normal 3' splice site and U-to-C and U-to-G replacements reduce its recognition (Merritt et al. 1997). In pea *rbcS3A* intron 1, U-to-A replacements of all nine uridines from −4 to −17 activate two cryptic 3' splice sites at significant distances (+62, +95) downstream from the normal 3' splice site (Baynton et al. 1996). Replacements between the normal 3' splice site and first cryptic 3' splice site have indicated that cryptic downstream sites are capable of competing with the normal 3' splice site at the AU transition only when 10–14 uridines precede the cryptic 3' splice site. The uridine composition of this region is especially important when nonconsensus nucleotides occur at −3 and −4 or when the 3' splice site is not located at a strong AU transition point. The fact that insertion of U-rich sequences immediately upstream from the normal 3' splice site in a poorly spliced version of maize *Adh1* intron 1 enhances its recognition (Luehrsen and Walbot 1994) provides further evidence that uridine tracts contribute to 3' splice site choice.

Compared to these many studies, the branchpoint and 3' splice site requirements of U12-dependent introns have been evaluated only in three *Arabidopsis* transcripts [CBP20 (20-kDa cap binding protein), GSH2 (glutathione synthetase), LD (LUMINIDEPENDENS protein)] expressed in tobacco protoplasts. All of these plant U12-type introns required an adenosine at the purine position immediately upstream of the branchpoint in the UCCUU(A/G)$\underline{A}$U(C/U) consensus sequence (Lewandowska et al. 2004). With few conserved nucleotides preceding the (C/U)AG/ or (C/U)AC/ sequences downstream from this branchpoint consensus, the /GU...AG/ splice sites of LD intron 10 utilize splicing enhancer elements positioned in the upstream exon to increase its splicing efficiency (Lewandowska et al. 2004). Also, the /AU...AC/ splice sites of CBP20 intron 4 and GSH2 intron 6, which differ in having A and G as their branchpoint nucleotides, depend on the AU-richness of their introns as well as exon-bridging interactions between the U12-dependent intron and adjacent U2-dependent introns (Kmieciak et al. 2002; Lewandowska et al. 2004).

Natural and Chemically Induced Splice Site Mutations

Naturally occurring and chemically induced mutations affecting splicing have been identified in a wide range of genes. In most of these mutants, the effects of splice site changes are more complex than in the single intron test constructs described above because of the possibility of exon skipping in multi-intron transcripts derived from full-length genes. The most recent review summarizing splice site changes in *Arabidopsis* mutants (Brown 1996) lays out the variety of alternative splicing options that can occur in multi-intron transcripts having defective splice sites.

As in the site-directed conversion of the first guanosine in pea *rbcS3A* intron 1, where multiple cryptic 5′ splice sites are activated (McCullough et al. 1993), a G-to-A change at +1 in the *rcd1-1* mutant of the *Arabidopsis RCD1* locus and the *irx4* mutant of the *CCR1* locus activate alternate /GU dinucleotides in the vicinity of the 5′ splice site and promote intron retention (Jones et al. 2001; Ahlfors et al. 2004). As summarized in Table 2, in other mutants such as the *cop1-2* mutant of the *Arabidopsis COP1* locus, this G-to-A change results in exon skipping and low level activation of an alternate canonical 5′ splice site (Simpson et al. 1998). In contrast, the *dwf5-3* mutant of the *Arabidopsis DWARF5* locus accumulates partially processed transcripts because it cannot recognize the defective intron (Choe et al. 2000). As noted above, the *rca* mutant of *Arabidopsis RCA* locus recognizes the mutant /AU 5′ splice site at +1 but blocks splicing after formation of the lariat-exon intermediate (Orozco et al. 1993; Liu and Filipowicz 1996). In the maize *bt2-7503* mutant of the *BRITTLE-2* locus, a G-to-A change at +1 activates a cryptic /GC dinucleotide upstream of the normal 5′ cleavage site of intron 3 and skipping of the entire preceding exon (exon 3) (Lal et al. 1999b). In addition, activation of a cryptic /GA at a significant distance (23 nt) upstream of the normal 5′ cleavage site results in the formation of a lariat-exon intermediate detectable by primer extension analysis but not by the RT-PCR analysis typically used in characterization of plant splicing products.

In contrast to these 5′ splice site changes, a G-to-U change at +1 in the rice Wx^b allele of the *waxy* locus (replacing the /GU at the 5′ splice site with /UU) allows for low-level excision of /UU...AG/ at the normal 5′ and 3′ cleavage sites and higher-frequency excision of /GU...AG/ introns at a cryptic 5′ splice site 1 nucleotide upstream from the normal 5′ cleavage site and another further upstream (Bligh et al. 1998; Cai et al. 1998; Isshiki et al. 1998). Larkin and Park (1999) have shown that usage of these sites is affected by temperature, with the noncanonical /UU at +1 being more efficiently recognized at 25°–30°C than at 18°C and the upstream canonical /GU at −93 being predominant at the higher temperature possibly because the mutation at +1 affects base pairing potential with U1 and/or U6 snRNAs.

At other positions near the 5′ splice site, a G-to-A change at +5 in the *det1-1* and *det1-3* mutants of the *Arabidopsis DET1* locus blocks splicing of the defective intron and partially processed transcripts accumulate (Pepper et al. 1994). Also, a G-to-A change at −1 of intron 8 (the position preceding the 5′ splice site) in the *spy-1* mutant of the *Arabidopsis SPINDLY* locus causes skipping of the preceding exon (exon 8) because the 5′ splice site is not correctly recognized (Jacobsen et al. 1996). Importantly, this same exon skipping event occurs in the *spy-2* mutant containing a G-to-A change at the last nucleotide of the preceding intron (intron 7), which converts the canonical AG/ 3′ splice site to AA/.

In the *Arabidopsis ap3-1* mutant of the *APETALA3* locus, a G-to-U mutation at −2 of the 5′ splice site of intron 5 disrupts recognition of the suboptimal splice site and results in skipping of the downstream exon at 28°C but not at

Table 2 Splice site mutations

Gene	Mutant	Intron	Mutation	Effect	Reference
Arabidopsis					
5′ splice site mutations					
APETALA3	*ap3-1*	5	**AG/GU** to **UG/GU** transition	Exons 4 and 6 are directly joined together	Jack et al. (1992); Sablowski and Meyerowitz (1998)
CCR1	*irx4*	2	**AG/GU** to **AG/AU** transition	Intron 2 retained (with in-frame stop codon in intron 2) Possible activation of upstream cryptic splice sites	Jones et al. (2001)
COP1	*cop1-2*	6	**AG/GU** to **AG/AU** transition	Exons 5 and 7 are directly joined together	McNellis et al. (1994); Simpson et al. (1998)
DWARF5	*dwf5-3*	12	**AG/GU** to **AG/AU** transition	Intron 12 retained (with in-frame stop codon in intron 12)	Choe et al. (2000)
RUBISCO ACTIVASE	*rca*	3	**AG/GU** to **AG/AU** transition	Accumulation of lariat-exon intermediate	Liu and Filipowicz (1996); Orozco et al. (1993)
SPINDLY	*spy-1*	8	**AG/GU** to **AA/GU** transition	Exons 7 and 9 are directly joined together	Jacobsen et al. (1996)
3′ splice site mutations					
AGAMOUS	*ag-4*	5	AG/ to AA/ transition	Exons 5 and 7 are directly joined together Activation of cryptic 3′ splice site in downstream exon 6	Sieburth et al. (1995)
APETALA2	*ap2-6*	6	AG/ to AA/ transition	Activation of cryptic splice site 2 nt downstream	Wakem and Kohalmi (2003)
CHS	*tt4(2YY6)*	1	AG/ to AA/ transition	Normal splicing reduced Activation of cryptic splice site 1 nt downstream	Burbulis et al. (1996)
COP1	*cop1-1*	5	CGCAG/ to CACAG/ transition	Exons 5 and 7 are directly joined together Activation of cryptic 5′ splice site in downstream intron	McNellis et al. (1994); Simpson et al. (1998)
COP1	*cop1-11*	12	AG/ to AA/ transition	Intron 12 retained Activation of cryptic splice site 1 nt downstream	McNellis et al. (1994)
COP1	*cop1-6*	4	AG/ to GG/ transition	Intron 4 retained	McNellis et al. (1994)

Splice Site Requirements and Switches in Plants

Gene	Allele	Position	Mutation	Effect	Reference
COP1	cop1–8	10	AG/ to AA/ transition	Activation of cryptic splice sites 16 nt downstream and 15 and 37 nt upstream	Simpson et al. (1998)
DWARF5	dwf5–2	8	AG/ to AA/ transition	Exon 11 is retained Activation of cryptic splice sites 25 and 62 nt downstream	Choe et al. (2000)
HY8	hy8–2	4	AG/ to AA/ transition	Unspliced	Dehesh et al. (1993)
SPINDLY	spy–2	7	AG/ to AA/ transition	Exons 7 and 9 are directly joined together	Jacobsen et al. (1996)
Branchpoint mutants					
DET3	det3–1	1	CUAA̱U to CAAA̱U transition	Possible activation of branchpoint sequence 10 nt upstream Transcript reduced twofold	Schumacher et al. (1999)
APETALA3	ap3–11	4	GUUGGA to GUUGA̱A transition at position –33	Activation of branchpoint sequence at –33	Yi and Jack (1998)
Other plants					
5' splice site mutations					
BRITTLE–2	bt2-7503	3	UG/GU to UG/AU transition	Intron 3 retained Activation of cryptic splice site 23 nt upstream results in accumulation of lariat intermediate Activation of cryptic splice site 9 nt upstream	Lal et al. (1999b)
Waxy Wx[b]		1	AG/GU to AG/UU transition	At higher temperatures (25°–32°C), nonconsensus 3' splice site and cryptic splice site 1 nt upstream are activated At lower temperatures (18°C), cryptic splice site 93 nt upstream is activated	Ayres et al. (1997); Bligh et al. (1998); Isshiki et al. (1998); Larkin and Park (1999); Wang et al. (1995)
3' splice site mutations					
Sh2 (shrunken 2)	sh2-i	2	GA/ to AA/ transition	Normal splicing Exons 2 and 4 are directly joined together	Lal et al. (1999a)
Sh2 (shrunken 2)	sh2–7460	12	GA/ to AA/ transition	Activation of cryptic splice site 22 nt downstream	Lal et al. (1999a)
PHYA fri		1	AG/ to UG/ transition	Exons 1 and 6 are directly joined together Intron 1 retained, some normal splicing Activation of cryptic sites 6, 21, and 99 nt downstream and 81 and 53 nt upstream	Lazarova et al. (1998)

16°C (Sablowski and Meyerowitz 1998). The temperature sensitivity of this particular mutant, which is opposite to that of the rice *Wx^b* allele, provides evidence of an interaction at position −2 of the 5′ splice site that is needed for proper splice site recognition. Mammalian and yeast studies of the basepairing interactions occurring in this region (Zhuang and Weiner 1986; Siliciano and Guthrie 1988; Newman and Norman 1992; Wyatt et al. 1992) suggest that the uridine substitution in the *ap3-1* mutant would disrupt interactions with both U1 and U5 snRNAs. Similar to the exon spanning effects of the *spy-1* and *spy-2* mutants mentioned above, this splicing defect is suppressed by a G-to-A change in the preceding intron (intron 4) that introduces an alternate branchpoint and enhances recognition of the preceding 3′ splice site and the intervening exon (Yi and Jack 1998).

By far, the most predominant change at the 3′ splice site disrupting intron splicing is a G-to-A change at the last position of the intron (−1) that converts the canonical AG/ splice site to AA/. In the presence of the U2-dependent 5′ splice sites found in naturally occurring introns, this mutation prevents recognition of the affected intron in the *hy8-2* mutant of the *Arabidopsis HY8* locus (Dehesh et al. 1993) and the *cop1-8* mutant of the *Arabidopsis COP1* locus (Simpson et al. 1998) and causes them to accumulate precursor transcripts without any apparent activation of adjacent AG/ dinucleotides. In other situations, downstream AG/ 3′ splice sites at significant distances are activated. In the *clpC2* mutant of the *ClpC2* locus, these occur at +9 and +13 (Park and Rodermel 2004) and, in the *dwf5-2* mutant of the *DWARF5* locus, these occur at +25 and +62 (Choe et al. 2000). In some, such as the *Arabidopsis cop1-11* mutant and the *Arabidopsis tt4* mutant of the *CHS* locus, these are just 1 nucleotide downstream (+1) of the normal 3′ cleavage site due to the presence of a fortuitous guanosine immediately downstream from the normal 3′ cleavage site (McNellis et al. 1994; Burbulis et al. 1996). Also, in the *ap2-6* mutant of the *APETALA2* locus, these are just 2 nucleotides downstream of the normal 3′ cleavage site (Wakem and Kohalmi 2003). These many examples documenting the activation of downstream canonical AG/ dinucleotides provide solid support for the notion that plant 3′ splice sites are selected based on scanning for suitable AG/ dinucleotides downstream from a branchpoint sequence (Smith et al. 1989, 1993; Simpson et al. 1996). However, additional factors clearly contribute to recognition of the proper 3′ cleavage site since, in some cases such as the *spy-2* mutant of the *Arabidopsis SPINDLY* locus (Jacobsen et al. 1996), the downstream exon is skipped probably because no alternate consensus 3′ splice sites are available. And, in the *ag-4* mutant of the *Arabidopsis AGAMOUS* locus, the downstream exon (exon 6) is partially included because of a cryptic 3′ splice site just 6 nucleotides upstream from the 5′ splice site of intron 6 or it is entirely skipped (Sieburth et al. 1995). As summarized in Brown (1996), many other *Arabidopsis* mutants carry this G-to-A change at −1 but, in most of these, the molecular effects of this replacement have not been identified.

In other plants, the G-to-A replacement at the last position of the intron, as in the maize *sh2-i* mutant of the *Sh2* (*shrunken 2*) locus, allows the AA/ site in intron 2 to be used at 10% efficiency while the remaining transcripts skip exon 3 because this noncanonical site is not efficiently recognized (Lal et al. 1999a). And, in the *sh2-7460* mutant of this same locus, this replacement at the 3' splice site of intron 12 blocks its recognition and activates a canonical AG/ dinucleotide in the downstream exon (Lal et al. 1999a).

Mutations at other positions near the 3' splice site can also dramatically change splice site usage. In the *cop1-6* mutant of the *Arabidopsis COP1* locus, an A-to-G mutation at −2 of intron 4 converting AG/ to GG/ activates cryptic canonical AG/ splice sites at significant distances upstream (−15, −37) and downstream (+16) of the normal 3' splice site position (McNellis et al. 1994). In the *fri* mutants of the tomato *PHYA* locus, an A-to-U mutation at this same position of intron 1 converting AG/ to UG/ leads to intron retention and activation of cryptic canonical 3' splice sites upstream (−53, −81) and downstream (+6, +21, +99) of the normal 3' cleavage site as well as the rare usage of the noncanonical UG/ at the normal 3' cleavage site and, occasionally, exon skipping (Lazarova et al. 1998). In the *cop1-1* mutant, a G-to-A mutation at −4 of intron 5 converting the optimal CGCAG/U 3' splice site to CACAG/U primarily causes skipping of downstream exon 6 and secondarily activates a cryptic 5' splice site in intron 6 that is 46 nucleotides downstream of the normal 5' splice site (Simpson et al. 1998). In support of plants sometimes using exon definition in the process of intron recognition, this is the same set of splicing events occurring when the 5' splice site of intron 6 is converted from /GU to /AU (the *cop1-2* mutant discussed above).

Exonic Elements

In addition to the splice site and intron characteristics mentioned above, exonic sequences can influence splice site recognition in plant cells. Competitions between two equal-strength 5' splice sites positioned at the normal AU transition boundary and in the downstream intron have indicated that, in the U2-type soybean β-conglycinin intron 4, the introduction of guanosines creating purine-rich sequences similar to those found in many mammalian exon splicing enhancers promotes recognition of the downstream 5' splice site normally hidden within the intron (McCullough and Schuler 1997). Likewise, site-directed replacements between two competing 5' splice sites in the U2-type *rbcS3A* intron 1 selectively activate a perfect, but normally silent, 5' splice site embedded in the intron when AG-rich enhancer-like elements as short as 9 nucleotides are placed upstream of it (Egoavil et al. 1997). Additional mutations have indicated that specificity exists in the purine-rich motifs acting as plant splicing enhancers since simple AG dinucleotide repeats are not capable of activating the normally silent 5' splice site. Similarly, deletion analyses of the exon sequences flanking the U12-type *Arabidopsis* LD intron have indicated that sequences between 3 and 38 nucleotides upstream from the 5' splice site contribute to its recognition (Lewandowska et al. 2004); specific elements participating in enhanced recognition of the downstream 5' splice site have not yet been identified.

Multi-intron Transcripts

Clearly, many of the splicing defects in naturally occurring multi-intron transcripts recapitulate the splicing defects of site-directed mutations seen in single intron test transcripts. In the few instances where multi-intron test transcripts have been analyzed, transcripts containing soybean β-conglycinin introns 3 and 4 showed splicing of the downstream intron at several sets of noncanonical 5′ splice sites and some exon skipping when the normal /GU 5′ splice site of the downstream intron (intron 4) was mutated to /AU and a functional upstream intron (intron 3) was present (McCullough et al. 1996). While the overall splicing efficiency of this mutant intron 4 is reduced relative to wild-type intron 4, the noncanonical 5′ and 3′ splice sites used (/AU...AC/, /AU...AC/, /UU...AG/) are the same as those used in the β-conglycinin single intron constructs containing this same mutation (Schuler 1998; C. E. Baynton and M. A. Schuler, unpublished). Contrary to the mode of exon definition used in vertebrate transcripts and some plant transcripts, splicing at these noncanonical sites is not dramatically affected in mutants containing uridine or cytosine substitutions in the 3′ splice site of the preceding intron (intron 3) that enhances its recognition and prevents exon skipping (McCullough et al. 1996). The relative autonomy of these intron splicing events contrasts with the behavior of the *spy-1* and *spy-2* mutants and the *ap3-1* and *ap3-11* mutants that are affected greatly by splice signals in the adjacent introns: *ap3-11* is an intragenic suppressor of *ap3-1* affecting the previous branchpoint (Yi and Jack 1998), and *spy-2* is an intragenic suppressor affecting the previous 3′ splice site (Jacobsen et al. 1996). Whether intron definition that seems to operate in these multi-intron test transcripts represents a major or minor mode affecting intron recognition is not yet clear since so few test transcripts other than these and the potato mini-exon transcript have been evaluated. Low frequency of exon skipping in many mutants of naturally occurring multi-intron plant transcripts suggests that intron definition may be prevalent and that the sequences within and surrounding individual introns are the most significant factors affecting recognition of U2-type plant introns. It is likely that the mode of recognition is much different for U12-type introns since, in the two cases analyzed, the presence of adjacent U2-type introns significantly increases the splicing efficiency of U12-type introns (Kmieciak et al. 2002; Lewandowska et al. 2004).

Conclusions

Usage of a wide variety of noncanonical dinucleotides at the 5′ and 3′ splice sites of plant introns is evident in a number of studies on site-directed and chemically induced mutants. As summarized in Schuler (1998), several of these alternative first and last nucleotides (/GU...AG/, /AU...AC/, /AU...AA/) are the same as those used in U2-type introns in mammals and yeast and U12-type introns in mammals (Parker and Siliciano 1993; Ruis et al. 1994; Scadden and Smith 1995). Others such as /UU...AG/

function only at low efficiency in yeast and plants (Chanfreau et al. 1994; Luukkonen and Seraphin 1997), while /CU...AG/ functions only in plants (Schuler 1998). Mutational analyses of the options for these sites indicates that, even with non-Watson–Crick base pairing allowing for usage of some dinucleotide pairings and not others, adjacent splice site sequences, intronic AU compositions, and exonic elements define the actual cleavage sites recognized in the plant splicing process. Within plant transcripts that often contain multiple constitutively spliced U2-type introns, single nucleotide changes in their splice sites have the potential to redirect splicing to alternate cryptic 5' and 3' sites containing the predominant U2-type dinucleotides and, to a lesser extent, to activate exon skipping. Understanding the role of plant-specific factors in discrimination of these sites should enable researchers to improve splicing of inefficiently spliced introns and enhance accumulation of proteins that are rate-limiting in biochemical pathways.

Acknowledgements The author thanks Dr. Wei Zhu for providing information on the U12-type introns in rice and Ms. Amy Dunlap for scientific input on splice site mutants.

References

Ahlfors R, Lang S, Overmyer K, Jaspers P, Brosche M, Tauriainen A, Kollist H, Tuominen H, Belles-Boix E, Piippo M, Inze D, Palva ET, Kangasjarvi J (2004) Arabidopsis RADICAL-INDUCED CELL DEATH1 belongs to the WWE protein-protein interaction domain protein family and modulates abscisic acid, ethylene, and methyl jasmonate responses. Plant Cell 16:1925–1937

Alexandrov NN, Troukhan ME, Brover VV, Tatarinova T, Flavell RB, Feldmann KA (2006) Features of *Arabidopsis* genes and genome discovered using full-length cDNAs. Plant Mol Biol 60:69–85

Ayres NM, McClung AM, Larkin PD, Bligh HFJ, Jones CA, Park WD (1997) Microsatellites and a single-nucleotide polymorphism differentiate apparent amylose classes in an extended pedigree of US rice germ plasm. *TAG Theor Appl Genet* 94:773–781

Baynton CE, Potthoff SJ, McCullough AJ, Schuler MA (1996) U-rich tracts enhance 3'' splice site recognition in plant nuclei. Plant J 10:703–711

Bligh HF, Larkin PD, Roach PS, Jones CA, Fu H, Park WD (1998) Use of alternate splice sites in granule-bound starch synthase mRNA from low-amylose rice varieties. Plant Mol Biol 38:407–415

Brown JWS (1996) *Arabidopsis* intron mutations and pre-mRNA splicing. Plant J 10:771–780

Burbulis IE, Iacobucci M, Shirley BW (1996) A null mutation in the first enzyme of flavonoid biosynthesis does not affect male fertility in Arabidopsis. Plant Cell 8:1013–1025

Burge CB, Padgett RA, Sharp PA (1998) Evolutionary fates and origins of U12-type introns. Mol Cell 2:773–785

Cai XL, Wang ZY, Xing YY, Zhang JL, Hong MM (1998) Aberrant splicing of intron 1 leads to the heterogeneous 5' UTR and decreased expression of waxy gene in rice cultivars of intermediate amylose content. Plant J 14:459–465

Chanfreau G, Legrain P, Dujon B, Jacquier A (1994) Interaction between the first and last nucleotides of pre-mRNA introns is a determinant of 3' splice site selection in *S. cerevisiae*. Nucleic Acids Res 22:1981–1987

Choe S, Tanaka A, Noguchi T, Fujioka S, Takatsuto S, Ross AS, Tax FE, Yoshida S, Feldmann KA (2000) Lesions in the sterol delta reductase gene of *Arabidopsis* cause dwarfism due to a block in brassinosteroid biosynthesis. Plant J 21:431–443

Dehesh K, Franci C, Parks BM, Seeley KA, Short TW, Tepperman JM, Quail PH (1993) Arabidopsis HY8 locus encodes phytochrome A. Plant Cell 5:1081–1088

Egoavil C, Marton HA, Baynton CE, McCullough AJ, Schuler MA (1997) Structural analysis of elements contributing to 5′ splice site selection in plant pre-mRNA transcripts. Plant J 12:971–980

Gniadkowski M, Hemmings-Mieszczak M, Klahre U, Liu H-X, Filipowicz W (1996) Characterization of intronic uridine-rich sequence elements acting as possible targets for nuclear proteins during pre-mRNA splicing in *Nicotiana plumbaginifolia*. Nucleic Acids Res 24:619–627

Goodall GJ, Filipowicz W (1989) The AU-rich sequences present in the introns of plant nuclear pre-mRNAs are required for splicing. Cell 58:473–483

Goodall GJ, Filipowicz W (1991) Different effects of intron nucleotide composition and secondary structure on pre-mRNA splicing in monocot and dicot plants. EMBO J 10:2635–2644

Hall SL, Padgett RA (1994) Conserved sequences in a class of rare eukaryotic nuclear introns with non-consensus splice sites. J Mol Biol 239:357–365

Hall SL, Padgett RA (1996) Requirement of U12 snRNA for in vivo splicing of a minor class of eukaryotic nuclear pre-mRNA introns. Science 271:1716–1718

Haseloff J, Siemering KR, Prasher DC, Hodge S (1997) Removal of a cryptic intron and subcellular localization of green fluorescent protein are required to mark transgenic *Arabidopsis* plants brightly. Proc Natl Acad Sci USA 94:2122–2127

Hong X, Scofield DG, Lynch M (2006) Intron size, abundance, and distribution within untranslated regions of genes. Mol Biol Evol 23:2392–2404

Hornig H, Aebi M, Weissmann, C (1986) Effect of mutations at the lariat branch acceptor site on beta-globin pre-mRNA splicing in vitro. Nature 324:589–591

Ibrahim AF, Watters JA, Clark GP, Thomas CJ, Brown JW, Simpson CG (2001) Expression of intron-containing GUS constructs is reduced due to activation of a cryptic 5′ splice site. Mol Genet Genomics 265:455–460

Isshiki M, Morino K, Nakajima M, Okagaki RJ, Wessler SR, Izawa T, Shimamoto K (1998) A naturally occurring functional allele of the rice waxy locus has a GT to TT mutation at the 5′ splice site of the first intron. Plant J:15:133–138

Jack T, Brockman, LL, Meyerowitz, EM (1992) The homeotic gene APETALA3 of *Arabidopsis thaliana* encodes a MADS box and is expressed in petals and stamens. Cell 68:683–697

Jackson IJ (1991) A reappraisal of non-consensus mRNA splice sites. Nucleic Acids Res 19:3795–3798

Jacobsen SE, Binkowski KA, Olszewski NE (1996) SPINDLY, a tetratricopeptide repeat protein involved in gibberellin signal transduction in *Arabidopsis*. Proc Natl Acad Sci USA 93:9292–9296

Jones L, Ennos AR, Turner SR (2001) Cloning and characterization of *irregular xylem4 (irx4)*: a severely lignin-deficient mutant of *Arabidopsis*. Plant J 26:205–216

Kitamura-Abe S, Itoh H, Washio T, Tsutsumi A, Tomita M (2004) Characterization of the splice sites in GT-AG and GC-AG introns in higher eukaryotes using full-length cDNAs. J Bioinform Comput Biol 2:309–331

Kmieciak M, Simpson CG, Lewandowska D, Brown JW, Jarmolowski A (2002) Cloning and characterization of two subunits of *Arabidopsis thaliana* nuclear cap-binding complex. Gene 283:171–183

Ko CH, Brendel V, Taylor RD, Walbot V (1998) U-richness is a defining feature of plant introns and may function as an intron recognition signal in maize. Plant Mol Biol 36:573–583

Kramer A (1996) The structure and function of proteins involved in mammalian pre-mRNA splicing. Annu Rev Biochem 65:367–409

Lal S, Choi J-H, Hannah LC. (1999a) The AG dinucleotide terminating introns is important but not always required for pre-mRNA splicing in the maize endosperm. Plant Physiol. 120:65–72

Lal S, Choi JH, Shaw JR, Hannah LC (1999b) A splice site mutant of maize activates cryptic splice sites, elicits intron inclusion and exon exclusion, and permits branch point elucidation. Plant Physiol 121:411–418

Larkin PD, Park WD (1999) Transcript accumulation and utilization of alternate and non-consensus splice sites in rice granule-bound starch synthase are temperature-sensitive and controlled by a single-nucleotide polymorphism. Plant Mol Biol 40:719–727

Lazarova GI, Kerckhoffs LH, Brandstadter J, Matsui M, Kendrick RE, Cordonnier-Pratt MM, Pratt LH (1998) Molecular analysis of *PHYA* in wild-type and phytochrome A-deficient mutants of tomato. Plant J 14:653–662

Levine A, Durbin R (2001) A computational scan for U12-dependent introns in the human genome sequence. Nucleic Acids Res 29:4006–4013

Lewandowska D, Simpson CG, Clark GP, Jennings NS, Barciszewska-Pacak M, Lin CF, Makalowski W, Brown JW, Jarmolowski A (2004) Determinants of plant U12-dependent intron splicing efficiency. Plant Cell 16:1340–1352

Liu H-X, Filipowicz W (1996) Mapping of branchpoint nucleotides in mutant pre-mRNAs expressed in plant cells. Plant J 9:381–389

Lou H, McCullough AJ, Schuler MA (1993a) 3′ splice site selection in dicot plant nuclei is position-dependent. Mol Cell Biol 13:4485–4493

Lou H, McCullough AJ, Schuler MA (1993b) Expression of maize *Adh1* intron mutants in tobacco nuclei. Plant J 3:393–403

Luehrsen KR, Walbot V (1994) Addition of A- and U-rich sequence increases the splicing efficiency of a deleted form of a maize intron. Plant Mol Biol 24:449–463

Luukkonen BGM, Seraphin B (1997) The role of branchpoint-3′ splice site spacing and interaction between intron terminal nucleotides in 3′ splice site selection in *Saccharomyces cerevisiae*. EMBO J 16:779–792

McCullough AJ, Lou H, Schuler MA (1991) In vivo analysis of plant pre-mRNA splicing using an autonomously replicating vector. Nucleic Acids Res 19:3001–3009

McCullough AJ, Schuler MA (1993) AU-rich intronic elements affect pre-mRNA 5′ splice site selection in *Drosophila melanogaster*. Mol Cell Biol 13:7689–7697

McCullough AJ, Baynton CE, Schuler MA (1996) Interactions across exons can influence splice site recognition in plant nuclei. Plant Cell 8:2295–2307

McCullough AJ, Schuler MA (1997) Intronic and exonic sequences modulate 5′ splice site selection in plant nuclei. Nucleic Acids Res 25:1071–1077

McNellis TW, von Arnim AG, Araki T, Komeda Y, Misera S, Deng XW (1994) Genetic and molecular analysis of an allelic series of cop1 mutants suggests functional roles for the multiple protein domains. Plant Cell 6:487–500

Merritt H, McCullough AJ, Schuler MA (1997) Internal AU-rich elements modulate activity of two competing 3′ splice sites in plant nuclei. Plant J 12:937–943

Moore MJ, Query CC, Sharp PA (1993) Splicing of precursors to messenger RNAs by the spliceosome. In: Gesteland R, Atkins J (eds) The RNA World. Cold Spring Harbor Laboratory Press, Cold Spring Harbor, NY, pp 303–357

Newman AJ (1997) The role of U5 snRNP in pre-mRNA splicing. EMBO J 16:5797–5800

Newman AJ, Norman C (1992) U5 snRNA interacts with exon sequences at 5′ and 3′ splice sites. Cell 68:743–754

Orozco BM, McClung CR, Werneke JM, Ogren WL (1993) Molecular basis of the ribulose-1,5-bisphosphate carboxylase/oxygenase activase mutation in *Arabidopsis thaliana* is a guanine-to-adenine transition at the 5′-splice junction of intron 3. Plant Physiol 102:227–232

Park S, Rodermel SR (2004) Mutations in ClpC2/Hsp100 suppress the requirement for FtsH in thylakoid membrane biogenesis. Proc Natl Acad Sci USA 101:12765–12770

Parker R, Siliciano PG (1993) Evidence for an essential non-Watson-Crick interaction between the first and last nucleotides of a nuclear pre-mRNA intron. Nature 361:660–662

Patel AA, Steitz JA (2003) Splicing double: insights from the second spliceosome. Nat Rev Mol Cell Biol 4:960–970

Pepper A, Delaney T, Washburn T, Poole D, Chory J (1994) *DET1*, a negative regulator of light-mediated development and gene expression in Arabidopsis, encodes a novel nuclear-localized protein. Cell 78:109–116

Query CC, Moore MJ, Sharp PA (1994) Branch nucleophile selection in pre-mRNA splicing: evidence for the bulged duplex model. Genes Dev 8:587–597

Ruis BL, Kivens WJ, Siliciano PG (1994) The interaction between the first and last intron nucleotides in the second step of pre-mRNA splicing is independent of other conserved intron nucleotides. Nucleic Acids Res 22:5190–5195

Ruskin B, Greene JM, Green MR (1985) Cryptic branch point activation allows accurate in vitro splicing of human β-globin intron mutants. Cell 41:833–844

Rymond BC, Rosbash M (1992) Yeast pre-mRNA splicing. In: Jones EW, Pringle JR, Broach JR (eds) The Molecular and Cellular Biology of the Yeast *Saccharomyces*: Gene Expression. Cold Spring Harbor Laboratory Press, pp 143–193

Sablowski RW, Meyerowitz EM (1998) Temperature-sensitive splicing in the floral homeotic mutant *apetala3-1*. Plant Cell 10:1453–1463

Scadden ADJ, Smith CWJ (1995) Interactions between the terminal bases of mammalian introns are retained in inosine-containing pre-mRNAs. EMBO J 14:3236–3246

Schuler MA (1998) Plant pre-mRNA splicing. In: Bailey-Serres J, Gallie Daniel RA (eds) Look Beyond Transcription: Mechanisms Determining mRNA Stability and Translation in Plants. American Society Plant Physiologists pp 1–19

Schumacher K, Vafeados D, McCarthy M, Sze H, Wilkins T, Chory J (1999) The *Arabidopsis* det3 mutant reveals a central role for the vacuolar H^+-ATPase in plant growth and development. Genes Dev 13:3259–3270

Segault V, Will CL, Polycarpou-Schwarz M, Mattaj IW, Branlant C, Luhrmann R (1999) Conserved loop I of U5 small nuclear RNA is dispensable for both catalytic steps of pre-mRNA splicing in HeLa nuclear extracts. Mol Cell Biol 19:2782–2790

Sheth N, Roca X, Hastings ML, Roeder T, Krainer AR, Sachidanandam R (2006) Comprehensive splice-site analysis using comparative genomics. Nucleic Acids Res 34:3955–3967

Sieburth LE, Running MP, Meyerowitz EM (1995) Genetic separation of third and fourth whorl functions of *AGAMOUS*. Plant Cell 7:1249–1258

Siliciano PG, Guthrie C (1988) 5′ splice site selection in yeast: genetic alterations in base pairing with U1 reveal additional requirements. Genes Dev 2:1258–1267

Simpson GG, Filipowicz W (1996) Splicing of precursors to messenger RNA in higher plants: mechanism, regulation and sub-nuclear organization of the spliceosomal machinery. Plant Mol Biol 32:1–41

Simpson CG, Clark G, Davidson D, Smith P, Brown JWS (1996) Mutation of putative branchpoint consensus sequences in plant introns reduces splicing efficiency. Plant J 9:369–380

Simpson CG, McQuade C, Lyon J, Brown JW (1998) Characterization of exon skipping mutants of the *COP1* gene from *Arabidopsis*. Plant J 15:125–131

Simpson CG, Hedley PE, Watters JA, Clark GP, McQuade C, Machray GC, Brown JW (2000) Requirements for mini-exon inclusion in potato invertase mRNAs provides evidence for exon-scanning interactions in plants. RNA 6:422–433

Simpson CG, Thow G, Clark GP, Jennings SN, Watters JA, Brown JW (2002) Mutational analysis of a plant branchpoint and polypyrimidine tract required for constitutive splicing of a mini-exon. RNA 8:47–56

Simpson CG. Jennings SN, Clark GP, Thow G, Brown JW (2004) Dual functionality of a plant U-rich intronic sequence element. Plant J 37:82–91

Smith CW, Chu TT, Nadal-Ginard B (1993) Scanning and competition between AGs are involved in 3′ splice site selection in mammalian introns. Mol Cell Biol 13:4939–4952

Smith CW, Porro EB, Patton JG, Nadal-Ginard B (1989) Scanning from an independently specified branch point defines the 3′ splice site of mammalian introns. Nature 342:243–247

Tarn W-Y, Steitz JA (1996a) A novel spliceosome containing U11, U12, and U5 snRNPs excises a minor class (AT-AC) intron in vitro. Cell 84:801–811

Tarn W-Y, Steitz JA (1996b) Highly diverged U4 and U6 small nuclear RNAs required for splicing rare AT-AC introns. Science 273:1824–1832

Wakem MP, Kohalmi SE (2003) Mutation in the *ap2-6* allele causes recognition of a cryptic splice site. J Exp Bot 54:2655–2660

Wang BB, Brendel V (2006) Genomewide comparative analysis of alternative splicing in plants. Proc Natl Acad Sci USA 103:7175–7180

Wang ZY, Zheng FQ, Shen GZ, Gao JP, Snustad DP, Li MG, Zhang JL, Hong MM (1995) The amylose content in rice endosperm is related to the post-transcriptional regulation of the waxy gene. Plant J 7:613–622

Wiebauer K, Herrero J-J, Filipowicz W (1988) Nuclear pre-mRNA processing in plants: Distinct modes of 3′-splice site selection in plants and animals. Mol Cell Biol 8:2042–2051

Wong GK, Wang J, Tao L, Tan J, Zhang J, Passey DA, Yu J (2002) Compositional gradients in *Gramineae* genes. Genome Res 12:851–856

Wyatt JR, Sontheimer EJ, Steitz JA (1992) Site-specific crosslinking of mammalian U5 snRNP to the 5′ splice site prior to the first step of premessenger RNA splicing. Genes Dev 6:2542–2553

Yi J, Jack T (1998) An intragenic suppressor of the Arabidopsis floral organ identity mutant apetala3-1 functions by suppressing defects in splicing. Plant Cell 10:1465–1477

Yu J, Hu S, Wang J et al (2002) A draft sequence of the rice genome (*Oryza sativa* L. ssp. *indica*). Science 296:79–92

Zhu W, Brendel V (2003) Identification, characterization and molecular phylogeny of U12-dependent introns in the *Arabidopsis thaliana* genome. Nucleic Acids Res 31:4561–4572

Zhuang Y, Weiner AM (1986) A compensatory base change in U1 snRNA suppresses a 5′ splice site mutation. Cell 46:827–835

U12-Dependent Intron Splicing in Plants

C.G. Simpson, J.W.S. Brown(✉)

Contents

Introduction 62
Properties and Splicing Signals of U12 Introns 62
   Conserved Splicing Signals of U12 Introns 62
   UA-Richness 65
U12-Type (Minor) Spliceosome 66
   U12-Type snRNAs 67
   Protein Components of the U12-Dependent Spliceosome in Animals 69
   Plant U12 Splicing Factors 69
   U12-Type Spliceosome Assembly 70
SR Proteins, Exon-Splicing Enhancers, and Intron-Exon
Definition in U12-Dependent Splicing 71
   SR Proteins and Exon-Splicing Enhancers 71
   Intron-Exon Definition and SR Proteins 72
   Alternative Splicing of U12 Introns 73
Role of U12 Introns in Gene Expression 76
U12 Intron Conservation and Conversion 77
Perspectives 78
References 79

Abstract U12-dependent (U12) introns have persisted in the genomes of plants since the ancestral divergence between plants and metazoans. These introns, which are rare, are found in a range of genes that include essential functions in DNA replication and RNA metabolism and are implicated in regulating the expression of their host genes. U12 introns are removed from pre-mRNAs by a U12 intron-specific spliceosome. Although this spliceosome shares many properties with the more abundant U2-dependent (U2) intron spliceosome, four of the five small nuclear RNAs (snRNAs) required for splicing are different and specific for the unique splic-

---

J.W.S. Brown
Genetics Programme, Scottish Crop Research Institute, Invergowrie, Dundee DD2 5DA, Scotland, UK
e-mail: john.brown@scri.ac.uk

A.S.N. Reddy, M. Golovkin (eds.), *Nuclear pre-mRNA Processing in Plants:*
Current Topics in Microbiology and Immunology 326.
© Springer-Verlag Berlin Heidelberg 2008

ing of U12 introns. Evidence in plants so far indicates that splicing signals of plant U12 introns and their splicing machinery are similar to U12 intron splicing in other eukaryotes. In addition to the high conservation of splicing signals, plant U12 introns also retain unique characteristic features of plant U2 introns, such as UA-richness, which suggests a requirement for plant-specific components for both the U2 and U12 splicing reaction. This chapter compares U12 and U2 splicing and reviews what is known about plant U12 introns and their possible role in gene expression.

Introduction

In higher eukaryotes there are two recognized types of nuclear pre-mRNA intervening sequence that must be removed to allow the correct assembly of mRNA exons to occur. The most common type is the U2-dependent canonical introns (U2 intron) (see the chapter by M. A. Schuler, this volume), and the second is the subject of this chapter, the rarer U12-dependent introns (U12 intron). U12 introns, identified over 15 years ago, contain splice sites and branchpoint sequences and are spliced by spliceosomes that differ from U2 splicing. They were originally called AT-AC introns, because the first such introns to be identified had the terminal dinucleotides 5'-AT and AC-3' (Jackson 1991). The first U12 introns to be described were from the human *P120* and chicken cartilage matrix protein genes and were identified on the basis of a lack of complementarity to the 5' end of the U1snRNA and the unexpected intron terminal sequences (Jackson 1991). U12-dependent introns are widespread in higher eukaryotes and have been identified in the genomes of plants, insects, and vertebrates but are absent from budding yeast, protists, and the nematode *Caenorhabditis elegans* (Alioto 2007; Sharp and Burge 1997; Tarn and Steitz 1997; Wu and Krainer 1997, 1999; Burge et al. 1998; Patel and Steiz, 2003). Initial computational searches for U12 introns within available genomic sequences identified 60 examples from plants, insects, and vertebrates (Burge et al. 1998). With the availability of completely sequenced genomes, 404 U12 introns were found in the human genome and, more recently, bioinformatic analysis of the sequenced genomes of human, mouse, *Drosophila melanogaster*, and *Arabidopsis thaliana* revealed that U12 introns are rare, making up only 0.35% of the total number of introns in human and mouse and 0.17% of the total numbers of introns in *Arabidopsis* (Sheth et al. 2006). They are found in any position within the coding region or UTRs and are usually in genes that contain multiple U2-dependent introns (Levine and Durbin 2001; Sheth et al. 2006).

Properties and Splicing Signals of U12 Introns

Conserved Splicing Signals of U12 Introns

Initial identification of U12 introns in a variety of eukaryotic genes revealed highly conserved 5' splice site and branchpoint sequences (Hall and Padgett 1994). The consensus sequences at the 5' splice site and branchpoint of U12 introns are

RTATCCTY and TTCCTTRAY, respectively, and are more highly conserved compared to the variable 5′ splice site and degenerate branchpoint of the U2-type introns (Hall and Padgett 1994; Sharp and Burge 1997; Burge et al. 1998; Patel and Steitz 2003). The underlined branchpoint purine and adenosine nucleotides identify branchpoint positions that have been mapped in vitro in human nuclear splicing extracts. The U12 branchpoint sequence (TTCCTTRAY) is related to that of U2 introns (YURAY), where the adenosine is invariant. However, some U12 introns contain a G at the analogous position, and vertebrate U12 intron branchpoints have been mapped to adenosines when present at either of the two purine positions, suggesting some flexibility in the selection of the branchpoint nucleotide (Tarn and Steitz 1996a; McConnell et al. 2002). In U12 introns, the 3′ splice site consensus sequence is YA(C/G) and is less informative than the 5′ splice site and branchpoint sequences. Although the first U12 introns had AT-AC terminal dinucleotides, it is now clear that they have either GT-AG or AT-AC as the most common terminal nucleotides. In a small number of cases, other non-canonical splice sites are observed where the terminal acceptor nucleotide is the most variable (AT-AN or GT-AT) while the highly conserved 5′ splice site and branchpoint sequences are maintained (Wu and Krainer 1999; Levine and Durbin 2000; Sheth et al., 2006). This suggests that the intron termini are not the defining feature of this class of introns, and a systematic mutational analysis of the terminal dinucleotides of U12 introns using the human nucleolar *P120* gene intron F showed that the penultimate nucleotides are more important for U12 splicing (Dietrich et al. 1997, 2005; Sharp and Burge 1997). In comparison to U2 introns, characteristic features of U12 introns are the lack of a polypyrimidine tract and the short distance between the branchpoint and 3′ splice site. This rarely exceeds 20 nt, with a clear preference for 12 nt in AT-AC introns and 12-16 nt in GT-AG U12 introns (Hall and Padgett 1994; Burge et al. 1998; Dietrich et al. 2001a, 2005; Zhu and Brendel 2003). Increasing the distance between the 3′ splice site and the branchpoint reduced U12 splicing efficiency and increased activation of cryptic splice sites. U12 introns, therefore, have a greater similarity to yeast U2 introns than to mammalian U2 introns. For example, positions from +2 to +4 of the 5′ splice site are identical to the yeast U2 intron, the branchpoint region is highly conserved, and the polypyrimidine tract is absent (Tarn and Steitz, 1997). Finally, the length of U12 introns ranges from less than 100 bases to more than 3,000 bases, which is significantly smaller than some human U2 intron sequences (Hall and Padgett 1994; Wu and Krainer 1997; Levine and Durbin 2001).

In plants, bioinformatic screens of the *Arabidopsis thaliana* genome have identified increasing numbers of plant U12 introns as different genome sequence releases and increasing numbers of ESTs and full-length cDNA sequences have become available (Burge et al. 1998; Zhu and Brendel 2003; Sheth et al. 2006; Alioto 2007). The consensus sequences for *Arabidopsis* U12 splicing signals are identical to those of human with GTATCCTT or: ATATCCTT at the 5′ splice site, TCCTT(A/G)AC at the branchpoint, and YAG: or YAC: at the 3′ splice site (colons denote the intron/exon border) (Fig. 1a). The latest catalogue of *Arabidopsis* U12 introns (http://genome.imim.es/cgi-bin/u12db/u12db.cgi) identified 186 GT-AG and 60 AT-AC *Arabidopsis* U12 introns (Alioto 2007). Previous analysis identified two introns

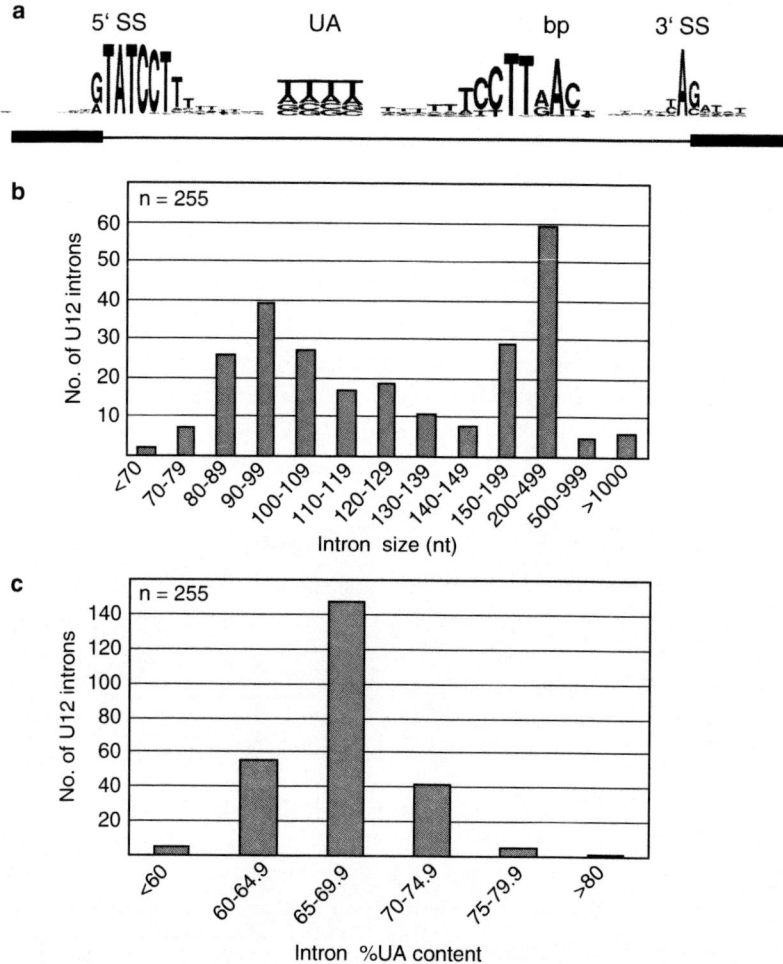

Fig. 1 Composition of *Arabidopsis* U12 introns. Sequence alignments of the three main splicing signals and composition of 255 of the 257 predicted *Arabidopsis* U12 introns are shown (Alioto, 2007; http://genome.imim.es/cgi-bin/u12db/u12db.cgi). (Two were removed from the analysis because one predicted 5′ splice site was part of the same intron's branchpoint sequence and the second was an intron that was only 30 nt in length.). **a** Sequence alignments were displayed as a sequence logo with WebLogo (http://weblogo.berkeley.edu/logo.cgi) (Crooks et al. 2004). The height of each nucleotide represents the relative frequency of the nucleotides at each position. *Boxes* represent the position of exon sequence, and the *line* indicates the intron sequence. **b** Distribution of *Arabidopsis* U12 intron sizes. Intron sizes ranged from 65 nt to 3731 nt. The majority of introns were found in the 80–140 size range. **c** Distribution of percentage UA content of *Arabidopsis* U12 introns. Most plant U12 introns fall within a narrow percentage range of 65%–70%

with non-canonical AT-AA and GT-AT (Zhu and Brendel 2003). As these screens used strict 5′ splice site and branchpoint consensus sequences, as described for their animal counterparts, they may be biased toward animal U12-like introns, and it is possible that U12 introns with more degenerate splice site signals exist in plants. Indeed, extending the distance between the branchpoint and the 3' splice site from a maximum distance of 21 nt to 39 nt identified a further 12 potential U12 introns, suggesting that, in plants, the local scanning process from the branchpoint to select its associated 3' splice site is less stringent (Zhu and Brendel 2003). The lengths of plant U12 introns, like human U12 introns, range from less than 100 nt to 3,731 nt found in a nodulin family protein (At1g11460), and over a third are found in the 80-110 nt size range (Fig. 1b).

The function of conserved splicing signals in U12 introns in plants has been investigated by analyzing splicing efficiency of mutated introns in vivo in tobacco protoplasts (Lewandowska et al. 2004). Substitution of the terminal dinucleotides at the 5' and 3' splice sites of three different *Arabidopsis* U12 introns revealed that GU-AG and AU-AC splice site combinations were interchangeable, but hybrid GU-AC and AU-AG combinations were unable to support plant U12 intron splicing (Lewandowska et al. 2004). The plant U12 branchpoint consensus sequence (TCCTTRAC) is the same as the human sequence, and branchpoint mutations were analyzed in three different U12 introns. In this study all three introns contained an adenosine at the −3 position (a purine in the consensus sequence, underlined). Site-specific mutation of this adenosine to a uridine resulted in severe inhibition of splicing irrespective of whether an adenosine was immediately downstream or not. The apparent inflexibility at this position in plants suggests either intolerance of a uridine at the −3 position or that, in contrast to human U12 introns, the adenosine at −3 is the preferred branchpoint (Lewandowska et al. 2004).

UA-Richness

A characteristic feature of plant U2 introns is a high UA content, which is a requirement for efficient splicing of introns particularly in dicots (Goodall and Filipowicz 1989; Simpson and Filipowicz 1996; Brown and Simpson 1998; Lorković et al. 2000a; Reddy 2001) (see the chapter by M. A. Schuler, this volume). The role of UA-richness in plant intron splicing is still poorly understood, but such sequences may minimize secondary structure in plant introns or they may be sites for binding of UA-rich binding proteins important in intron recognition and/or early events in spliceosome assembly (Goodall and Filipowicz 1989, 1991; Gniadkowski et al. 1996; Brown and Simpson 1998). Possible candidates include RNA-binding proteins that have been isolated from *Arabidopsis* called UBP1, RBP45, and RBP47, which have affinity for U-rich sequences (Lambermon et al. 2000; Lorković et al. 2000b).

Plant U12 introns also have high UA contents, and a screen of 255 predicted *Arabidopsis* U12 introns (http://genome.imim.es/cgi-bin/u12db/u12db.cgi) showed that, with the exception of 6 introns, plant U12 introns had a UA content of >60%, similar to U2 introns (Simpson, unpublished) (Fig. 1c). An effect of UA-rich sequences on U12 splicing was shown by sequentially adding U-rich sequence elements to the poorly spliced CBP20 U12 intron. Insertion of two heptamer U-rich elements, which increased UA content from 64% to 67.5%, were required to stimulate CBP20 U12 splicing from 10% to 35% in the transient protoplast system. Although this indicated that U-rich elements increase the splicing efficiency of U12 introns (Kmieciak et al. 2002; Lewandowska et al. 2004), the generally poor improvement in splicing suggests that other factors are also important for U12 splicing.

Overexpression of UBP1, an RNA binding protein with affinity for U-rich sequences, enhanced the splicing of UA-poor and inefficiently spliced U2 introns but did not affect splicing efficiency of U2 introns with normal UA content (Lambermon et al. 2000). When UBP1 was overexpressed along with inefficiently spliced U12 intron constructs in tobacco protoplasts, there was no stimulatory effect on U12 splicing efficiency (Lewandowska et al. 2004). Nevertheless, the relatively high UA-richness of plant U2 and U12 introns suggests that both classes have similar characteristics that could rely on specific U-rich binding proteins with affinity for U12 introns or that interact with U12 splicing factors to determine U12 intron splicing efficiency. In addition, the context of U12 introns in multi-intron transcripts may be important and U12 introns may depend on interactions with flanking U2 introns to promote efficient splice site definition and U12 removal (Lewandowska et al. 2004).

U12-Type (Minor) Spliceosome

The U12 spliceosome resembles the major U2-dependent class of spliceosome in its assembly and mode of action. It contains U11, U12, U4atac, and U6atac snRNP particles that are functionally equivalent to U1, U2, U4, and U6 of the U2-type spliceosome, with U5snRNP common to both (Tarn and Steitz 1996a, 1997; Burge et al. 1999, Wu and Krainer 1999; Patel and Steitz 2003). The U2-type spliceosome contains up to 300 proteins, including snRNP proteins and proteins that transiently associate with the spliceosome during its assembly (Rappsilber et al. 2002; Zhou et al 2002). Similarly, the U12-type spliceosome is a complex RNA-protein machine consisting of numerous proteins (Will et al. 1999, 2004).

Use of in vitro splicing extracts characterized splicing of U12 introns and the discovery that U12 introns are excised from the pre-mRNA by the U12 spliceosome. In addition, the U12 splicing reaction, like U2 splicing, proceeds via a two-step transesterification catalytic pathway that involves the generation of a reactive nucleophile at the labile 2′OH group on the ribose of the branchpoint nucleotide, leading to the formation of an intron-lariat intermediate before complete removal of the intron and ligation of the exons (Tarn and Steitz 1996a). Thus there are many similarities and parallels between the two types of spliceosome in terms of their composition and chemistry, but clear differences exist that distinguish their specificity for their cognate introns.

U12-Type snRNAs

U11, U12, U4atac, and U6atac snRNAs are unique to the U12-type spliceosome and are absolutely required for U12 splicing. These snRNAs do not show an extensive sequence homology with the major snRNAs, but their predicted secondary structures are similar to their U2-type counterparts. For example, U11, U12, and U4atac have Sm-binding sites located in single-stranded regions of the snRNA. In addition, the positions of essential snRNA sequences that take part in the intermolecular RNA interactions in the spliceosome are nearly identical in both groups of snRNAs (Yu et al. 1996; Tarn and Steitz 1996b). During the biogenesis of snRNPs, the major snRNAs are post-transcriptionally modified to include, for example, $2'$-O-ribose methylation and pseudouridylation at specific positions. The minor snRNAs in animals show a lower level of pseudouridylation, which may contribute to the differences between the two splicing systems. However, a key pseudouridine in U12 snRNA, at the position where U12 snRNA base pairs with the branchpoint sequence, is conserved at the analogous position in U2snRNA, suggesting a similar function in both spliceosomes (Massenet and Branlant 1999). Finally, the biogenesis pathways of the snRNP components for both U2 and U12 spliceosomes are assumed to be similar (Will and Lührmann 2001).

U11 and U12 snRNAs

Human U11 and U12 snRNAs were discovered before their role in U12-dependent splicing was identified (Montzka and Steitz 1988). It is now known from in vivo and in vitro studies that U12 snRNA base pairs with the highly conserved branchpoint region (Tarn and Steitz 1996a; Hall and Padgett 1996) and U11 snRNA interacts with the 5' splice site of U12 introns by base pairing in an analogous way to the interaction of U2snRNA and U1snRNA, respectively, in U2 splicing (Yu and Steitz 1997; Kolossova and Padgett 1997). However, a major difference in the establishment of these interactions is that U11 and U12 snRNAs enter the splicing reaction as a stable di-snRNP, while U1 and U2 snRNAs enter separately as discrete particles (Wassarman and Steitz 1992).

U4atac and U6atac snRNAs in Humans

Human U4atac and U6atac snRNAs were coprecipitated from affinity-purified U12-dependent spliceosomes by immunoprecipitation with an antibody specific to the U6snRNA 5' γ-methylated guanosine cap structure (Tarn and Steitz 1996b). Both U4atac and U6atac snRNAs are essential for splicing of *P120* and *SCN4A* U12 introns in vitro (Tarn and Steitz 1996b; Wu and Krainer 1997). In the major U2-type spliceosome, U4 and U6 are initially base paired, and during assembly of the spliceosome this interaction is unwound, allowing U6snRNA to interact with the 5' splice site and U2snRNA (Brow 2002). Although the sequences of human

U4atac and U6atac snRNAs show only ~40% identity with human U4 and U6snRNAs, the secondary structure of U4atac-U6atac snRNA complex is remarkably similar to that of U4-U6 (Tarn and Steitz 1996b; Yu et al. 1999). For example, the sequence of U6atac, which base pairs with the 5′ splice site during splicing, is located upstream of Stem I of U4atac-U6atac and is identical to the position of the analogous U6 snRNA sequence in U4-U6 (Tarn and Steitz 1996b; Incorvaia and Padgett 1998). In addition, the region of highest homology between human U6atac and U6 (>80%) contains the sequences involved in base-pairing interactions thought to form the spliceosomal catalytic core, highlighting the conservation of sequence function of these snRNAs in the different spliceosomes (Tarn and Steitz 1996b; Incorvaia and Padgett 1998; Otake et al., 2002). Thus U4atac and U6atac have functions analogous to those of U4 and U6, where U4atac has a role as chaperone for U6atac snRNA that functions at the catalytic core of the spliceosome. Activation of the U12 spliceosome is likely to occur through unwinding of U4atac-U6atac base pairing (Tarn and Steitz 1997).

Plant U12-Type snRNAs

The similarity between splicing signals of plant U12 introns and those of other eukaryotes suggests that the components of the plant U12 spliceosome will also be similar. *Arabidopsis* and *Drosophila* U6atac genes were identified by a BLASTN search with human U6atac (Burge et al. 1998). However, identification of the other U12 spliceosomal snRNAs was difficult because of limited sequence homology with human and *Drosophila* snRNAs (Shukla and Padgett 1999). The U6atac and U12 snRNAs plant orthologs were cloned from *Arabidopsis* by identifying the snRNA regions that are involved in conserved base-pairing interactions (Shukla and Padgett 1999). U11snRNA was identified in a cDNA library of small non-messenger RNAs from *Arabidopsis*. It was similar in size and secondary structure to human U11, and the gene contained promoter elements (USE and TATA) characteristic of UsnRNAs (Marker et al. 2002; Schneider et al. 2004). With the use of a sequence complementary to stem II of plant U6atac, a single candidate U4atac was identified, which had the expected secondary structure and complementarity to U6atac, and again contained promoter features characteristic for snRNAs (Lorković et al. 2005). The U6atac gene from *Arabidopsis* is probably transcribed by RNA polymerase III, similar to U6 snRNA genes (Shukla and Padgett 1999; Schneider et al. 2004; Lorković et al. 2005).

The degree of divergence of plant U12 spliceosomal snRNAs from their human and fly counterparts was unexpected. For example, U6atac, U12, U11, and U4atac from *Arabidopsis* show 65%, 55%, 54%, and 45% identity, respectively, with their human counterparts (Simpson, unpublished). In both U6atac and U12, the regions of highest similarity are in the 5′ regions, which are essential for RNA-RNA interactions with pre-mRNA. Despite the low sequence homology, plant U12 spliceosomal snRNAs have similar secondary structures including the base pairing of U6atac and U4atac, such that despite the sequence divergence, the function of plant U12 spliceosomal snRNAs is likely to be conserved (Shukla and Padgett 1999; Lorković et al. 2005).

Protein Components of the U12-Dependent Spliceosome in Animals

A proteomic analysis of the major and minor spliceosome revealed that many of the protein components of both spliceosomes are shared (Will et al. 1999). Nevertheless, one of the main differences between the major and minor spliceosomes is that U11 and U12 snRNPs form highly stable 18S U11/U12 di-snRNP complexes, which interact with pre-mRNA during the early steps of U12 spliceosome assembly (Wassarman and Steitz 1992; Frilander and Steitz 1999). Of the 24 distinct proteins identified in the 18S U11/U12 di-snRNP, seven were unique to the U12-type spliceosome: 65K, 59K, 48K, 35K, 31K, 25K, and 20K (Will et al. 1999; 2004). siRNA knockdowns of four of these genes severely inhibited cell growth, demonstrating the importance of U12 splicing in cell viability (Will et al. 2004). One of these proteins, the U11/U12-35K protein, exhibits features of the U1 snRNP-specific 70K protein, containing a similar RNA recognition motif and a glycine-rich region and being associated with the U11snRNP-rich fraction (Will et al. 2004). The 20K and 65K proteins have structures that are similar to U1-C and U1A/U2B" (Will et al. 2004). It is possible that these proteins function in a comparable way to their U1 counterparts and may be involved in 5′ splice site recognition in U12-dependent splicing. Proteins common to the U1 and U2snRNPs and the 18S U11/U12 di-snRNP include the Sm proteins, B/B′, D3, D2, D1, E, F, and G, and subunits of the essential splicing factor SF3b, SF3b155, SF3b145, SF3b130, and SF3b49, initially discovered in U2 snRNPs. However, proteins constituting the essential U2 snRNP SF3a were absent (Will et al. 1999, 2004).

The 14-kDa protein (p14), which is also part of the U2 snRNP SF3b component and which interacts with the bulged adenosine residue at the branchpoint sequence of U2-type introns early in spliceosome assembly, is also present in the U12 spliceosome (Will et al. 2001). Furthermore, electron microscopy and X-ray crystallography indicate that p14 is positioned within a pocket presented on the outer domains of the U11/U12 di-snRNP (Golas et al. 2005; Schellenberg et al. 2006), suggesting a similar role in branchpoint formation. In U2-dependent splicing, the U4/U6snRNP and U5 snRNP form a tri-snRNP particle. Similarly, human U4atac/U6atac di-snRNPs associate with U5 snRNP, forming 25S tri-snRNP particles, and immunoprecipitation experiments show that U4/U6 di-snRNP- and tri-snRNP-specific proteins are present in the U4atac/U6atac·U5 tri-snRNP (Schneider et al. 2002). Similar functions in the different snRNPs are exemplified by the U4/U6 snRNP-specific 61K protein, which facilitates the association of U4atac/U6atac snRNP with U5, and the 15.5-kDa protein, which binds to the 5′ stem-loop of the both U4 and U4atac snRNAs in vitro (Luo et al. 1999; Nottrott et al. 1999; Schneider et al. 2002).

Plant U12 Splicing Factors

Orthologs of all of the unique U11/U12 di-snRNP protein genes, with the exception of a clear example for the 20K (U1C like) protein, have been found in the *Arabidopsis* and rice genomes with the highest degree of conservation found within the RRMs of the 65K protein (Will et al. 2004; Lorković et al. 2005). Some of the plant proteins vary in size from their human counterparts, for example, the U12-48K protein has a 260-amino acid N-terminal extension (Lorković et al. 2005). There is little biochemical characterization of the plant U12 spliceosome or its components, but *Arabidopsis* U11-35K was isolated in a yeast two-hybrid screen with the SR domain-containing cyclophilin protein, CypRS92. The AtU11-35K protein localized to the nucleus, had a similar gene structure to U1-70K, and was isolated by affinity selection with oligonucleotides specific to U11 and U12 snRNAs (Lorković et al. 2005). Furthermore, U11 and U12snRNAs sedimented in glycerol gradients with similar size coefficients to those of mono-snRNP and U11/U12 di-snRNP particles found in human extract. Further similarities in protein composition were the presence of SF3b49 and the absence of SF3a components in both human and *Arabidopsis* U2 snRNP and the U11/U12 di-snRNP (Will et al. 1999, 2004; Lorković et al. 2005).

U12-Type Spliceosome Assembly

The stepwise assembly of the major U2-dependent spliceosome involves an early E complex, precommitment complex H, and the presplicing complexes A, B, and C. In U12-dependent spliceosome assembly, the E complex containing early splicing factors prior to the incorporation of the snRNP particles has not been detected (Tarn and Steitz 1996a). This suggests differences in early intron recognition and commitment to splicing, and the incorporation of U11 and U12 as a U11/U12 di-snRNP also implies differences in early spliceosome assembly. For example, the 5' splice site and branchpoint in U12 introns are recognized cooperatively by the U11/U12 di-snRNP particle during formation of the A complex. Moreover, the U11 snRNP/5' splice site interaction is strongly stimulated by the binding of U12 snRNA to the branchpoint, which contrasts with the U2 spliceosome, where the 5' splice site is recognized by the U1 snRNA independently of U2 snRNA (Madhani and Guthrie 1994). In the absence of the E complex, it was suggested that the 5' splice site and branchpoint sequence were the only *cis*-elements required for U12 spliceosomal A complex formation (Frilander and Steitz 1999). However, an in vitro U12-dependent spliceosome assembly assay supports a need for an active 3' splice site for the formation of the U12-type A complex (Dietrich et al. 2001a), indicating a requirement for all of the major U12-dependent intron splicing signals.

Further differences in spliceosome assembly are also found in later spliceosomal complexes. For example, a novel intermediate, containing both U11snRNP and U4atac/U6atac·U5 tri-snRNP, has been identified in Complex B, which suggests an alternative pathway for catalytic core formation in the U12 spliceosome (Frilander and Steitz, 1999, 2001). In particular, U4atac/U6atac and U12/U6atac snRNA interactions can occur without displacement of U11 snRNA by U6atac snRNA at the 5' splice site contrasting the displacement of U1 snRNP by U6 in the major spliceosome. Despite these differences the formation of the active site has clear parallels between the two spliceosomes. For example, the U5 and U6atac snRNAs can interact simultaneously with the pre-mRNA near the 5′ splice site, even in the absence of U12 snRNA and is dependent on sequences within the 5′ exon (Frilander and Steitz 2001). The sequence and structure of U6snRNA is the key to the splicing reaction by the major spliceosome, requiring the resolution of U4/U6 base pairing. A similar stem-loop structure in U6atac, which is proposed to be at or near the catalytic center of the U2 spliceosome, is required for *in vitro* splicing within the U12-dependent spliceosome (Shukla and Padgett 2001). Finally, the sequence requirements of U4atac snRNA are similar to those described previously for U4snRNA, supporting a chaperone role for U4atac (Shukla et al. 2002).

In the absence of plant in vitro splicing extracts, there is no information on the assembly of either the U2- or U12-dependent spliceosomes. However, the conservation of both snRNAs and protein components and the evidence that plants form U11/U12 disnRNP particles (see Sect. 3.3 above) is highly indicative that the assembly and splicing of U12 introns in plants will be similar (see the chapter by Y. Ru et al., this volume). One potential difference, however, lies in the sequence composition of plant introns. UA-richness is required in early intron recognition for U2 introns in plants, potentially involving specific proteins that recognize UA-rich sequences. It is also possible that such sequences are important in U12 intron splicing and presumably would also function early in spliceosome assembly.

SR Proteins, Exon-Splicing Enhancers, and Intron-Exon Definition in U12-Dependent Splicing

SR Proteins and Exon-Splicing Enhancers

SR proteins play essential roles in constitutive and alternative U2-dependent splicing pathways (Kalyna and Barta 2004; Graveley 2000) (see chapters by A. Barta et al. and G. S. Ali and A. S. N. Reddy, this volume). They are also required for U12-dependent splicing as shown by the rescue of splicing of U12 introns in HeLa S100 extracts after the addition of a nuclear fraction enriched with SR proteins. Individual recombinant SR proteins were able to activate splicing of U12 introns from the *P120* and *SCN4A* genes but showed some substrate preferences (Hastings and Krainer 2001). SR proteins interact with sequences in flanking exons called exon-splicing enhancers (ESEs), which are associated with both types of splicing pathways (Wu and Krainer 1998; Hastings and Krainer 2001; Dietrich et al. 2001b; Graveley 2000; Scamborova et al. 2004) (Fig. 2). ESEs are usually purine-rich sequences, and heterologous purine-rich ESEs placed in the exon downstream of the U12-dependent intron from the human *SCN4A* gene strongly stimulated splicing in vitro. A natural purine-rich ESE has been identified in the exon downstream of the U12 intron in *P120* pre-mRNA. This ESE stimulated in vitro splicing of the U12 intron, and its effect was mediated through the action of SR proteins (Hastings

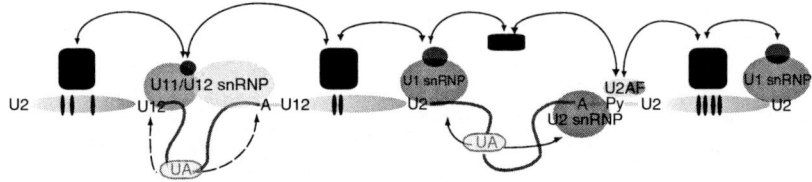

Fig. 2 Interactions between U2- and U12-dependent introns (adapted from Hastings and Krainer 2001). SR proteins (*black boxes*), exon splicing enhancer sequence (ESEs) (*black lines*), and splicing factors (snRNP labeled *ovals*) promote exon and intron defining interactions (*arrowed*) that identify the correct splice sites. These occur between U2 (bordered by U2) and U12 (bordered by U12) introns. Plant U2 and U12 introns have high (>60%) UA content, which may be recognized by UA-binding proteins (*oval* labeled UA). Other factors may also be more important for plant U12 splicing such as intron definition by the U11/U12 di-snRNP, the context of U12 introns with their U2 intron neighbors, and exon definition interactions

and Krainer 2001). Similarly, ESE-like splicing elements were discovered in the exon upstream of the *Arabidopsis LUMINIDEPENDENS* U12 intron and exon deletions, and an exon splicing enhancer trap has demonstrated that these sequences are required for increased splicing efficiency of the associated downstream U12 intron (Lewandowska et al. 2004; Simpson, unpublished; S. Mount, University of Maryland, USA, personal communication). Plants have a large family of SR proteins, and many have been shown to be important in U2 splicing (Lazar et al. 1995; Lopato et al. 1996a, 1996b, 1999a, 1999b, 2002; Kalyna et al. 2003). Many plant SR proteins have been shown to interact with the U11-35K protein, suggesting that SR proteins are involved in plant U12 splicing (Lorković et al. 2005), but their role in plant U12 splicing and the sequences (ESEs) that they interact with remain to be discovered.

Intron-Exon Definition and SR Proteins

Animal genes have introns that are often considerably larger than their associated exons. This basic structural difference has consequences for the recognition of exons and for spliceosome assembly. In animals, exon definition, where splicing factors associated with the upstream 3' splice site and the downstream 5' splice site interact across an exon, identifies an exon within very long transcripts and often involves SR proteins (Berget 1995; Black 1995). Plant and animal genes have similar average sizes for exons, but plant U2 introns are consistently smaller than those of animals (over 80% are between 80 and 120 nt long) and plant introns also contain high intronic UA-rich sequences. These characteristics have suggested that splice sites are selected predominantly by intron definition where splicing factors at each end of the intron can interact across the intron (see Brown and Simpson

1998; Lorković et al. 2000a; Reddy 2001 for reviews). However, many mutations in plant splice sites lead to exon skipping, which is a characteristic feature that exon-defining interactions have been disrupted (Brown 1996), and exon definition is also necessary for the constitutive splicing of a 9-nt mini-exon (Simpson et al. 2000). In addition, efficiently spliced introns enhance the splicing of adjacent poorly spliced introns (Simpson et al. 1999). Thus interactions across exons in multiple intron pre-mRNA transcripts in plants not only determine splice site selection but also influence the efficiency of splicing (Simpson et al. 1999, 2000, 2002).

U12-dependent introns are situated in multi-intron transcripts, and in most cases the U12 introns are flanked by U2-dependent introns. When expressed individually in tobacco protoplasts, the splicing of U12 introns was inefficient but increased when flanked by U2 introns, such that it is likely that there is communication and cooperation in the assembly of U2 and U12 spliceosomes on adjacent introns (Lewandowska et al. 2004). In animal systems, a downstream conventional 5′ splice site was found to stimulate U12 splicing in vitro and was dependent on the presence of U1 snRNP. Exon enhancer sequences were also found to promote U12 splice site usage through interactions with SR proteins, suggesting that exon defining interactions occur between U12 and U2 introns and that SR proteins have an important role in uniting both splicing mechanisms (Fig. 2) (Wu and Krainer 1996; Hastings and Krainer 2001). Different SR proteins were also found to maximize splicing of single presumably intron-defined U12 intron constructs in vitro (Hastings and Krainer 2001). For example, U11-35K protein is thought to have a function analogous to U1-70K by facilitating 5′ splice selection through interactions with SR proteins (Will et al. 1999, 2004) (Fig. 2). In plants, the U11-35K ortholog has been shown to interact with variable affinity for different SR proteins (Lorković et al. 2005), suggesting that plant U12 intron definition and exon defining interactions between U12 and U2 introns occur through SR proteins to help determine splice site selection between the two splicing systems.

There are no natural examples of two adjacent U12 introns; however, some genes contain more than one U12 intron. For example, 15 human genes contain two U12 introns, and a Na^+/H^+ antiporter gene contains three U12 introns (Levine and Durbin 2001). In *Arabidopsis* two orthologous Na^+/H^+ antiporter genes contain three U12 introns along with 18 U2 introns (Zhu and Brendel 2003). It is likely in these cases that communication between U12 and U2 introns is important in determining antiporter gene expression, which suggests a reliance on efficient spliceosome assembly on flanking U2 introns to support U12 spliceosome assembly.

Alternative Splicing of U12 Introns

Alternative splicing increases the protein-coding capacity of genomes by producing multiple protein forms through, in many cases, highly regulated alternative selection of splice sites (see the chapter by B. J. Haas, this volume). For many alternative splicing events, *cis*-acting enhancer or suppressor sequence elements promote or inhibit splice site selection, respectively, such that genes can be regulated by complex coordinated alternative splicing networks (Blencowe 2006). The stringent requirement of the conserved 5′ splice site and branchpoint in U12 intron splicing makes these introns less likely to be involved in alternative splicing events. However, alternative splicing has been induced in the human *P120* U12 intron F, which normally uses the 3′ splice site adjacent to the branchpoint, by the addition of a purine-rich enhancer sequence downstream of this splice site, causing selection of a cryptic splice site (Dietrich et al. 2001b). Computational analysis of 404 human U12-dependent introns identified 13 cases with evidence of alternative splicing, 11 of which used 3′ alternative splice sites (Levine and Durbin 2001). In plants, there are four GT-AG U12 introns that are alternatively spliced with alternative 3′ splice sites (Zhu and Brendel 2003). Two have alternative 3′ splice sites in close proximity to each other, suggesting selection by local scanning (Fig. 3a). The third has an alternative branchpoint/3′ splice site combination that precedes the preferred U12 branchpoint sequence/3′ splice site (Fig. 3b). The fourth example utilizes an unusual alternative AU dinucleotide as a 3′ splice site (Fig. 3c). Thus some U12 introns undergo alternative splicing and, as in U2 intron alternative splicing, enhancer or suppressor elements may be involved in the determination of which splice sites are used.

Alternative Splicing of U2 and U12 Introns

U2 and U12 introns provide a unique form of alternative splicing, whereby the U12 and U2 spliceosomes compete to splice out an intron positioned in the same region of a pre-mRNA transcript. In plants, two examples of alternative 5′ splice site selection and one example of exon skipping was observed that involved alternative selection of U12 and U2 splice sites (Zhu and Brendel 2003). In the examples of alternative 5′ splice site usage, both involved alternative selection of the U12 splice site with a U2-type GC 5′ splice site in paralogous genes (Fig. 3d and e). The limited number of ESTs suggests that there may be splice site selection preferences, one case showing preference for the U2 GC site and the other a preference for U12 5′ splice site. In the example of exon skipping, alternative splicing occurred between a U12-type GT-AG intron and use of the upstream 5′ splice site of a U2 type GT-AG intron (Zhu and Brendel 2003) (Fig. 3f). Commitment to splicing by one or the other spliceosome is likely to be due to the early recognition of these splicing signals that leads to a switch between the two spliceosomes.

Nested U2 and U12 introns have also been identified in animal introns, whereby the U12 and U2 spliceosomes also compete to splice out an intron positioned in the same region of a pre-mRNA transcript. The second intron of the *Drosophila Prospero* (*pros*) protein gene contains U2-type splice sites located within an intron bordered by U12-type splice sites (Fig. 3g). Splicing of the two introns is mutually exclusive and developmentally regulated. U2 splicing preferentially occurs during early embryogenesis, while U12 splicing is more abundant at the late embryogenic stage. A purine-rich element positioned downstream of the U2 5′ splice site preferentially promotes U2 splicing, and disruption of this

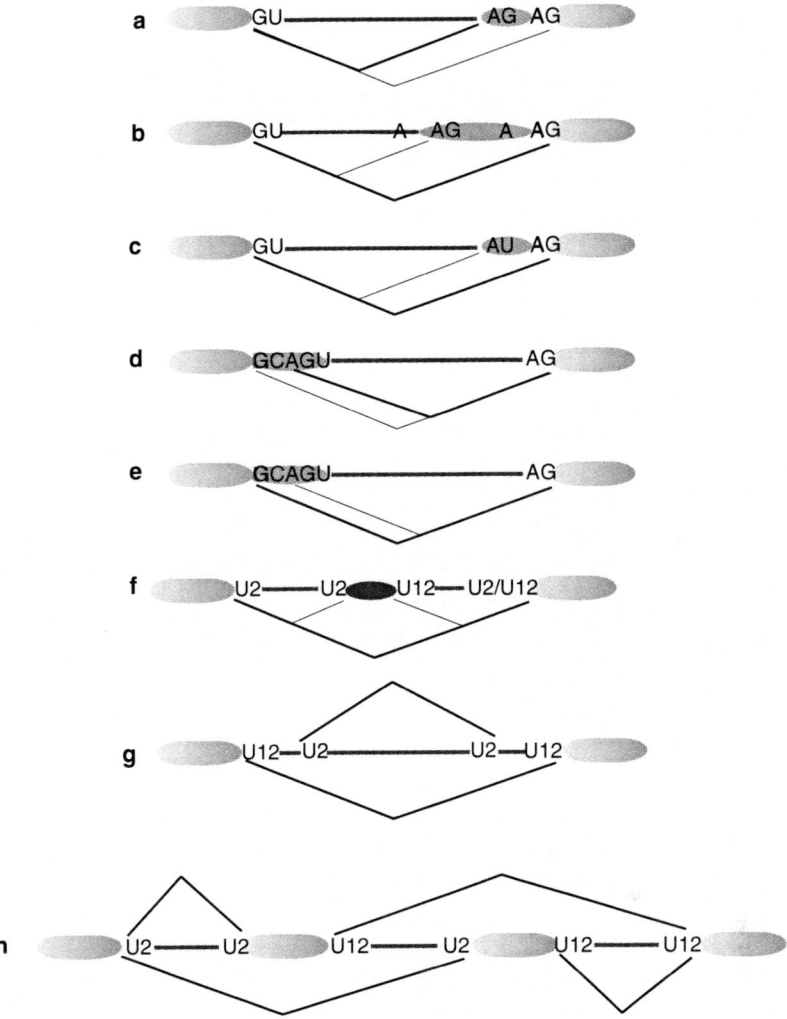

Fig. 3 U12 Alternative splicing. U12 alternative splicing events in *Arabidopsis* are shown by *lines* linking 5′ splice sites (GU) to alternative 3″ splice sites (AG), or where use of alternative spliceosomes are used, splice sites are indicated by *U2* or *U12*. The intensity of the *lines* indicates the preferred splicing selection. **a** Selection of alternative U12 3′ splice sites are found <10 nt apart in At2g26430 and At3g13460. **b** Selection of alternative U12 branchpoint and 3′ splice site in At3g52180. **c** Selection of an alternative nonconsensus 3′ splice site in At4g09720. **d** and **e** two paralogous genes, At2g44680 and At3g60250, show alternative U2 and U12 5′ splice site selection 3 nt apart with different preferences for U2 and U12 splicing. **f** Exon skipping in At1g49160 through activation of a U12 intron 5′ splice site. **g** Nested U2 and U12 introns. Splice sites are shown as *U2* or *U12*. **h** Mutually exclusive selection of exons through alternative spliceosome selection. U2 and U12 splice sites are shown. (From Zhu and Brendel 2003; Scamborova et al. 2004; Chang et al. 2007)

element has a greater consequential effect on U2 splicing, leading to a preference for U12 splicing (Otake et al. 2002; Scamborova et al. 2004). Plants may also undergo this type of alternative splicing. In the *Arabidopsis* gene of unknown function (At1g73970) a potential U12 intron is positioned within the second U2 intron, reminiscent of the situation in the *Drosophila pros* gene.

Finally, exons 6a and 6b of the human JNK2 gene are bordered by U2 and U12 splice sites that lead to mutually exclusive selection of these exons, dependent on whether the intron is spliced out by the U2 or the U12 spliceosome. U2-dependent splicing predominates in nonneuronal cells because of a strong polypyrimidine tract sequence upstream of the exon 6b 3′ splice site (Fig. 3h). Expression of neuron-specific Nova proteins may block use of this site and activate U12-dependent splicing that leads to inclusion of exon 6a in this cell type (Chang et al. 2007). There are no examples of this form of mutually exclusive selection by U2 and U12 spliceosomes described in plants, and considering the rarity in plants of mutually exclusive splicing by U2 splicing it is unlikely that this form of alternative splicing will be found.

Role of U12 Introns in Gene Expression

U12 introns are found in important information processing genes such as those for DNA replication/repair, RNA processing, and translation (Burge et al. 1998; Chang et al. 2007). In *Drosophila*, it was also shown that the U12-dependent spliceosome was essential for cell viability and development (Otake et al. 2002) such that U12 intron splicing has an important role in cell survival. It has been postulated that the presence of U12 introns in genes may help regulate expression of the genes in which they are found. Human U12 introns have been shown to splice significantly more slowly in vivo and in vitro than U2 introns, and this has been described as a posttranscriptional bottleneck, preventing the overexpression of U12 intron-containing genes that may be harmful to the cell (Tarn and Steitz 1996b; Wu and Krainer 1996; Patel et al. 2002; Hirose et al. 2004). The generally low efficiency of U12 intron splicing was also observed in plant U12 introns in vivo when processed in tobacco protoplasts either as single U12 intron constructs or as part of multiple intron constructs flanked by U2 introns (Kmieciak et al. 2002; Lewandowska et al. 2004). In this system the high level of expression from the 35S CaMV promoter may explain the apparent low levels of splicing; however, unspliced U12 intron transcripts have been detected in endogenous *LUMINIDEPENDENS* transcripts in different organs with an RT-PCR panel designed to monitor changes in alternative splicing, suggesting that some limitation in splicing of these introns occurs in endogenous genes (unpublished data).

The lower level of U12 splicing activity was originally thought to be the result of ~100-fold lower abundance of U12 spliceosomal components (Montzka and Steitz 1988; Tarn and Steitz 1996a; Yu et al. 1999). However, recent evidence suggests that U12 spliceosome abundance is not a limiting factor for efficient U12 splicing. In a mouse cell line with reduced U4atac expression, which is the least abundant U12 snRNA, the rate of U12 intron excision was the same in cell lines with normal levels of U4atac (Pessa et al. 2006). In addition, upregulation of U12 intron-containing genes is compensated by increased activity of the U12 spliceosome during replication in rapidly growing cells (Patel et al. 2002; Scamborova et al. 2004). Thus the spliceosome is abundant enough to splice the relatively low number of U12 introns and is actively regulated. In plants, some U12 introns are found in genes that are upregulated in response to salt stress, suggesting a role in regulation of abiotic stress response, but whether there is any concomitant increase in U12 spliceosome activity or differential activity of U12 and U2 spliceosomes under these conditions is unknown (Zhu and Brendel 2003).

Regulation of expression of U12 intron-containing genes has also been linked to nonsense-mediated decay (NMD). If U12 intron removal is a rate-limiting step, then partially spliced transcripts could be targeted for degradation through NMD (Patel and Steitz 2002; Hirose et al. 2004). NMD is activated when a premature stop codon (PTC) is identified over 55 nt upstream of an exon junction and occurs through interaction between a ribosome stalled at the PTC and an exon junction complex (EJC) deposited 20-25 nt upstream of the exon junction (Maquat 2004). A PTC positioned upstream of a U12 intron led to lower levels of observable spliced RNA, and immunoprecipitation of FLAG-tagged EJC proteins on a U12 intron spliced substrate found many of the components of the EJC normally deposited on a spliced U2-intron transcript. Thus U12 intron-containing transcripts can be subject to NMD and EJCs may be deposited by both spliceosomes (Hirose et al. 2004). In plants, a pre-mRNA with a retained U12 intron has been identified, which is a potential target for NMD (Zhu and Brendel 2003).

U12 Intron Conservation and Conversion

U12 introns have been identified in a wide range of distantly related eukaryotes (Russell et al. 2006). The high identity of U12 splicing signals and the high degree of conservation of the plant and human U12 splicing factors suggests that the two types of spliceosome have existed side by side since the divergence of plants and metazoans 1.2 billion years ago (Feng et al. 1997). There are two models for the origin of the two intron and spliceosome types. In the first model, the presence of conserved proteins within U2-type and U12-type spliceosomes supports a common origin for both spliceosomes derived from early Group II-like introns (Burge et al. 1998; Lynch and Richardson 2002). In the second model, the presence of unique proteins within the U11/U12 disnRNP, the absence of a number of essential U2-type protein components, the variation shown between the U2 and U12snRNAs, the rarity of the U12 introns, and the presence of genes with multiple U12 introns support an endosymbiotic fusion of organisms with the different splicing types (the fission/fusion model) that resulted in a progenitor with both spliceosomes (Burge et al. 1998).

The multi-U12 intron-containing Na$^+$/H$^+$ antiporter genes from plants and humans described above have maintained the positions of two of the three U12 introns in the same position, which supports an ancient origin for these gene sequences and maintenance of these introns (Zhu and Brendel 2003). However, mutations in the stringent U12 splice sites are prone to conversion to the more variable U2 splice sites and splicing by the U2 spliceosome. Polymorphisms in the U12 5' splice site of the human WDFY1 led to activation of U2 splicing and selection of cryptic splice sites. On the other hand, mutations that weakened the branchpoint also led to activation of cryptic U2 splice sites, but in this case different transcripts were spliced by either U2 or the U12 spliceosome. The U12 5' splice site, therefore, has an essential role in committing the intron to U12 splicing (Chang et al. 2007). Considering the relative rarity of U12 introns, the similarity of the U2 and U12 splicing signals, and the more stringent U12 splicing signal requirements compared to U2 type introns, the number of mutations required to convert functional U2-intron 5' splice site and branchpoint sequences from a U12 splice site is much lower than the number of mutations required to convert U2 sites to U12 (Dietrich et al. 1997).

In plants, the *Arabidopsis* acidic ribosomal protein P1 gene family provides an example of conversion of a U12 intron to a U2 intron. All four gene family members have intron 1 positioned in virtually the same location, but one of these genes (At4g00810) contains a U12 intron with a U12 5' splice site (:GTATCCTT) and a branchpoint sequence (ATCCTTAAC) located 16 nt upstream of the 3' splice site (TAG:). The other genes have typical U2 intron 5' splice sites (5/9 match with U2 - CAG:GTAAGT 5' splice site consensus) and branchpoints (U2 branchpoint-CURAY). Two of these genes (At5g47700 and At1g01100) contain 5' splice sites that are identical to the U12 intron of At4g00810 except for a C to G at position +5 or a C to U at position +6, respectively. Both of these changes would disrupt the U12 5' splice site and generate acceptable U2-type intron 5' splice sites. Neither of these genes retained the highly conserved U12 branchpoint. The presence of U2-type introns in identical positions to U12-type introns in orthologous genes supports the conversion of U12- to U2-type introns and the gradual reduction in the number of U12 introns (Burge et al. 1998; Zhu and Brendel 2003).

Perspectives

Whatever the evolutionary history of the two types of nuclear pre-mRNA introns and spliceosomes in eukaryotes, there are still a number of key questions regarding plant U12 intron splicing that remain to be addressed. Much of what we know about plant U12 introns and the U12 spliceosome is based on bioinformatic analyses of the introns, snRNAs, and protein components. There have been some analyses of U12 intron mutants and initial biochemical studies, but our knowledge is still heavily reliant on comparison with animal systems. Further concerted efforts to generate in vitro plant splicing systems are needed to open up the field and allow detailed analysis of intron recognition, spliceosome assembly, and composition. Alternatively, genetic and molecular analyses may begin to address questions such as why some plant U12 introns are maintained, whether plant U12 introns confer regulation of the genes containing them, whether the activity of the plant U12 spliceosome is regulated, and how the different spliceosomes behave in different conditions?

References

Alioto TS (2007) U12DB: a database of orthologous U12-type spliceosomal introns. Nucl Acid Res 35:D110–D115
Berget SM (1995) Exon recognition in vertebrate splicing. J Biol Chem 27:2411–2414
Black DL (1995) Finding splice sites within a wilderness of RNA. RNA 1:763–771
Blencowe BJ (2006) Alternative splicing: New insights from global analysis. Cell 126:37–47
Brow DA (2002) Allosteric cascade of spliceosome activation. Annu Rev Genet 36:333–360
Brown JWS (1996) *Arabidopsis* intron mutations and pre-mRNA splicing. Plant J 10:771–780
Brown JWS, Simpson CG (1998) Splice site selection in plant pre-mRNA splicing. Annu Rev Plant Physiol Plant Mol Biol 49:77–95
Burge CB, Padgett RA, Sharp PA (1998) Evolutionary fates and origins of U12-type introns. Mol Cell 2:773–785
Burge CB, Tuschl T, Sharp PA (1999) Splicing of precursors to mRNAs by the spliceosomes. In: RNA World, Gesteland R, Cech TR, Atkins JF eds. (New York: Cold Spring Harbor Laboratory Press, Cold Spring Harbor), pp. 525–560
Chang W-C, Chen Y-C, Lee K-M, Tarn W-Y (2007) Alternative splicing and bioinformatics analysis of human U12-type introns. Nucl Acid Res. 35:1833–1841
Crooks GE, Hon G, Chandonia J-M and Brenner SE. (2004) WebLogo: A sequence logo generator. Genome Res 14:1188–1190
Dietrich RC, Incorvaia R, Padgett RA (1997) Terminal intron dinucleotide sequences do not distinguish between U2- and U12-dependent introns. Mol Cell 1:151–160
Dietrich RC, Peris MJ, Seyboldt AS, Padgett RA (2001a) Role of the 3′ splice site in U12-dependent intron splicing. Mol Cell Biol 21:1942–1952
Dietrich RC, Shukla GC, Fuller JD, Padgett RA (2001b) Alternative splicing of U12-dependent introns in vivo responds to purine-rich enhancers. RNA 7:1378–1388
Dietrich RC, Fuller JD, Padgett RA (2005) A mutational analysis of U12-dependent splice site dinucleotides. RNA 11:1430–1440
Feng D-F, Cho G, Doolittle RF (1997) Determining divergence times with a protein clock: Update and re-evaluation. Proc Natl Acad Sci USA 94:13028–13033
Frilander MJ, Steitz JA (1999) Initial recognition of U12-dependent introns requires both U11/5′ splice-site and U12/branchpoint interactions. Genes Dev 13:851–863
Frilander MJ, Steitz JA (2001) Dynamic exchanges of RNA interactions leading to catalytic core formation in the U12-dependent spliceosome. Mol Cell 7:217–226
Gniadkowski M, Hemmings-Mieszczak M, Klahre U, Liu HX, Filipowicz W (1996) Characterization of intronic uridine-rich elements acting as possible targets for nuclear proteins during pre-mRNA splicing in *Nicotiana plumbaginifolia*. Nucl Acid Res 24:619–627
Golas MM, Sander B, Will CL, Lührmann R, Stark H (2005) Major conformational change in the complex SF3b upon integration into the spliceosomal U11/U12 di-snRNP as revealed by electron cryomicroscopy. Mol Cell 17:869–883
Goodall GJ, Filipowicz W (1989) The AU-rich sequences present in the introns of plant nuclear mRNAs are required for splicing. Cell 58:473–483

Goodall GJ, Filipowicz W (1991) Different effects of intron nucleotide composition and secondary structure on pre-mRNA splicing in monocot and dicot plants. EMBO J 10:2635–2644

Graveley BR (2000) Sorting out the complexity of SR protein functions. RNA 6:1197–1211

Hall SL, Padgett RA (1994) Conserved sequences in a class of rare eukaryotic nuclear introns with non-consensus splice sites. J Mol Biol 239:357–365

Hall SL, Padgett RA (1996) Requirement of U12 snRNA for in vivo splicing of a minor class of eukaryotic nuclear pre-mRNA introns. Science 271:1716–1718

Hastings ML, Krainer AR (2001) Functions of SR proteins in the U12-dependent AT-AC pre-mRNA splicing pathway. RNA 7:471–482

Hirose T, Shu M-D, Steitz JA (2004) Splicing of U12-type introns deposits an exon junction complex competent to induce nonsense-mediated mRNA decay. Proc Natl Acad Sci USA 101:17976–17981

Incorvaia R, Padgett RA (1998) Base pairing with U6atac snRNA is required for 5′ splice site activation of U12-dependent introns in vivo. RNA 4:709–718

Jackson IJ (1991) A reappraisal of non-consensus mRNA splice sites. Nucl Acid Res 19:3795-3798

Kalyna M, Lopato S, Barta A (2003) Ectopic expression of atRSZ33 reveals its function in splicing and causes pleiotropic changes in development. Mol Biol Cell 14:3565–3577

Kalyna M, Barta A (2004) A plethora of plant serine/arginine-rich proteins: Redundancy or evolution of novel gene functions? Biochem Soc Trans 32:561–564

Kolossova I, Padgett RA (1997) U11 snRNA interacts in vivo with the 5′ splice site of U12-dependent (AU-AC) pre-mRNA introns. RNA 3:227–233

Kmieciak M, Simpson CG, Lewandowska D, Brown JWS, Jarmowlowski A (2002) Cloning and characterization of two sub-units of *Arabidopsis thaliana* nuclear cap-binding complex. Gene 283:171–183

Lambermon MH, Simpson GG, Wieczorek Kirk DA, Hemmings-Mieszczak M, Klahre U, Filipowicz W (2000) UBP1, a novel hnRNP-like protein that functions at multiple steps of higher plant nuclear pre-mRNA maturation. EMBO J 19:1638–1649

Lazar G, Schaal T, Maniatis T, Goodman HM (1995) Identification of a plant serine-arginine-rich protein similar to the mammalian splicing factor SF2/ASF. Proc Natl Acad Sci USA 92:7672–7676

Levine A, Durbin R (2001) A computational scan for U12-dependent introns in the human genome sequence. Nucl Acid Res 29:4006–4013

Lewandowska D, Simpson CG, Clark GP, Jennings NS, Barciszewska-Pacak M, Lin CF, Makalowski W, Brown JWS, Jarmolowski A (2004) Determinants of plant U12-dependent intron splicing efficiency. Plant Cell 16:1340–1352

Lopato S, Mayeda A, Krainer AR, Barta A (1996a) Pre-mRNA splicing in plants: characterization of Ser/Arg splicing factors. Proc Natl Acad Sci USA 93:3074–3079

Lopato S, Waigmann E, Barta A (1996b) Characterization of a novel arginine/serine-rich splicing factor in *Arabidopsis*. Plant Cell 8:2255–2264

Lopato S, Gattoni R, Fabini G, Stevenin J, Barta A (1999a) A novel family of plant splicing factors with a Zn knuckle motif: examination of RNA binding and splicing activities. Plant Mol Biol 39:761–773

Lopato S, Kalyna M, Dorner S, Kobayashi R, Krainer AR, Barta A (1999b) atSRp30, one of two SF2/ASF-like proteins from *Arabidopsis thaliana*, regulates splicing of specific plant genes. Genes Dev 13:987–1001

Lopato S, Forstner C, Kalyna M, Hilscher J, Langhammer U, Indrapichate K, Lorković ZJ, Barta A (2002) Network of interactions of a novel plant-specific Arg/Ser-rich protein, atRSZ33, with atSC35-like splicing factors. J Biol Chem 277:39989–39998

Lorković ZJ, Wieczorek Kirk DA, Lambermon MH, Filipowicz W (2000a) Pre-mRNA splicing in higher plants. Trends Plant Sci 5:160–167

Lorković ZJ, Wieczorek Kirk DA, Klahre U, Hemmings-Mieszczak M, Filipowicz W (2000b) RBP45 and RBP47, two oligouridylate-specific hnRNP-like proteins interacting with poly(A)+ RNA in nuclei of plant cells. RNA 6:1610–1624

Lorković ZJ, Lehner R, Forstner C, Barta A (2005) Evolutionary conservation of minor U12-type spliceosome between plants and humans. RNA 11:1095–1107

Luo HR, Moreau GA, Levin N, Moore MJ (1999) The human Prp8 protein is a component of both U2- and U12-dependent spliceosomes. RNA 5:893–908

Lynch M, Richardson AO (2002) The evolution of spliceosomal introns. Curr Opin Genet Dev 12:701–710

Madhani HD, Guthrie C (1994) Dynamic RNA-RNA interactions in the spliceosome. Annu Rev Genet 28:1–26

Maquat LE (2004) Nonsense-mediated mRNA decay: splicing, translation and mRNP dynamics. Nat Rev Mol Cell Biol 5:89–99

Marker C, Zemann, A, Terhorst T, Kiefmann M, Kastenmayer JP, Green P, Bachellerie JP, Brosius J, Huttenhofer A (2002) Experimental RNomics: identification of 140 candidates for small non-messenger RNAs in the plant *Arabidopsis thaliana*. Curr Biol 12:2002–2013

Massenet S, Branlant C (1999) A limited number of pseudouridine residues in the human atac spliceosomal UsnRNAs as compared to human major spliceosomal UsnRNAs. RNA. 5:1495–1503

McConnell TS, Cho SJ, Frilander MJ, Steitz JA (2002) Branchpoint selection in the splicing of U12-dependent introns in vitro. RNA 8:579–586

Montzka KA, Steitz JA (1988) Additional low-abundance human small nuclear ribonucleoproteins: U11, U12, etc. Proc Natl Acad Sci USA 85:8885–8889

Nottrott S, Hartmuth K, Fabrizio P, Urlaub H, Vidovic I, Ficner R, Lührmann R (1999) Functional interaction of a novel 15.5kD [U4/U6.U5] tri-snRNP protein with the 5' stem-loop of U4 snRNA. EMBO J 18:6119–6133

Otake LR, Scamborova P, Hashimoto C, Steitz JA (2002) The divergent U12-type spliceosome is required for pre-mRNA splicing and is essential for development in *Drosophila*. Mol Cell 9:439–446

Patel AA, McCarthy M, Steitz JA (2002) The splicing of U12-type introns can be a rate-limiting step in gene expression. EMBO J 21:3804–3815

Patel AA, Steitz JA (2003) Splicing double: Insights from the second spliceosome. Nat Rev Mol Cell Biol 4:960–970

Pessa HK, Ruokolainen A, Frilander MJ (2006) The abundance of the spliceosomal snRNPs is not limiting the splicing of U12-type introns. RNA 12:1883–1892

Rappsilber J, Ryder U, Lamond AI, Mann M. (2002) Large-scale proteomic analysis of the human spliceosome. Genome Res 12:1231–1245

Reddy AS (2001) Nuclear pre-mRNA splicing in plants. Crit Rev Plant Sci 20:523–571

Russell AG, Charette JM, Spencer DF, Gray MW (2006) An early evolutionary origin for the minor spliceosome. Nature 443:863–866

Scamborova P, Wong A, Steitz JA (2004) An intronic enhancer regulates splicing of the twintron of *Drosophila melanogaster* prospero pre-mRNA by two different spliceosomes. Mol Cell Biol 24:1855–1869

Schellenberg MJ, Edwards RA, Ritchie DB, Kent OA, Golas MM, Stark H, Lührmann R, Glover JNM, MacMillan AM (2006) Crystal structure of a core spliceosomal protein interface. Proc Natl Acad Sci USA 103:1266–1271

Schneider C, Will CL, Makarova OV, Makarov EM, Lührmann R (2002) Human U4/U6.U5 and U4atac/U6atac.U5 tri-snRNPs exhibit similar protein compositions. Mol Cell Biol 22:3219–3229

Schneider C, Will CL, Brosius J, Frilander MJ, Lührmann R (2004) Identification of an evolutionarily divergent U11 small nuclear ribonucleoprotein particle in Drosophila. Proc Natl Acad Sci USA 101:9584–9589

Sharp PA, Burge CB (1997) Classification of introns: U2-type or U12-type. Cell 91:875–879

Sheth N, Roca X, Hastings ML, Roeder T, Krainer AR, Sachidanandam R (2006) Comprehensive splice-site analysis using comparative genomics. Nucl Acid Res 34:3955–3967

Shukla GC, Padgett RA (1999) Conservation of functional features of U6atac and U12 snRNAs between vertebrates and higher plants. RNA 5:525–538

Shukla GC, Padgett RA (2001) The intramolecular stem-loop structure of U6 snRNA can functionally replace the U6atac snRNA stem-loop. RNA 7:94–105

Shukla GC, Cole AJ, Dietrich RC, Padgett RA (2002) Domains of human U4atac snRNA required for U12-dependent splicing in vivo. Nucl Acid Res 30:4650–4657

Simpson CG, Clark GP, Lyon J, Watters J, McQuade C, Brown JWS (1999) Interactions between introns via exon definition in plant pre-mRNA splicing. Plant J 18:293–302

Simpson CG, Hedley PE, Watters JA, Clark GP, McQuade C, Machray GC, Brown JWS (2000) Requirements for mini-exon inclusion in potato invertase mRNAs provides evidence for exon-scanning interactions in plants. RNA 6:422–433

Simpson CG, Thow G, Clark GP, Jennings SN, Watters JA, Brown JWS (2002) Mutational analysis of a plant branchpoint and polypyrimidine tract required for constitutive splicing of a mini-exon. RNA 8:47–56

Simpson GG, Filipowicz W (1996) Splicing of precursors to mRNA in higher plants: mechanism, regulation and sub-nuclear organisation of the spliceosomal machinery. Plant Mol Biol 32:1–41

Tarn WY, Steitz JA (1996a) A novel spliceosome containing U11, U12, and U5 snRNPs excises a minor class (AT-AC) intron in vitro. Cell 84:801–811

Tarn WY, Steitz JA (1996b) Highly diverged U4 and U6 small nuclear RNAs required for splicing rare AT-AC introns. Science 273:1824–1832

Tarn WY, Steitz JA (1997) Pre-mRNA splicing: the discovery of a new spliceosome doubles the challenge. Trends Biochem Sci 22:132–137

Wassarman KM, Steitz JA (1992) The low-abundance U11 and U12 small nuclear ribonucleoproteins (snRNPs) interact to form a two-snRNP complex. Mol Cell Biol 12:1276–1285

Will CL, Schneider C, Reed R, Lührmann R (1999) Identification of both shared and distinct proteins in the major and minor spliceosomes. Science 284:2003–2005

Will CL, Lührmann R (2001) Spliceosomal UsnRNP biogenesis, structure and function. Curr Opin Cell Biol 13:290–301

Will CL, Schneider C, MacMillan AM, Katopodis NF, Neubauer G, Wilm M, Lührmann R, Query CC (2001) A novel U2 and U11/U12 snRNP protein that associates with the pre-mRNA branch site. EMBO J 20:4536–4546

Will CL, Schneider C, Hossbach M, Urlaub H, Rauhut R, Elbashir S, Tuschl T, Lührmann R (2004) The Human 18S U11/U12 snRNP contains a set of novel proteins not found in the U2-dependent spliceosome. RNA 10:929–941

Wu Q, Krainer AR (1996) U1-mediated exon definition interactions between AT-AC and GT-AG introns. Science 274:1005–1008

Wu Q, Krainer AR (1997) Splicing of a divergent subclass of AT-AC introns requires the major spliceosomal snRNAs. RNA 3:586–601

Wu Q, Krainer AR (1998) Purine-rich enhancers function in the AT-AC pre-mRNA splicing pathway and do so independently of intact U1 snRNP. RNA 4:1664–1673

Wu Q, Krainer AR (1999) AT-AC pre-mRNA splicing mechanisms and conservation of minor introns in voltage-gated ion channel genes. Mol Cell Biol 19:3225–3236

Yu YT, Tarn WY, Yario TA, Steitz JA (1996) More Sm snRNAs from vertebrate cells. Exp Cell Res 229:276–281

Yu YT, Steitz JA (1997) Site-specific crosslinking of mammalian U11 and u6atac to the 5' splice site of an AT-AC intron. Proc Natl Acad Sci USA 94:6030–6035

Yu YT, Scharl EC, Smith CM, Steitz JA (1999) The growing world of small nuclear ribonucleoproteins. In RNA World, R.Gesteland, T.R.Cech, and J.F.Atkins, eds. (New York: Cold Spring Harbor Laboratory Press), pp 487–524

Zhou Z, Licklider LJ, Gygi SP, Reed R (2002) Comprehensive proteomic analysis of the human spliceosome. Nature 419:182–185

Zhu W, Brendel V (2003) Identification, characterization and molecular phylogeny of U12-dependent introns in the *Arabidopsis thaliana* genome. Nucl Acid Res 31:4561–4572

Plant SR Proteins and Their Functions

A. Barta(✉), M. Kalyna, and Z.J. Lorković

Contents

Introduction 84
    General 84
    Plant Splicing 85
Plant SR Proteins 85
    A Plenitude of Plant SR Proteins 85
    Binding Specificity of Plant SR Proteins 89
    Complex Network of Interactions of Plant SR Proteins 90
SR Proteins and Alternative Splicing 92
    Alternative Splicing 92
    Regulation of Alternative Splicing by Plant SR Proteins
    and Consequences for Plant Growth and Development 93
    Conserved Alternative Splicing of SR Genes 95
Conclusions 96
References 97

Abstract SR proteins are a family of splicing factors important for splice site recognition and spliceosome assembly. Their ability to bind to RNA and to interact with proteins as well identifies them as important players in splice site choice and alternative splicing. Plants possess twice as many SR proteins as animals, and some of the subfamilies are plant specific. Arabidopsis SR proteins are involved in different aspects of plant growth and development as well as in responses to environmental cues. The plant-specific subfamilies have been shown to be regulated by alternative splicing events, which are highly conserved in evolution. The tight regulation of splicing factors by alternative splicing might allow coordinated responses of their target genes.

A. Barta
Max F. Perutz Laboratories, Medical University of Vienna, Dr. Bohrgasse 9/3, A-1030,
Vienna, Austria
e-mail: andrea.barta@meduniwien.ac.at

Introduction

General

SR (serine/arginine-rich) proteins are evolutionarily highly conserved splicing factors. They are characterized by the presence of one or two RNA binding domains of the RRM type and by a reversibly phosphorylated arginine/serine-rich (RS) domain (Bourgeois et al. 2004; Fu 1995; Graveley 2000). A closer look at the splicing process demonstrates the importance of SR proteins for gene regulation.

Splicing of pre-mRNA introns involves one of the largest RNP complexes of the cell, the spliceosome. This multifaceted machinery consists of five small nuclear RNPs (termed U1, U2, U4/U6, and U5 snRNP) and about 150 additional proteins (Behzadnia et al. 2007; Deckert et al. 2006). Spliceosome assembly on a pre-mRNA molecule depends on the recognition of specific intronic sequences by individual snRNPs such that the two ends of an intron are in close proximity for the two transesterification reactions to take place. This is achieved by base pairing interactions of U1 and U2 snRNAs with the 5′ splice site and the branchpoint region, respectively. These interactions are aided by RS domain-containing proteins like U2AF (U2 snRNA auxiliary factor) and the SR protein SF2/ASF, which supports binding of U1 snRNP to the 5′ splice site. In addition, another SR protein, SC35, was shown to interact with other RS-containing proteins, U1-70K (U1 snRNP-specific protein 70K) and U2AF35, thus enhancing spliceosome assembly (Cao and Garcia Blanco 1998; Kohtz et al. 1994; Wu and Maniatis 1993). Addition of the tri-snRNP complex U4/U6/U5 and many proteins, which include again SR proteins, allows spliceosome assembly and the splicing reaction to occur (for review see Hastings and Krainer 2001b). Therefore, the most critical step in spliceosome assembly is the recognition of the 5′ and 3′ splice sites, and it is certain that SR proteins play a vital role in this process.

Binding of SR proteins to pre-mRNA is mediated by RNA recognition motifs (RRMs), which recognize rather short, degenerate, and in most cases purine-rich sequences. RS domains are mainly responsible for protein-protein interactions, although they seem to be able to modulate RNA binding as well. By binding specific RNA sequences on the pre-mRNA with the RRM, the protein-protein interactions mediated by RS domains are important for the formation of cross-intron and cross-exon protein networks necessary for splice site selection and spliceosome assembly (Black 2003; Maniatis and Tasic 2002; Smith and Valcarcel 2000). In contrast to the highly structured RRM, the RS domain is in essence disordered and can be highly phosphorylated by SR protein-specific kinases. Phosphorylation/dephosphorylation of SR proteins is important for their ability to interact with RNA and other splicing factors, as well as for their localization within the nucleus (Blencowe et al. 1999; Stojdl and Bell 1999; Tenenbaum and Aguirre-Ghiso 2005). The functions of SR proteins seem to be redundant in splicing of some introns but have unique properties in others. Although SR proteins were originally identified as essential factors for constitutive and alternative splicing, recent studies on shuttling SR proteins revealed additional roles, including mRNA export, RNA stability, mRNA quality control (nonsense-mediated mRNA decay), and translation (Huang et al. 2004; Reed and Cheng 2005 and references therein). In addition, a new role for SF2/ASF in the maintenance of genome stability has been proposed recently (Li and Manley 2005, 2006).

Plant Splicing

First insights into plant splicing came from early observations that animal introns could not be processed in plants, albeit plant introns were spliced in HeLa cell nuclear extracts (Barta et al. 1986; Brown et al. 1986; Hartmuth and Barta 1986). Plant introns contain similar *cis*-acting sequences as metazoans but are on average shorter (for review see Brown and Simpson 1998; Lorković et al. 2000b). The intron recognition process in plants seems to differ, as splicing of plant introns requires U-rich sequences within the intron. This could be the reason that in general metazoan introns are not spliced in plants. However, it is not known how these U-rich intronic sequences contribute to intron recognition and splicing at the molecular level. This has stimulated research on proteins that bind U-rich sequences. Several proteins binding to intronic U-rich sequences in vitro have been identified; however, only UBP1 increased splicing efficiency of poorly spliced introns when overexpressed in plant protoplasts (Lambermon et al. 2000, 2002; Lorković et al. 2000a).

The existence of plant SR proteins has been known for a decade, but biochemical analysis has been hampered by the lack of in vitro splicing extracts. Despite many efforts with several different plant tissues, no functional splicing extracts were obtained. Although the protein and RNA contents of these extracts were reasonable, no splicing activity could be observed. The likely reason was that exogenous RNAs formed large complexes in these extracts that could not be analyzed on glycerol gradients or on native gels (our unpublished observations). However, the nature of these complexes (proteins or vacuole compounds) has not been resolved to date. Plant alternative splicing has gained more recognition recently as an important regulatory process for plant development and for environmental responses. As SR proteins contribute significantly to intron recognition and are equally important for constitutive and alternative splicing, our review focuses on what is known about plant SR proteins to date.

Plant SR Proteins

A Plenitude of Plant SR Proteins

Plant SR proteins were originally identified with antibodies against a conserved serine phosphoepitope located in their RS domains (Lazar et al. 1995; Lopato et al. 1996a). These proteins can be isolated by two salt precipitations (Zahler et al. 1992) and are able to complement inactive HeLa cell cytoplasmic splicing extracts. Furthermore, they are active in heterologous alternative splicing assays, suggesting conserved functions of plant SR proteins in splicing. Nevertheless, further research revealed a different complexity of plant SR proteins as well as some differences of their domain structure in comparison to metazoan proteins (Fig. 1) (Kalyna and Barta 2004). Analysis of the fully sequenced *Arabidopsis* and rice genomes has shown that they encode 19 and 24 SR proteins, respectively (Table 1). This is more than found in humans (10 SR proteins) (Bourgeois et al. 2004) or *Caenorhabditis elegans* (7 SR proteins) (Longman et al. 2000). However, only true orthologs of human SF2/ASF, SC35, and 9G8 are unambiguously identified in *Arabidopsis* and rice (Golovkin and Reddy 1998; Isshiki et al. 2006; Lazar et al. 1995; Lopato et al. 1999a, 1999b, 2002; Lorković and Barta 2002). In contrast to humans, plants possess multiple homologs of these proteins and thus form subfamilies. Interestingly, proteins with long RS domains like human SRp55 and SRp75 are not encoded in either the *Arabidopsis* or rice genome (Isshiki et al. 2006; Kalyna and Barta 2004; Lorković and Barta 2002; Reddy 2004).

The other plant SR proteins have a unique domain organization not found in any metazoan organism and are good candidates for proteins with plant-specific functions (Fig. 1). There are several plant-specific subfamilies (Table 1). Proteins of the RS subfamily were identified quite early in *Arabidopsis* (Lopato et al. 1996b), and their name comes from an RS domain highly enriched in arginines rather than serine-arginine dipeptides. These proteins have two typical RRM domains, however

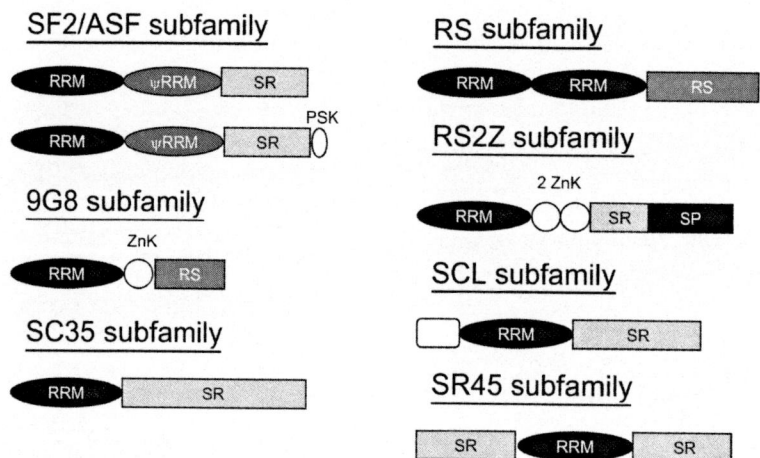

Fig. 1 Schematic representation of plant SR proteins. Subfamilies in the right column are specific for the plant kingdom. *RRM*, RNA recognition motif; *ψRRM*, ψRRM (contains the SWQDLKD motif, which is present in all SF2/ASF homologs); *SR*, domain rich in serine-arginine dipeptides; *RS*, domain rich in arginines and serines; *ZnK*, zinc knuckle of CCHC type; *SP*, domain rich in serines and prolines; *PSK*, region rich in proline, serine, and lysine. SCL proteins have an N-terminal extension rich in arginines, prolines, serines, glycines, and tyrosines.

Table 1 Plant SR proteins

Subfamily	Name	Accession	Reference
SF2/ASF	atSRp30	At1g09140	Lopato et al. 1999a
	atSRp34/SR1	At1g02840	Lazar et al. 1995; Lopato et al. 1999a
	atSRp34a	At3g49430	Lorković and Barta 2002
	atSRp34b	At4g02430	Lorković and Barta 2002
	osSRp32	Os03g22380	Isshiki et al. 2006
	osSRp33a	Os05g30140	Isshiki et al. 2006
	osSRp33b	Os07g47630	Isshiki et al. 2006
	osSRp20*	Os01g21420**	Isshiki et al. 2006*; Iida and Go. 2006**
	zmSRp30	AY649842	Gao et al. 2004
	zmSRp31	AY649843	Gao et al. 2004
	zmSRp32	AY649841	Gao et al. 2004
	taSRp30	DQ019630	Lopato et al. 2006
	taSRp30a	DQ019639	Lopato et al. 2006
9G8	atRSZp21/SRZ21	At1g23860	Golovkin and Reddy 1998; Lopato et al. 1999b
	atRSZp22/SRZ22	At4g31580	Golovkin and Reddy 1998; Lopato et al. 1999b
	atRSZp22a	At2g24590	Lopato et al. 1999b
	osRSZp21a	Os06g08840	Isshiki et al. 2006
	osRSp21b	Os02g54770	Isshiki et al. 2006
	osRSZp23	Os02g39720	Isshiki et al. 2006
	taRSZ22	DQ019626	Lopato et al. 2006
	taRSZ22a	DQ019627	Lopato et al. 2006
SC35	atSC35	At5g64200	Lopato et al. 2002
	osSC35a	Os08g37960	Isshiki et al. 2006
	osSC35b	Os07g43050	Isshiki et al. 2006
	osSC35c	Os03g27030	Isshiki et al. 2006
RS	atRSp31a	At2g46610	Kalyna and Barta 2004
	atRSp31	At3g61860	Lopato et al. 1996
	atRSp40	At4g25500	Lopato et al. 1996
	atRSp41	At5g52040	Lopato et al. 1996
	osRSp29	Os04g02870	Isshiki et al. 2006
	osRSp33	Os02g03040	Isshiki et al. 2006
	zmRSp31A	AY616013	Gupta et al. 2005
	zmRSp31B	AY616024	Gupta et al. 2005
RS2Z	atRSZ32	At3g53500	Lopato et al. 2002
	atRSZ33	At2g37340	Lopato et al. 2002
	osRSZ36	Os05g02880	Isshiki et al. 2006
	osRSZ37a	Os01g06290	Isshiki et al. 2006
	osRSZ37b	Os03g17710	Isshiki et al. 2006
	osRSZ39	Os05g07000	Isshiki et al. 2006
	taRSZ38	DQ019628	Lopato et al. 2006
SCL	atSCL28	At5g18810	Lopato et al. 2002
	atSCL30	At3g55460	Lopato et al. 2002
	atSCL30a	At3g13570	Lopato et al. 2002

(continued)

Table 1 (continued)

Subfamily	Name	Accession	Reference
	atSCL33/SR33	At1g55310	Golovkin and Reddy 1999
	osSCL25	Os07g43950	Isshiki et al. 2006
	osSCL26	Os03g25770	Isshiki et al. 2006
	osSCL30a	Os02g15310	Isshiki et al. 2006
	osSCL30b	Os12g38430	Isshiki et al. 2006
	–	Os03g24890	Iida and Go 2006
	–	Os11g47830	Iida and Go 2006
SR45	atSR45	At1g16610	Golovkin and Reddy 1999
	–	Os05g01540	Iida and Go 2006
	–	Os01g72890	Iida and Go 2006

at, *Arabidopsis thalina*; os, *Oryza sativa*; zm, *Zea mays*; ta, *Triticum aestivum*.
* Shortened version of the protein; **full-length version of the protein.
The subfamilies RS, RS2Z, SCL, and SR45 are specific to plants. In *Pinus taeda*, five proteins in the RS subfamily and one protein in the RS2Z subfamily have been identified (Kalyna et al. 2006). In *Physcomitrella patiens*, one protein in the RS subfamily, three proteins in the RS2Z subfamily, and three proteins in the SCL subfamily have been found (Iida and Go, 2006; Kalyna et al. 2006; our unpublished data). One protein in the RS subfamily has been identified in *Chlamydomonas reinhardtii* (Kalyna et al. 2006). Sequence information is available in the respective publications

without the SWQDLKD signature in their second RRM, which is characteristic for SF2/ASF-like proteins. Different plant species show different rates of expansion within this subfamily. There are four genes in the *Arabidopsis* RS subfamily (Kalyna and Barta 2004; Lopato et al. 1996b) but only two in rice (Isshiki et al. 2006). Interestingly, in gymnosperms (*Pinus taeda*) there are at least five proteins from this subfamily, and a single protein was detected both in moss *Physcomitrella patens* and in unicellular green alga *Chlamydomonas reinhardtii* (Kalyna et al. 2006).

Of particular interest are the members of the RS2Z family. In contrast to the homologs of the human 9G8, these proteins have an RRM, two Zn knuckles, and a SR domain followed by a domain rich in serines and prolines (Fig. 1). *Arabidopsis* has two members, atRSZ32 and atRSZ33 (Lopato et al. 2002), rice has four (Isshiki et al. 2006), wheat has two (Lopato et al. 2006), and at least three proteins are detected in *Physcomitrella* (Iida and Go 2006; Kalyna et al. 2006; our unpublished data); however, no true homolog was found in green algae (Kalyna et al. 2006).

The members of the SCL (SC35-like) protein family are most similar to SC35 and have a single RRM followed by an RS domain, but in addition they possess a short N-terminal extension with several RS and SP dipeptides (Golovkin and Reddy 1999; Isshiki et al. 2006; Lopato et al. 2002). These proteins show also some similarity to human SRp38, which in contrast to the majority of SR proteins acts as a splicing repressor (Cowper et al. 2001; Shin et al. 2004, 2005).

SR45 proteins have an atypical structure for SR proteins with two RS domains separated by the RRM (Golovkin and Reddy 1999). There is one SR45 protein in *Arabidopsis* and two genes in rice, and it is not present in animals and algae, indicating that it evolved in flowering plants later in evolution (Ali et al. 2007; Iida and Go 2006).

Within an *Arabidopsis* SR family there are closely related protein pairs (atSRp34/SR1 and atSRp34b, atRSp31 and atRSp31a, atRSp40 and atRSp41, atRSZ32 and atRSZ33, atSCL33 and atSCL30a, atRSZ22 and atRSZ22a) that are encoded by genes that arose through large interchromosomal duplications found in the *Arabidopsis* genome, explaining the higher SR protein complexity in plants (Kalyna and Barta 2004). This observation raises the question of whether these proteins have acquired different activities or are redundant in their functions. In general, a duplicate gene can potentially evolve new functions or become a pseudogene. In addition, fusions to new genomic loci can change the expression pattern of a gene, while functional constraints will preserve protein sequence. It has been shown that most *C. elegans* SR proteins are functionally redundant (Kawano et al. 2000; Longman et al. 2000). In plants, functional redundancy of the SR genes has not been addressed. For some of these protein pairs evidence exists that they are expressed in different organs or different cell types with little overlap, indicating that they may regulate splicing of different pre-mRNAs during plant development (M. Kalyna, unpublished results). Individual activities of paralogous pairs will only be evident from analysis of *Arabidopsis* mutant lines lacking individual SR proteins together with double mutant lines of the close homologs.

Binding Specificity of Plant SR Proteins

Plant SR proteins have been shown to regulate splice site choices *in planta* (Ali et al. 2007; Isshiki et al. 2006; Kalyna et al. 2003; Lazar and Goodman 2000; Lopato et al. 1999b). However, so far, only a few splicing regulatory *cis*-acting elements have been described in plants (Lewandowska et al. 2004; Simpson et al. 2000; Yoshimura et al. 2002). Most functions of SR proteins require the presence of specific RNA binding sites, usually very degenerate, purine-rich sequences often located in exons (Blencowe 2000; Bourgeois et al. 2004; Graveley 2000; Tacke and Manley 1999). As in general SR proteins stimulate splicing reactions, their binding sequences are termed splicing enhancers. Indeed, many mammalian constitutive exons contain functional splicing enhancers, many of them related to SR proteins (for review see Bourgeois et al. 2004; Goren et al. 2006).

The observation that plant SR proteins can functionally substitute mammalian SR proteins in an in vitro alternative splicing assay argues for similar RNA binding sequences in plant genes (Lazar et al. 1995; Lopato et al. 1996a, 1999a). Indeed, *Arabidopsis* RSZp22 and the closely related human SR proteins, 9G8 and hSRp20, exhibit similar RNA-binding specificities in vitro (Lopato et al. 1999a). Therefore, purine-rich exonic sequences might also serve as binding sites for plant SR proteins and could act as exonic splicing enhancers or silencers. A general survey for splicing enhancer sequences in the *Arabidopsis* genes has revealed 35 oligomeric sequences showing alternative splicing activity (Pertea et al. 2007). Nevertheless, for none of the plant SR proteins have such sequences been directly identified and characterized. Information about binding sites for individual plant SR proteins will either come from in vitro RNA selection procedures, like conventional or genomic SELEX (Cavaloc et al. 1999; Kim et al. 2003; Singer et al. 1997; Tacke and Manley 1995), or from selection of RNA sequences in vivo by cross-linking and immunopreciptation methods (Keene et al. 2006; Niranjanakumari et al. 2002; Ule et al. 2003, 2005a).

A critical question for plant splicing is the role of SR proteins, in particular the plant specific SR proteins, in the recognition process of the U-rich elements in pre-mRNA splicing. However, plant-specific SR proteins might not bind directly to the U-rich sequences but might exert their activity indirectly by binding to proteins that recognize the intronic U-rich sequences. These specific protein-protein interactions might be required to stimulate the plant spliceosome assembly process and might be the reason why animal introns in general are not recognized in plants (as most of them lack U-rich intronic elements). Therefore, efforts will concentrate on the sequence specificity of members of the RS and RS2Z subfamilies to determine in vivo complex formation. This could potentially help to answer the obvious question of whether interactions/functions of *Arabidopsis* SR proteins are intron- and/or exon dependent.

Complex Network of Interactions of Plant SR Proteins

One way to establish activities for SR proteins is to investigate complex formation in vitro and in vivo. Indeed, protein-protein interaction studies with *Arabidopsis* SR proteins not only revealed interactions found between human SR proteins and other splicing factors but also established some novel interactions.

In human cells U1-70K interacts with SF2/ASF, thereby stabilizing U1 snRNP binding to the 5′ splice site. Concurrently, U1–70K interacts with SC35, which also binds near the 3′ splice site, thereby helping spliceosome assembly (Cao and Garcia Blanco 1998; Kohtz et al. 1994; Wu and Maniatis 1993). A similar picture arises in plants, as immunoprecipitation experiments performed with the SF2/ASF homolog atSRp34/SR1 and atU1-70K protein coexpressed in protoplasts clearly revealed an interaction between them (Lorković et al. 2004). However, in plants atU1-70K was found to interact also with other SR proteins including atRSZp21, atRSZp22 and the plant-specific atSR33/SCL33, and atSR45 (Golovkin and Reddy 1998, 1999). The interaction with atRSZp21 and atRSZp22, which are *Arabidopsis* homologs of the human 9G8 protein, is of particular interest, as 9G8 was not found to interact with human U1-70K. In addition, it has been found that some plant SR proteins interact with the U2AF65 and U2AF35, two proteins important for recognition of the polypyrimidine tract and the 3′ splice site (our unpublished results). These interactions would argue that plant SR proteins perform at least some functions similar to their metazoan counterparts, that is, by stabilizing components bound at the 5′ and 3′ splice sites during early steps of the spliceosomal assembly.

Most *Arabidopsis* SR proteins were found to interact with U11-35K (U11 snRNP-specific protein 35K), a component of the minor spliceosome (Lorković et al. 2004, 2005). This, together with the previously reported involvement of SR proteins in splicing of minor AT-AC introns in vitro (Hastings and Krainer 2001a), indicates that plant SR proteins have similar functions in major and minor spliceosomes. Moreover, these results also indicate that U11-35K protein plays a role in splicing of minor introns similar to that of U1-70K in splicing of major introns, namely, interacting with SR proteins bound to splicing enhancers, which in turn stabilizes U11 snRNP at the 5′ splice site (Graveley 2000; Hastings and Krainer 2001b).

Extensive yeast two-hybrid screens with selected *Arabidopsis* SR proteins revealed many interactions with themselves and with other SR proteins, most of which have been confirmed by in vitro binding assays or coprecipitation (Lopato et al. 2002; our unpublished observation). Of particular interest is the interaction of atRSZ33 with all members of the SCL subfamily and with atSC35, atSRp34, atRSZp21, and atRSZp22 (Lopato et al. 2002). In addition, the wheat homolog taRSZ38 interacted with several splicing-related proteins among them, taU1-70K, taU2AF65, taU2AF35, and with the transportin TRN-SR (Lopato et al. 2006).

In the course of the yeast two-hybrid screening in *Arabidopsis* several additional non-SR proteins with either RS domains or with RD/ KE-rich domains have been identified. These included novel proteins potentially involved in splicing regulation, like kinases, helicases, cyclins, and cyclophilins (de la Fuente van Bentem et al. 2006; Lorković et al. 2004; Gullerova et al. 2006, 2007).

One of the kinases interacting with atSR33/SCL33 is atSRPK-4, a member of a family of SRPK kinases in plants (de la Fuente van Bentem et al. 2006). As shown for the metazoan SRPKs, atSRPK-4 phosphorylates SR proteins in vitro (our unpublished results), and it has been shown that the phosphorylation sites of atRSp31 in vivo are identical to the phosphorylation sites of atRSp31 in vitro (de la Fuente van Bentem et al. 2006). One of these sites, RpSP, is also found in other SR proteins in vivo, suggesting that SRPK-4 is one of the kinases regulating SR protein function.

The group of cyclophilins that are peptidyl-prolyl *cis-trans* isomerases (PPIase) are particularly interesting as they have the potential to regulate activities by changing protein structure, thus influencing phosphorylation/dephosphorylation events. The RS-containing cyclophilins, atCypRS64 and atCypRS92, have been investigated in more detail and have indeed been proposed to regulate interactions between SR proteins and the spliceosomal proteins U1-70K and U11-35K through changing their conformation and/or phosphorylation status (Lorković et al. 2004). In addition, the highly conserved nuclear cyclophilin, atCyp59, which contains an RRM domain, was also found to interact with most *Arabidopsis* SR proteins in vitro (Gullerova et al. 2006). As atCyp59 and its ortholog in *Schizosaccharomyces pombe* were also found to strongly influence the phosphorylation status of the carboxy terminal domain (CTD) of RNA Pol II, this protein is implicated in regulation of both transcription and splicing activities (Gullerova et al. 2006, 2007). These findings together with data on an U4/U6 snRNP-associated cyclophilin in humans indicate that PPIase activities are important for pre-mRNA splicing (Horowitz et al.

2002). This can be further supported by observations that several PPIases copurified with the in vitro assembled spliceosome from HeLa nuclear extracts (Deckert et al. 2006; Rappsilber et al. 2002; Zhou et al. 2002).

From these analyses it seems that a much more complex network of SR protein interactions exists in plants compared to metazoans. However, no such extensive yeast two-hybrid analysis of metazoan SR proteins has been performed to date. In any case, these data must be treated with caution as RS domains are readily interacting with other RS domains. Thus an important task will be to verify these interactions by identification of complexes in vivo. It seems possible that at least some of these interactions are responsible for differences observed between plant and metazoan splicing.

SR Proteins and Alternative Splicing

Alternative Splicing

Intron-containing genes are one of the hallmarks of modern eukaryotes, and it was estimated that introns occur in about 80% of plant genes (Alexandrov et al. 2006). Introns are implicated in the design of newly assembled genes and are therefore a driving force for extending gene diversity (see the chapter by B. J. Haas, this volume). In addition, introns allow for alternative splicing pathways, which are an important means to regulate gene expression and to enlarge the eukaryotic proteome. As an outcome, alternative splicing may create transcripts with various protein domains and hence varying activities but might also target the mRNA for nonsense-mediated decay (NMD) by creating a premature termination codon (PTC). In mammalian genomes about 60% of the genes are alternatively spliced, and this number rises to 74% for human multi-intron-containing genes (Johnson et al. 2003; Modrek and Lee 2003). In plants, the number of known cases of alternative splicing is constantly increasing, and current analysis of *Arabidopsis* and rice EST and full-length cDNA sequences are estimating at least 30% of genes with transcript support to undergo alternative splicing (Campbell et al. 2006; Wang and Brendel 2006; Xiao et al. 2005). Although the numbers might still increase, alternative splicing appears to occur less frequently in plants than in animal systems; this might be the consequence of smaller intron sizes in plants (Lorković et al. 2000b). With up to 9,000 Arabidopsis genes undergoing alternative splicing, these events will have a significant impact on gene expression, although in most cases the biological consequences of such events are unknown (Lorković et al. 2000b; Reddy 2007).

The mechanisms of selection of alternative splice sites involve the recognition of *cis*-acting splicing signals by RNA binding factors to enhance or suppress use of a particular splice site (Black 2003; Maniatis and Tasic 2002; Matlin et al. 2005; Smith and Valcarcel 2000). Beside the classical splice site consensus sequences and the polypyrimidine tract, sequences within exons or introns called exonic/intronic enhancers or suppressors (ESE/ISE and ESS/ISS, respectively) are important. Their position relative to competing splice sites and their interaction or interference with proteins or complexes determine splice site choice.

Regulated splicing is mainly controlled by changing the levels of protein factors and by their ability to act antagonistically as exemplified by the antagonism between SR proteins and hnRNP A1 (Mayeda and Krainer 1992). As proteins highly similar to metazoan hnRNP A/B proteins have been identified in plants (Lorković and Barta 2002; Lorković et al. 2000b), this kind of regulation, although not experimentally confirmed, is likely to occur in plants as well. This can be further supported by the fact that plant SR proteins change splice site usage of competing 5' splice sites in a heterologous mammalian splicing extract (Lazar et al. 1995; Lopato et al. 1996a). In addition to the action of more general splicing factors, genes are often regulated in cell-, tissue-, or developmental stage-specific manner by more specific proteins whose expressions are restricted and which might coordinate alternative splicing events of genes with related function (Blencowe 2006; Matlin et al. 2005; Ule et al. 2005b).

Regulation of Alternative Splicing by Plant SR Proteins and Consequences for Plant Growth and Development

In plants, alternatively spliced genes are involved in plant growth and development, signal transduction, responses to stresses, disease resistance, flowering, circadian rhythm, and metabolism, with regulatory and stress-response genes well represented (Dinesh-Kumar and Baker 2000; Egawa et al. 2006; Iida et al. 2004; Jia et al. 2004; Jordan et al. 2002; Kazan 2003; Larkin and Park 1999; Ner-Gaon et al. 2004; Quesada et al. 2003; Zhou et al. 2003) (see the chapters by W. Gassmann and G. S. Ali and A. S. N. Reddy, this volume). In particular, genes from the plant splicing machinery itself have been found to be subject to complex regulation involving alternative splicing (Iida and Go 2006; Iida et al. 2004; Kalyna et al. 2006; Palusa et al. 2007). For example, plant SR proteins have been shown to be regulated by developmental or stress cues on the transcriptional, posttranscriptional, and posttranslational levels (Gao et al. 2004; Gupta et al. 2005; Iida and Go 2006; Iida et al. 2004; Isshiki et al. 2006; Kalyna and Barta 2004; Kalyna et al. 2003, 2006; Lazar and Goodman 2000; Lopato et al. 1999b, 2002; Palusa et al. 2007). That means that plant SR proteins are targets of several signaling pathways and might serve as central players in the coordination of responses to developmental and environmental signals.

The first hint that plant SR proteins might regulate splicing in a similar manner to the mammalian SR proteins came from in vitro results in a heterologous HeLa cell splicing extract where SR protein preparations from plants or individual recombinant SR proteins (atSRp34/SR1, atRSp31, atRSZ22, atSR45) were able to complement SR deficient extracts and/or change splice site selection in a quantitative manner

(Ali et al. 2007; Lazar et al. 1995; Lopato et al. 1996a, 1996b, 1999a). More revealing information has been obtained from *Arabidopsis* plants ectopically expressing additional copies of a particular SR protein, leading to elevated protein levels in all tissues (Kalyna et al. 2003; Lopato et al. 1999b). Overexpression of atSRp30, one of the *Arabidopsis* SF2/ASF homologs, resulted in morphological and developmental changes displaying mostly a late flowering phenotype (Lopato et al. 1999b). More interesting, however, was the observation that alternative splicing patterns of several genes, *atRSp31*, *atU1-70K*, and *atSRp34/SR1*, were changed and atSRp30 also regulated splicing of its own pre-mRNA. In particular, elevated levels of atSRp30 changed the splicing pattern in *atSRp34/SR1*, another *Arabidopsis* SF2/ASF homolog, in a way that mRNA1 encoding the full-length protein was decreased but mRNA3, which encodes a protein with a shorter RS domain, strongly increased. Thus the level of atSRp34 protein was downregulated, whereas these plants accumulated the shorter version of the atSRp34 protein (Lopato et al. 1999b).

Ectopic expression of the plant-specific SR protein atRSZ33 was hampered by a tight regulation of its protein level (Kalyna et al. 2003). As a consequence only transgenic plants transformed with the genomic clone could be recovered. Molecular analysis showed only a small increase in atRSZ33 protein levels in these plants. Nevertheless, this caused severe pleiotropic changes in plant development resulting from increased cell expansion and changed polarization of cell elongation and division (Kalyna et al. 2003). In addition to changes in the splicing of *atSRp30*, *atSRp34*, and *atRSp31*, atRSZ33 was found to autoregulate both exogenous and endogenous splicing of its own pre-mRNA, leading to a large increase in alternatively spliced transcript (with no protein product detectable) and only a small increase in full-length mRNA (Kalyna et al. 2003, 2006). This indicates a feedback control loop of atRSZ33 on its own splicing pattern. As rice osRSZ36 overexpression also caused changes in splicing of its own pre-mRNA (Isshiki et al. 2006), this autoregulatory loop seems to be conserved in the plant-specifc RS2Z subfamily and is likely to have a role in the regulation of RS2Z protein levels. In contrast to atRSZ33, neither atRSp31 nor its rice homologs, osRSp29 and osRSp33, regulate alternative splicing in the long introns of their own pre-mRNAs (Isshiki et al. 2006; Kalyna et al. 2006). Similar feedback-regulatory networks among alternatively spliced SR proteins have been described in animal systems as well (Jumaa and Nielsen 1997; Kumar and Lopez 2005).

Along these lines are the results of recent experiments with *atSR45* knockout mutants, which also displayed late flowering by influencing the autonomous flowering pathway and altered leaf and flower morphology. The observed phenotypic changes are likely due to altered splicing patterns of *atSRp30*, *atRSp31*, *atRSp31a*, *atSRp34*, and *atSRp34b* genes in *atSRp45* mutant background (Ali et al. 2007). The involvement of SR proteins in the regulation of flowering time is further supported by changes in splicing patterns during thermal induction of flowering of the flowering regulators *FCA*, *MAF2*, and *FLM* and the expression profiles of several SR genes, including *atSRp30*, *atRSZp22a*, and *atSCL33/SR33* (Balasubramanian et al. 2006). Together, these results pinpoint the importance of a tight control of SR protein levels in particular cell or tissue types.

Why do plants ectopically expressing SR proteins show developmental phenotypes? By binding to splicing enhancer sequences, they modulate splice site selection in a concentration-dependent manner (reviewed by Smith and Valcarcel 2000). Some alternative splicing events are restricted to particular tissue or cell types and/ or developmental stages. These events are of great importance as they contribute to major developmental decisions, thereby generating species-specific traits and functions. This is best illustrated by examples of alternative splicing regulating nervous system differentiation and sex determination in *Drosophila* (Black and Grabowski 2003; Grabowski and Black 2001; Graveley 2002; Lopez 1998; Maniatis and Tasic 2002). As the expression patterns of particular plant SR proteins show tissue- and cell type specificity one can speculate that ectopic and constitutive expression in whole plants results in misregulation of splicing of specific pre-mRNAs required for normal plant development ((Fang et al. 2004; Gao et al. 2004; Golovkin and Reddy 1998, 1999; Kalyna et al. 2003; Lazar and Goodman 2000; Lazar et al. 1995; Lopato et al. 1996b, 1999b, 2002, 2006). However, to get a more global picture of changes of gene expression by SR genes it will be necessary to analyze mutant transcriptomes by microarray technology. In addition, changes in alternative splicing pattern should be monitored by splicing-sensitive, high-throughput methods.

Conserved Alternative Splicing of SR Genes

SR proteins are evolutionarily conserved. In addition, alternative splicing in SR genes is well documented in various species, both in metazoans and in plants. However, only recently have evolutionary conservation and cross-species comparison of alternative splicing events in SR genes received attention. In fact, similarities in alternative splicing events of different *Arabidopsis* SR genes were recognized quite early. It has been shown that *atSRp30* and *atSRp34/SR1* have similar gene structure; both possess quite long introns in their RS domain. In both genes these introns are alternatively spliced resulting in proteins with a truncated RS domain due to the presence of premature termination codon (PTC) (Lopato et al. 1999b). Similar alternative splicing events were described later also in maize SF2/ASF genes (Gao et al. 2004). Alternative splicing in *atRSp31* (Lopato et al. 1999b), *atSR33/atSCL33* (Golovkin and Reddy 1999), and *atRSZ33* (Lopato et al. 2002) occurs also in relatively long introns, which are, however, situated within the RRM. These introns are more than 400 nt, long while the average plant introns are shorter than 150 nt (Lorković et al. 2000b). Further analysis revealed that the majority of *Arabidopsis* SR genes that belong to the plant-specific subfamilies RS, RS2Z, and SCL have these alternatively spliced long introns in their RRMs. Due to the presence of a PTC, these alternative splice variants potentially encode extremely truncated proteins containing only a part of the RRM (Kalyna and Barta 2004). Indeed, no protein encoded by the alternative splice forms of *atRSp31* could be detected (Kalyna et al. 2006).

In the RS and SCL subfamilies an intriguing feature of these PTC+ alternative transcripts is that they seem to be quite stable and are easily detected by Northern blot analysis. In some tissues these transcripts make up a significant portion of the transcript population of a given gene. Interestingly, these alternative transcripts are regulated during development, which indicates their importance for SR protein expression (Golovkin and Reddy 1999; Lopato et al. 1996b).

The functional significance of these alternative splicing events was also implicated by cross-species analyses of plant-specific SR genes (Iida and Go 2006; Kalyna et al. 2006). This revealed the presence of alternatively spliced introns at a conserved position between RNP2 and RNP1 of the RRM in distant taxa, from dicots and monocots to the unicellular green alga *C. reinhardtii*. The same types of alternative splicing events are conserved in different species. Mostly, either alternative acceptor or donor splice sites are being used; however, simultaneous usage of both alternative acceptor and donor splice sites creating a cassette exon is also a frequent event. Although the overall conservation of the intronic sequences is not very high, the sequences of alternative acceptor and/or donor splice sites are well conserved in the RS2Z subfamily from *Physcomitrella* to *Arabidopsis* during at least 700 MY (Kalyna et al. 2006). Interestingly, not every alternative splicing event in these introns uses conserved alternative splice sites. Moreover, in maize both conserved and nonconserved alternative splice sites can be used in the same transcript and some alternative splice variants preserve the reading frame (Gupta et al. 2005; Kalyna et al. 2006). It is interesting that not only sequence features but also the regulation of the alternative splicing event is conserved in the orthologs of the SR protein genes. In the RS2Z subfamily, both atRSZ33 and osRSZ36 have been demonstrated to autoregulate splicing of their own pre-mRNAs (Isshiki et al. 2006; Kalyna et al. 2003). However, in the RS subfamily, neither *atRSp31* nor *osRSp29* and *osRSp33* are subjected to autoregulation (Isshiki et al. 2006; Kalyna et al. 2006).

Interestingly, in contrast to the plant SR genes, alternative splicing events in animal SR genes are located in long ultraconserved sequences, whereby the relative abundance of PTC+ isoforms is quite low because of the regulation by nonsense-mediated decay (NMD) (Lareau et al. 2007; Ni et al. 2007) (see the chapter by D. A. Belostotsky, this volume). This differs from the situation in plant SR genes with their high abundance of alternative spliced transcripts with PTC, which argues against being regulated by an NMD-dependent mechanism.

Taken together, these results show that both plant and animal SR genes utilize highly conserved alternative splicing events for gene regulation, although the specific mechanisms might be different. Further studies will be needed to elucidate the cellular compartmentalization of the conserved alternative transcripts and to evaluate the impact of NMD on the regulation of plant SR genes.

Conclusions

SR proteins are one of the most important splicing factor families. They are highly conserved in all organisms that undergo alternative splicing and are therefore one of the key elements for enlarging genome diversity. Plants do possess a multitude of SR proteins due to several genome duplication events giving rise to several homologous SR gene pairs. Therefore, an interesting line to pursue is the question of whether their activities are redundant or whether new functions have evolved. Several of the plant SR protein families are specific for plants, suggesting that they evolved to carry out activities specific to the plant kingdom. Most plant SR genes are alternatively spliced, and their splicing patterns can be regulated by developmental and environmental signals. This regulation might be central for coordinated gene responses to these signals, as SR proteins are essential for proper gene expression of most of the eukaryotic genes. Consequently, deregulation of some plant SR proteins leads to distinct developmental changes. An exciting phenomenon is the discovery that three of the plant-specific families display highly conserved splicing events interrupting the RRM and with no indication for protein production. As this event is evolutionarily conserved even in unicellular green algae, it will be an important task to uncover the significance of these alternative splicing events for the regulation of plant-specific SR genes.

Acknowledgments Research in the author laboratory is supported by grants from the Austrian National Science Foundation (FWF: SFB-017), from the Austrian GENAU initiative, form the WWTF, and from the European Network of Excellence on Alternative Splicing (EURASNET).

References

Alexandrov NN, Troukhan ME, Brover VV, Tatarinova T, Flavell RB, Feldmann KA (2006) Features of Arabidopsis genes and genome discovered using full-length cDNAs. Plant Mol Biol 60:69–85
Ali GS, Palusa SG, Golovkin M, Prasad J, Manley JL, Reddy AS (2007) Regulation of plant developmental processes by a novel splicing factor. PLoS ONE 2: e471
Balasubramanian S, Sureshkumar S, Lempe J, Weigel D (2006) Potent induction of *Arabidopsis thaliana* flowering by elevated growth temperature. PLoS Genet 2: e106
Barta A, Sommergruber K, Thompson D, Hartmuth K, Matzke MA, Matzke AJM (1986) The expression of a nopaline synthase—human growth hormone chimaeric gene in transformed tobacco and sunflower callus tissue. Plant Mol Biol 6:347–357
Behzadnia N, Golas MM, Hartmuth K, Sander B, Kastner B, Deckert J, Dube P, Will CL, Urlaub H, Stark H, Luhrmann R (2007) Composition and three-dimensional EM structure of double affinity-purified, human prespliceosomal A complexes. EMBO J 26:1737–1748
Black DL (2003) Mechanisms of alternative pre-messenger RNA splicing. Annu Rev Biochem 72:291–336
Black DL, Grabowski PJ (2003) Alternative pre-mRNA splicing and neuronal function. Prog Mol Subcell Biol 31:187–216
Blencowe BJ (2000) Exonic splicing enhancers: mechanism of action, diversity and role in human genetic diseases. Trends Biochem Sci 25:106–10
Blencowe BJ (2006) Alternative splicing: new insights from global analyses. Cell 126:37–47
Blencowe BJ, Bowman JA, McCracken S, Rosonina E (1999) SR-related proteins and the processing of messenger RNA precursors. Biochem Cell Biol 77:277–291
Bourgeois CF, Lejeune F, Stevenin J (2004) Broad specificity of SR (serine/arginine) proteins in the regulation of alternative splicing of pre-messenger RNA. Prog Nucleic Acid Res Mol Biol 78:37–88
Brown JW, Feix G, Frendewey D (1986) Accurate in vitro splicing of two pre-mRNA plant introns in a HeLa cell nuclear extract. EMBO J. 5:2749–2758

Brown JWS, Simpson CG (1998) Splice site selection in plant pre-mRNA splicing. Annu Rev Plant Physiol Plant Mol Biol 49:77–95

Campbell MA, Haas BJ, Hamilton JP, Mount SM, Buell CR (2006) Comprehensive analysis of alternative splicing in rice and comparative analyses with Arabidopsis. BMC Genomics 7:327

Cao W, Garcia Blanco MA (1998) A serine/arginine-rich domain in the human U1 70k protein is necessary and sufficient for ASF/SF2 binding. J Biol Chem 273:20629–20635

Cavaloc Y, Bourgeois CF, Kister L, Stevenin J (1999) The splicing factors 9G8 and SRp20 trans-activate splicing through different and specific enhancers. RNA 5:468–483

Cowper AE, Caceres JF, Mayeda A, Screaton GR (2001) Serine-arginine (SR) protein-like factors that antagonize authentic SR proteins and regulate alternative splicing. J Biol Chem 276:48908–48914

de la Fuente van Bentem S, Anrather D, Roitinger E, Djamei A, Hufnagl T, Barta A, Csaszar E, Dohnal I, Lecourieux D, Hirt H (2006) Phosphoproteomics reveals extensive in vivo phosphorylation of Arabidopsis proteins involved in RNA metabolism. Nucleic Acids Res 34:3267–3278

Deckert J, Hartmuth K, Boehringer D, Behzadnia N, Will CL, Kastner B, Stark H, Urlaub H, Luhrmann R (2006) Protein composition and electron microscopy structure of affinity-purified human spliceosomal B complexes isolated under physiological conditions. Mol Cell Biol 26:5528–5543

Dinesh-Kumar SP, Baker BJ (2000) Alternatively spliced N resistance gene transcripts: their possible role in tobacco mosaic virus resistance. Proc Natl Acad Sci USA 97:1908–1913

Egawa C, Kobayashi F, Ishibashi M, Nakamura T, Nakamura C, Takumi S (2006) Differential regulation of transcript accumulation and alternative splicing of a DREB2 homolog under abiotic stress conditions in common wheat. Genes Genet Syst 81:77–91

Fang Y, Hearn S, Spector DL (2004) Tissue-specific expression and dynamic organization of SR splicing factors in Arabidopsis. Mol Biol Cell 15:2664–2673

Fu X-D (1995) The superfamily of arginine/serine-rich splicing factors. RNA 1:663–680

Gao H, Gordon-Kamm WJ, Lyznik LA (2004) ASF/SF2-like maize pre-mRNA splicing factors affect splice site utilization and their transcripts are alternatively spliced. Gene 339:25–37

Golovkin M, Reddy AS (1998) The plant U1 small nuclear ribonucleoprotein particle 70K protein interacts with two novel serine/arginine-rich proteins. Plant Cell 10:1637–1648

Golovkin M, Reddy AS (1999) An SC35-like protein and a novel serine/arginine-rich protein interact with Arabidopsis U1-70K protein. J Biol Chem 274:36428–36438

Goren A, Ram O, Amit M, Keren H, Lev-Maor G, Vig I, Pupko T, Ast G (2006) Comparative analysis identifies exonic splicing regulatory sequences—The complex definition of enhancers and silencers. Mol Cell 22:769–781

Grabowski PJ, Black DL (2001) Alternative RNA splicing in the nervous system. Prog Neurobiol 65:289–308

Graveley BR (2000) Sorting out the complexity of SR protein functions. RNA 6:1197–1211

Graveley BR (2002) Sex, AGility, and the regulation of alternative splicing. Cell 109:409–412

Gullerova M, Barta A, Lorković ZJ (2006) AtCyp59 is a multidomain cyclophilin from *Arabidopsis thaliana* that interacts with SR proteins and the C-terminal domain of the RNA polymerase II. RNA 12:631–643

Gullerova M, Barta A, Lorković ZJ (2007) Rct1, a nuclear RNA recognition motif-containing cyclophilin, regulates phosphorylation of the RNA polymerase II C-terminal domain. Mol Cell Biol 27:3601–3611

Gupta S, Wang BB, Stryker GA, Zanetti ME, Lal SK (2005) Two novel arginine/serine (SR) proteins in maize are differentially spliced and utilize non-canonical splice sites. Biochim Biophys Acta 1728:105–114

Hartmuth K, Barta A (1986) In vitro processing of a plant pre-mRNA in a HeLa cell nuclear extract. Nucleic Acids Res 14:7513–7528

Hastings ML, Krainer AR (2001a) Functions of SR proteins in the U12-dependent AT-AC pre-mRNA splicing pathway. RNA 7:471–482

Hastings ML, Krainer AR (2001b) Pre-mRNA splicing in the new millennium. Curr Opin Cell Biol 13:302–309

Horowitz DS, Lee EJ, Mabon SA, Misteli T (2002) A cyclophilin functions in pre-mRNA splicing. EMBO J 21:470–480

Huang Y, Yario TA, Steitz JA (2004) A molecular link between SR protein dephosphorylation and mRNA export. Proc Natl Acad Sci USA 101:9666–9670

Iida K, Go M (2006) Survey of conserved alternative splicing events of mRNAs encoding SR proteins in land plants. Mol Biol Evol 23:1085–1094

Iida K, Seki M, Sakurai T, Satou M, Akiyama K, Toyoda T, Konagaya A, Shinozaki K (2004) Genome-wide analysis of alternative pre-mRNA splicing in *Arabidopsis thaliana* based on full-length cDNA sequences. Nucleic Acids Res 32:5096–5103

Isshiki M, Tsumoto A, Shimamoto K (2006) The serine/arginine-rich protein family in rice plays important roles in constitutive and alternative splicing of pre-mRNA. Plant Cell 18:146–158

Jia Y, del Rio HS, Robbins AL, Louzada ES (2004) Cloning and sequence analysis of a low temperature-induced gene from trifoliate orange with unusual pre-mRNA processing. Plant Cell Rep 23:159–166

Johnson JM, Castle J, Garrett-Engele P, Kan Z, Loerch PM, Armour CD, Santos R, Schadt EE, Stoughton R, Shoemaker DD (2003) Genome-wide survey of human alternative pre-mRNA splicing with exon junction microarrays. Science 302:2141–4

Jordan T, Schornack S, Lahaye T (2002) Alternative splicing of transcripts encoding Toll-like plant resistance proteins—what's the functional relevance to innate immunity? Trends Plant Sci 7:392–398

Jumaa H, Nielsen PJ (1997) The splicing factor SRp20 modifies splicing of its own mRNA and ASF/SF2 antagonizes this regulation. EMBO J 16:5077–5085

Kalyna M, Barta A (2004) A plethora of plant serine/arginine-rich proteins: redundancy or evolution of novel gene functions? Biochem Soc Trans 32:561–564

Kalyna M, Lopato S, Barta A (2003) Ectopic expression of atRSZ33 reveals its function in splicing and causes pleiotropic changes in development. Mol Biol Cell 14:3565–3577

Kalyna M, Lopato S, Voronin V, Barta A (2006) Evolutionary conservation and regulation of particular alternative splicing events in plant SR proteins. Nucleic Acids Res 34:4395–4405

Kawano T, Fujita M, Sakamoto H (2000) Unique and redundant functions of SR proteins, a conserved family of splicing factors, in *Caenorhabditis elegans* development. Mech Dev 95:67–76

Kazan K (2003) Alternative splicing and proteome diversity in plants: the tip of the iceberg has just emerged. Trends Plant Sci 8:468–471

Keene JD, Komisarow JM, Friedersdorf MB (2006) RIP-Chip: the isolation and identification of mRNAs, microRNAs and protein components of ribonucleoprotein complexes from cell extracts. Nat Protoc 1:302–307

Kim S, Shi H, Lee DK, Lis JT (2003) Specific SR protein-dependent splicing substrates identified through genomic SELEX. Nucleic Acids Res 31:1955–1961

Kohtz JD, Jamison SF, Will CL, Zuo P, Luhrmann R, Garcia-Blanco MA, Manley JL (1994) Protein-protein interactions and 5′-splice-site recognition in mammalian mRNA precursors. Nature 368:119–124

Kumar S, Lopez AJ (2005) Negative feedback regulation among SR splicing factors encoded by Rbp1 and Rbp1-like in Drosophila. EMBO J 24:2646–2655

Lambermon MH, Fu Yu L, Wieczorek Kirk DA, Dupasquier M, Filipowicz W, Lorković Z-J (2002) UBA1 and UBA2, two proteins that interact with UBP1, a multifunctional effector of pre-mRNA maturation in plants. Mol Cell Biol 22: 4346–4357

Lambermon MH, Simpson GG, Wieczorek Kirk DA, Hemmings-Mieszczak M, Klahre U, Filipowicz W (2000) UBP1, a novel hnRNP-like protein that functions at multiple steps of higher plant nuclear pre-mRNA maturation. EMBO J 19:1638–1649

Lareau LF, Inada M, Green RE, Wengrod JC, Brenner SE (2007) Unproductive splicing of SR genes associated with highly conserved and ultraconserved DNA elements. Nature 446:926–929

Larkin PD, Park WD (1999) Transcript accumulation and utilization of alternate and non-consensus splice sites in rice granule-bound starch synthase are temperature-sensitive and controlled by a single-nucleotide polymorphism. Plant Mol Biol 40:719–727

Lazar G, Goodman HM (2000) The Arabidopsis splicing factor SR1 is regulated by alternative splicing. Plant Mol Biol 42:571–581

Lazar G, Schaal T, Maniatis T, Goodman HM (1995) Identification of a plant serine-arginine-rich protein similar to the mammalian splicing factor SF2/ASF. Proc Natl Acad Sci USA 92:7672–7676

Lewandowska D, Simpson CG, Clark GP, Jennings NS, Barciszewska-Pacak M, Lin CF, Makalowski W, Brown JW, Jarmolowski A (2004) Determinants of plant U12-dependent intron splicing efficiency. Plant Cell 16:1340–1352

Li X, Manley JL (2005) New talents for an old acquaintance: the SR protein splicing factor ASF/SF2 functions in the maintenance of genome stability. Cell Cycle 4:1706–1708

Li X, Manley JL (2006) Cotranscriptional processes and their influence on genome stability. Genes Dev 20:1838–1847

Longman D, Johnstone IL, Caceres JF (2000) Functional characterization of SR and SR-related genes in *Caenorhabditis elegans*. EMBO J 19:1625–1637

Lopato S, Borisjuk L, Milligan AS, Shirley N, Bazanova N, Langridge P (2006) Systematic identification of factors involved in post-transcriptional processes in wheat grain. Plant Mol Biol 62:637–653

Lopato S, Forstner C, Kalyna M, Hilscher J, Langhammer U, Indrapichate K, Lorković ZJ, Barta A (2002) Network of interactions of a novel plant-specific Arg/Ser-rich protein, atRSZ33, with atSC35-like splicing factors. J Biol Chem 277:39989–39998

Lopato S, Gattoni R, Fabini G, Stevenin J, Barta A (1999a) A novel family of plant splicing factors with a Zn knuckle motif: examination of RNA binding and splicing activities. Plant Mol Biol 39:761–773

Lopato S, Kalyna M, Dorner S, Kobayashi R, Krainer AR, Barta A (1999b) atSRp30, one of two SF2/ASF-like proteins from *Arabidopsis thaliana*, regulates splicing of specific plant genes. Genes Dev 13:987–1001

Lopato S, Mayeda A, Krainer AR, Barta A (1996a) Pre-mRNA splicing in plants: characterization of Ser/Arg splicing factors. Proc Natl Acad Sci USA 93:3074–3079

Lopato S, Waigmann E, Barta A (1996b) Characterization of a novel arginine/serine-rich splicing factor in Arabidopsis. Plant Cell 8:2255–2264

Lopez AJ (1998) Alternative splicing of pre-mRNA: developmental consequences and mechanisms of regulation. Annu Rev Genet 32:279–305

Lorković ZJ, Barta A (2002) Genome analysis: RNA recognition motif (RRM) and K homology (KH) domain RNA-binding proteins from the flowering plant *Arabidopsis thaliana*. Nucleic Acids Res 30:623–635

Lorković ZJ, Lehner R, Forstner C, Barta A (2005) Evolutionary conservation of minor U12-type spliceosome between plants and humans. RNA 11:1095–1107

Lorković ZJ, Lopato S, Pexa M, Lehner R, Barta A (2004) Interactions of Arabidopsis RS domain containing cyclophilins with SR proteins and U1 and U11 small nuclear ribonucleoprotein-specific proteins suggest their involvement in pre-mRNA splicing. J Biol Chem 279:33890–33898

Lorković ZJ, Wieczorek Kirk DA, Klahre U, Hemmings Mieszczak M, Filipowicz W (2000a) RBP45 and RBP47, two oligouridylate-specific hnRNP-like proteins interacting with poly(A)+ RNA in nuclei of plant cells. RNA 6:1610–1624

Lorković ZJ, Wieczorek Kirk DA, Lambermon MH, Filipowicz W (2000b) Pre-mRNA splicing in higher plants. Trends Plant Sci 5:160–167

Maniatis T, Tasic B (2002) Alternative pre-mRNA splicing and proteome expansion in metazoans. Nature 418:236–243

Matlin AJ, Clark F, Smith CW (2005) Understanding alternative splicing: towards a cellular code. Nat Rev Mol Cell Biol 6:386–398

Mayeda A, Krainer AR (1992) Regulation of alternative pre-mRNA splicing by hnRNP A1 and splicing factor SF2. Cell 68:365–375

Modrek B, Lee CJ (2003) Alternative splicing in the human, mouse and rat genomes is associated with an increased frequency of exon creation and/or loss. Nat Genet 34:177–180

Ner-Gaon H, Halachmi R, Savaldi-Goldstein S, Rubin E, Ophir R, Fluhr R (2004) Intron retention is a major phenomenon in alternative splicing in Arabidopsis. Plant J 39:877–885

Ni JZ, Grate L, Donohue JP, Preston C, Nobida N, O'Brien G, Shiue L, Clark TA, Blume JE, Ares M, Jr. (2007) Ultraconserved elements are associated with homeostatic control of splicing regulators by alternative splicing and nonsense-mediated decay. Genes Dev 21:708–718

Niranjanakumari S, Lasda E, Brazas R, Garcia-Blanco MA (2002) Reversible cross-linking combined with immunoprecipitation to study RNA-protein interactions in vivo. Methods 26:182–190

Palusa SG, Ali GS, Reddy AS (2007) Alternative splicing of pre-mRNAs of Arabidopsis serine/arginine-rich proteins: regulation by hormones and stresses. Plant J 49:1091–1107

Pertea M, Mount SM, Salzberg SL (2007) A computational survey of candidate exonic splicing enhancer motifs in the model plant *Arabidopsis thaliana*. BMC Bioinformatics 8:159

Quesada V, Macknight R, Dean C, Simpson GG (2003) Autoregulation of FCA pre-mRNA processing controls Arabidopsis flowering time. EMBO J 22:3142–3152

Rappsilber J, Ryder U, Lamond AI, Mann M (2002) Large-scale proteomic analysis of the human spliceosome. Genome Res 12:1231–1245

Reddy AS (2004) Plant serine/arginine-rich proteins and their role in pre-mRNA splicing. Trends Plant Sci 9:541–547

Reddy AS (2007) Alternative splicing of pre-messenger RNAs in plants in the genomic era. Annu Rev Plant Biol 58:267–294

Reed R, Cheng H (2005) TREX, SR proteins and export of mRNA. Curr Opin Cell Biol 17:269–273

Shin C, Feng Y, Manley JL (2004) Dephosphorylated SRp38 acts as a splicing repressor in response to heat shock. Nature 427:553–558

Shin C, Kleiman FE, Manley JL (2005) Multiple properties of the splicing repressor SRp38 distinguish it from typical SR proteins. Mol Cell Biol 25:8334–8343

Simpson CG, Hedley PE, Watters JA, Clark GP, McQuade C, Machray GC, Brown JW (2000) Requirements for mini-exon inclusion in potato invertase mRNAs provides evidence for exon-scanning interactions in plants. RNA 6:422–433

Singer BS, Shtatland T, Brown D, Gold L (1997) Libraries for genomic SELEX. Nucleic Acids Res 25:781–786

Smith CW, Valcarcel J (2000) Alternative pre-mRNA splicing: the logic of combinatorial control. Trends Biochem Sci 25:381–388

Stojdl DF, Bell JC (1999) SR protein kinases: the splice of life. Biochem Cell Biol 77:293–298

Tacke R, Manley JL (1995) The human splicing factors ASF/SF2 and SC35 possess distinct, functionally significant RNA binding specificities. EMBO J 14:3540–3551

Tacke R, Manley JL (1999) Determinants of SR protein specificity. Curr Opin Cell Biol 11:358–362

Tenenbaum SA, Aguirre-Ghiso J (2005) Dephosphorylation shows SR proteins the way out. Mol Cell 20:499–501

Ule J, Jensen K, Mele A, Darnell RB (2005a) CLIP: a method for identifying protein-RNA interaction sites in living cells. Methods 37:376–386

Ule J, Jensen KB, Ruggiu M, Mele A, Ule A, Darnell RB (2003) CLIP identifies Nova-regulated RNA networks in the brain. Science 302:1212–1215

Ule J, Ule A, Spencer J, Williams A, Hu JS, Cline M, Wang H, Clark T, Fraser C, Ruggiu M, Zeeberg BR, Kane D, Weinstein JN, Blume J, Darnell RB (2005b) Nova regulates brain-specific splicing to shape the synapse. Nat Genet 37:844–852

Wang BB, Brendel V (2006) Genomewide comparative analysis of alternative splicing in plants. Proc Natl Acad Sci USA 103:7175–7180

Wu JY, Maniatis T (1993) Specific interactions between proteins implicated in splice site selection and regulated alternative splicing. Cell 75:1061–1070

Xiao YL, Smith SR, Ishmael N, Redman JC, Kumar N, Monaghan EL, Ayele M, Haas BJ, Wu HC, Town CD (2005) Analysis of the cDNAs of hypothetical genes on Arabidopsis chromosome 2 reveals numerous transcript variants. Plant Physiol 139:1323–1337

Yoshimura K, Yabuta Y, Ishikawa T, Shigeoka S (2002) Identification of a cis element for tissue-specific alternative splicing of chloroplast ascorbate peroxidase pre-mRNA in higher plants. J Biol Chem 277:40623–40632

Zahler AM, Lane WS, Stolk JA, Roth MB (1992) SR proteins: a conserved family of pre-mRNA splicing factors. Genes Dev 6:837–847

Zhou Y, Zhou C, Ye L, Dong J, Xu H, Cai L, Zhang L, Wei L (2003) Database and analyses of known alternatively spliced genes in plants. Genomics 82:584–595

Zhou Z, Licklider LJ, Gygi SP, Reed R (2002) Comprehensive proteomic analysis of the human spliceosome. Nature 419:182–185

Spatiotemporal Organization of Pre-mRNA Splicing Proteins in Plants

G.S. Ali, A.S.N. Reddy(✉)

Contents

Introduction 104
Experimental Approaches in Studying Cell Biology
of Proteins Associated with Pre-mRNA in Plants 105
Subcellular Organization of Proteins Involved in
Pre-mRNA Splicing in Plant Cells. 108
Regulation of Spatiotemporal Organization and Dynamics
of Pre-mRNA Splicing Proteins. 110
Mobility Analyses of Proteins Associated with Pre-mRNA
Splicing at the Molecular Level 112
Model for the Distribution and Kinetics of Plant SR Proteins. 114
Conclusions and Future Directions 116
References 117

Abstract The general organization of eukaryotic nuclei, including plant nuclei, into functional domains is now widely recognized. Conventional immunocytochemistry and visualization of proteins fused to fluorescent proteins (FP) have revealed that in plants, RNA and protein components of pre-mRNA splicing are spatially organized depending on the stage of cell cycle, development, and the cell's physiological state. Application of some of the latest microscopy techniques, which reveal biophysical properties such as diffusion and interaction properties of proteins, has begun to provide important insights into the functional organization of spliceosomal proteins in plants. Although some progress has been made in understanding the spatial and temporal organization of splicing machinery in plants, the mechanisms that regulate this organization and its functional consequences remain unresolved.

A.S.N. Reddy
Department of Biology and Program in Molecular Plant Biology, Colorado State University,
Fort Collins, CO 80523, USA
e-mail: reddy@colostate.edu

Introduction

Advances in cell biological tools have allowed detailed analysis of the subcellular organization and dynamics of pre-mRNA splicing factors. Questions such as how and under what physiological conditions, and in what tissues and developmental states, plant splicing machinery is regulated are being vigorously pursued. Motivations for these investigations had primarily come from the availability of fluorescence microscopes combined with more efficient lasers and high-speed cameras and detectors that are capable of recording ultra-fast movements. So far these studies in plants are limited. Nevertheless, an analysis of existing literature indicates that, as in metazoans, plant nuclei are also intricately organized into functional domains. The most prominent of these are the nucleoli, chromosome territories, Cajal bodies, and speckles that harbor a variety of nuclear proteins (Lorković and Barta 2004; Shaw and Brown 2004). Experimental and comparative analyses based on sequence similarity searches have revealed an overall conservation of core and other major spliceosomal proteins between plants and animals (Wang and Brendel 2004), suggesting that the general mechanism of splicing is conserved across a broad spectrum of eukaryotes (see the chapter by V. Brendel and colleagues, this volume). These studies have also revealed important differences in spliceosomal proteins.

SR splicing factors, a family of non-snRNP proteins in the spliceosome, are known to play important roles in pre-mRNA splicing, mRNA export, RNA stability, protein translation, and genome stability (Huang and Steitz 2005; Li and Manley 2005, 2006). Several of these proteins are conserved between plants and metazoans (Kalyna and Barta 2004; Reddy 2004) (see the chapter by A. Barta et al., this volume). In contrast to 10 SR splicing factors in metazoans, in the dicot and monocot model plants, Arabidopsis and rice, there are 19 and 24 SR splicing factors, respectively, including several plant-specific members (Lorković and Barta 2002; Sanford et al. 2003; Kalyna and Barta 2004; Reddy 2007), indicating that certain aspects of RNA metabolism and its regulation in plants are likely to be different from those in animals. In the interphase nucleus of animal cells, several of these SR proteins are present in a diffuse nucleoplasmic pool and in concentrated areas termed speckles (Lamond and Spector 2003; Shaw and Brown 2004). Studies with animal SR proteins have provided useful information about their structure and the regulation of their cellular mobility (Misteli 2001). However, the mechanisms that regulate subcellular distribution and mobility of plant SR proteins are largely unknown.

Over the past several years considerable efforts have been devoted to understanding the spatial organization of pre-mRNA splicing in pants. Here we review (a) the function and cell cycle-dependent changes in the localization pattern of splicing-related proteins and (b) the dynamics of these proteins as a function of the transcription and phosphorylation state of the cells. First we give a brief introduction to techniques used in studying spatiotemporal dynamics of splicing machinery in plants including some of the latest microscopic techniques such as fluorescence recovery after photobleaching (FRAP) and bimolecular fluorescence complementation (BiFC), and then we provide an insight into the dynamics of these components revealed by these techniques. Since most of these studies focus on the SR proteins, a discussion of these proteins constitutes a major portion of this article.

Experimental Approaches in Studying Cell Biology of Proteins Associated with Pre-mRNA in Plants

Understanding the dynamic organization of pre-mRNA splicing and its relationship with other steps in gene expression, such as transcription, polyadenylation, 5'-capping, export to the cytoplasm, and translation requires the use of modern microscopic techniques. Although they are used extensively for studying the dynamics of a variety of proteins in animals, the use of these techniques in plant is limited but is gaining momentum. To orient the reader toward the use of these techniques, here we briefly introduce several of these techniques that have been used in plants or would be worth using in the future. For details on each of these techniques, the reader is referred to excellent reviews and articles (Reits and Neefjes 2001; Sprague and McNally 2005). The ability to genetically tag virtually any protein with a fluorescent protein such as green fluorescent protein (GFP) has revolutionized the way in which cell biology is studied. In addition to its use as a localization tool, the ability to photobleach GFP makes GFP fusion protein, in combination with high-speed image acquisition technology, an essential tool for studying the dynamics of proteins in living plant cells.

Initial analyses with plant mRNA splicing used mostly immunocytochemistry and electron microscopy with antibodies raised against mammalian splicing-related proteins and RNA (Testillano et al. 1993; Beven et al. 1995; Glyn and Leitch 1995). These analyses indicated that the structural properties of major splicing components are conserved between animals and plants. These analyses, however, are not a substitute for the use of native plant splicing components. Befittingly, plant researchers have begun to use plant splicing-related proteins for addressing issues about the spatiotemporal dynamics of splicing. In addition, plant biologists have also adopted a variety of microscopy techniques originally developed for studies in animal cells. Among these techniques, several photobleaching techniques are rapidly gaining popularity for studying the dynamics of plant proteins. Fluorescence recovery after photobleaching (FRAP) was first developed and used in the 1970s by a limited number of laboratories that had the technical expertise for studying the dynamics of membrane proteins (Sprague and McNally 2005). Since the mid 1990s, continuous progress in refining optical microscopy together with the advent of genetically encoded fluorescent tags have led to an explosion of commercially available systems making FRAP accessible to almost all cell biologists. In fact, FRAP and other complementary techniques have turned into indispensable tools in current cell biology research. FRAP has been extensively used in metazoans, with very few studies in plants. During the last several years, however, several studies in the literature have utilized FRAP for studying a variety of plant proteins. These articles have focused primarily on SR proteins. In FRAP an artificial optical gradient of the fluorescent protein is generated by selectively bleaching a small area with intense laser pulses. The extent of recovery of fluorescence to the bleached area from the surrounding unbleached areas provides a measure of mobility of a protein of interest. A plot of the recovery data against time provides a qualitative glimpse into the mobility of the protein of interest. FRAP analysis yields information about biophysical parameters such as mobile/immobile fractions and diffusion coefficients. The mobile fraction provides information about how much of the total available protein is free to diffuse in live cells and how much is anchored to fixed locations or in higher-order immobile complexes. Similarly, diffusion coefficients and the shape of the recovery curves of the mobile fraction reveal information about whether a protein is freely diffusing or is restricted in its mobility by confinement to domains, probably by interacting with other components. More detailed mathematical analysis of the recovery data and application of various diffusion models yield valuable information about association and dissociation constants of proteins (Sprague and McNally 2005). Figure 1 illustrates this technique, where a GFP-tagged Arabidopsis splicing factor (GFP-SR45) is subjected to FRAP analysis.

In addition to FRAP, fluorescence loss in photobleaching (FLIP), a technique similar and complementary to FRAP, has also been used in studying plant proteins. In FLIP, a region outside the region of interest is repeatedly bleached. A loss of fluorescence intensity in the unbleached region indicates that the protein is mobile. Mathematical analysis of the kinetics of fluorescence loss provides measures of several biophysical properties such as the diffusion coefficient. Given the importance of pre-mRNA splicing, several other fluorescence-based techniques such as fluorescence correlation spectroscopy (FCS) and bimolecular fluorescence complementation (BiFC) will be applied to studying pre-mRNA splicing in plants. In FCS, a beam of laser is focused on a small volume of fluorescent proteins. Fluctuations in fluorescent signals, which are recorded over time, can be used to determine the diffusion coefficients and binding constants of the labeled protein (Schmiedeberg et al. 2004). BiFC is based on the reconstitution of yellow fluorescent protein (YFP) fluorescence when the two nonfluorescent halves of YFP fused separately to two putative interacting proteins are brought together in close proximity when the two tagged proteins interact (Walter et al. 2004) (Fig. 2). A reasonably elaborate interactome of various splicing-related proteins has been established (reviewed in Reddy 2007). However, this interactome is primarily based on yeast two-hybrid and in vitro immunoprecipitation approaches and, therefore, might not reflect the in vivo situation. Besides, these interactions are dependent on the cell type and its physiological state. For any meaningful understanding of the relevance of these interactions to physiological responses, it is therefore important to elucidate interactions of these proteins in vivo. BiFC and fluorescence resonance energy

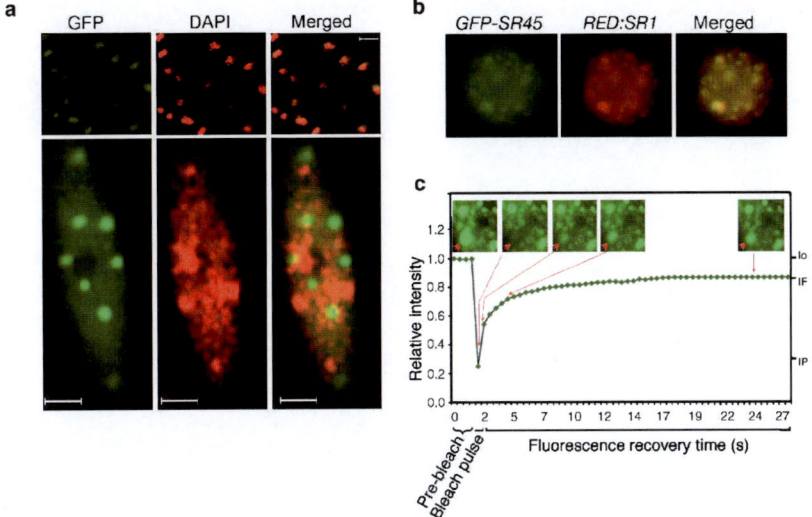

Fig. 1 Localization of GFP-SR45, RED:SR1 and Optimization of FRAP. **a** The Arabidopsis root epidermal cells stained with DAPI (*middle column*, pseudocolored as *red*) reveal that GFP-SR45 localizes to nuclei (*top panel*) and that the subnuclear distribution of GFP-SR45 (bottom panel) does not colocalize to the brightly stained (*red*) heterochromatin domains. Bars, 50 μm (*upper panel*), 5 μm (*lower panel*). **b** GFP-SR45 colocalizes with dsRED:SR1 in a well-characterized speckled and diffused pattern characteristic of splicing factors. **c** FRAP of GFP-SR45 nuclei. A small nuclear region of interest (ROI) in living cells was bleached. The recovery of fluorescence in the bleached ROI was monitored by time-lapse microscopy until there was no further recovery. Fluorescence signal in the bleached and unbleached regions at each time point was measured and normalized (from 0 to 1, with 1 representing the initial fluorescence before bleach) for loss in fluorescence during the bleach pulse and subsequent imaging. Shown is a plot of normalized intensity versus time. *Insets* show the fluorescence of a spot at pre- and postbleach times. The shape and extent of the recovery curves and mathematical curve-fitting reveals information about the kinetic and binding properties of the GFP-tagged SR45. *Io*, prebleach initial fluorescence intensity; *IF*, maximum final intensity after bleaching when the recovery curve has reached a plateau (the difference between the Io and IF represents the immobile fraction of the GFP-SR45); *IP*, fluorescence intensity immediately after bleaching representing the depth of bleaching. The difference between IP and IF corresponds to the mobile fraction

transfer (FRET) will undoubtedly contribute tremendously toward addressing these issues. Recently, we have demonstrated the suitability of BiFC for mapping domains of the Arabidopsis SR45 protein that interact with U1-70K (unpublished data). We suggest that combining BiFC with FRAP will be useful in studying the dynamics of spliceosomal complexes. A complete understanding of splicing-related proteins will require a system-level approach, encompassing several different disciplines such as microscopy, molecular biology, bioinformatics, and mathematical modeling.

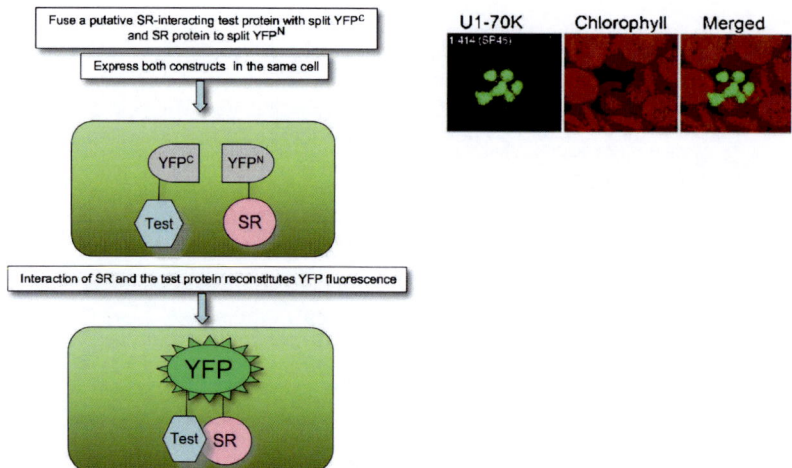

Fig. 2 The bimolecular fluorescence complementation (BiFC) system. *Left panel*: Outline of BiFC system. Putative interacting proteins of interests are fused to split nonfluorescent parts of YFP separately. The interaction of proteins brings the split parts of YFP in close proximity. resulting in fluorescence. *Right panel*: BiFC images of Arabidopsis protoplasts cotransfected with U1-70K-YFP^N and a full-length (1-414) SR45-YFP^C as indicated on each panel show that SR45 interacts with U1-70K in vivo

Subcellular Organization of Proteins Involved in Pre-mRNA Splicing in Plant Cells

Studies have shown that nuclei are organized into distinct subdomains/structures likely playing a role in regulating various nuclear processes. For example, chromosomes occupy specific locations in nuclei called "chromosomal territories" and rRNA processing and ribosome biogenesis take place in nucleoli, whereas several splicing-related proteins localize to Cajal bodies and speckles.

Several recent studies employing biochemical analysis, fluorescence and electron microscopy, and immunofluorescence techniques have clearly demonstrated that proteins associated with pre-mRNA splicing are present in both a diffused nucleoplasmic pool and in a variety of nuclear bodies such as Cajal bodies, speckles, and nucleoli in plants (Beven et al. 1995; Glyn and Leitch 1995; Fisher et al. 1998; Saudan et al. 1998; Shaw et al. 1998; Ali et al. 2003; Cui and Moreno Diaz de la Espina 2003; Docquier et al. 2004; Fang et al. 2004; Lorković et al. 2004; Tillemans et al. 2005, 2006; Ali and Reddy 2006; Riera et al. 2006). Several studies have verified the presence of SR proteins in nucleoli under certain conditions, suggesting that these bodies may also be instrumental in certain aspects of RNA metabolism (Tillemans et al. 2005) (see the chapter by J. W. S. Brown and P. J.Shaw, this volume). These observations were strengthened by proteomic analyses of nucleoli in both Arabidopsis and humans (Pendle et al. 2005), which showed the presence of several proteins associated with pre-mRNA splicing. In Arabidopsis, subsequent verification of the proteomic data by visualization with GFP-tagged fusion proteins revealed that these proteins include several SR proteins (Pendle et al. 2005), which are involved in pre-mRNA splicing and also in mRNA export, suggesting that the nucleolus likely serves as a processing or transit center in assembling splicing machinery (see the chapter by Brown and Shaw, this volume). Direct evidence for the involvement of the nucleolus in pre-mRNA splicing still remains to be determined.

Initial studies in monocot and dicot plants, using electron microscopy and immunocytochemistry with antibodies raised against core mammalian spliceosomal proteins and in situ hybridization with U snRNAs, revealed the presence and distribution of spliceosomal components in speckles, fibrils, and other nuclear bodies in a pattern very similar to that in mammalian cells. This suggested a high degree of similarity in nuclear architecture across phylogenetically different organisms (Testillano et al. 1993; Beven et al. 1995; Glyn and Leitch 1995). Immunofluorescence studies showed that the spliceosomal U2 snRNP-specific U2B protein in peas and onions localizes to interchromatin fibrils and Cajal bodies, some of which were closely associated with the nucleolus (Beven et al. 1995; Cui and Moreno Diaz de la Espina 2003). Similarly, using antibodies raised against small nuclear ribonucleoprotein particles (snRNP) and Sm proteins and mAb KSm2 antibody, which detects the "D" polypeptide of the core snRNP complex, immunofluorescence microscopy revealed a speckled and diffused nucleoplasmic pattern in wheat, barley, and onion cells (Testillano et al. 1993; Beven et al. 1995; Cui and Moreno Diaz de la Espina 2003). In addition, in pea cells, U1, U2, and U6 snRNAs also displayed a similar pattern (Beven et al. 1995). Together these studies provided the initial framework for investigation into the subcellular organization of splicing machinery in plant cells. Later, with the use of GFP-tagged U2B and U2A' proteins, similar results were reproduced in model plant systems including Arabidopsis and tobacco (Boudonck et al. 1998; Lorković et al. 2004). Additionally, several other spliceosomal proteins such as U1-70K and smB were also shown to localize to speckles.

More recent studies have focused on studying the subcellular organization and dynamics of SR proteins. Most of the Arabidopsis SR proteins, representing all major groups including plant-specific ones (RSZ33, RSp31, and SR45), have been shown to display a characteristic distribution pattern consisting of areas of high concentrations called speckles embedded in a diffused nucleoplasmic pool (Lorković and Barta 2004; Shaw and Brown 2004). Most of them display similar distribution patterns, but there are obvious differences among them in regard to their number and sizes. With transmission electron microscopy (TEM), the speckles of some of the Arabidopsis SR proteins (SR1, SR30, and SR33) have been shown to correspond to the interchromatin granule clusters (IGC) (Fang et al. 2004), structures that consist of 20- to 30-nm-size particles and have been shown previously to correspond to SR protein speckles in mammalian cells (Lamond and Spector 2003). This global conservation of the distribution pattern of SR proteins across phylogenetically diverse organisms underscores their importance in the spatial regulation of nuclear activities. The sizes and numbers of speckles vary considerably with development and in different tissues (Ali et al. 2003; Fang et al. 2004). Whether these differences have any functional significance remains to be shown, but they are likely to regulate the splicing and alternative splicing of tissue and cell type-specific gene expression.

The nuclear localization and the speckle targeting/retention signals seem to reside in the serine/arginine-rich (RS) domains of plant SR proteins but not in their RNA-recognition motifs (RRM). For example, the GFP-RRM fusions of SR45, atRSp31 and atRSZp22, displayed a nucleoplasmic and/or cytoplasmic distribution with no targeting to speckles (Tillemans et al. 2005; Ali and Reddy 2006). In contrast, speckle targeting signals in the human SF2/ASF were also localized to the RRM, which suggests that the targeting of plant and animal SR proteins might be regulated differently.

Regulation of Spatiotemporal Organization and Dynamics of Pre-mRNA Splicing Proteins

Interest in investigating the functional significance of nuclear domains in pre-mRNA has increased steadily. Most of these studies were conducted in metazoan systems. Given the importance of SR proteins in pre-mRNA splicing, it is natural that the dynamics of these proteins in plants have also been attracting considerable attention. Over the past several years we and several other groups have analyzed the regulation of their mobility in live plant cells (Ali et al. 2003; Fang et al. 2004; Tillemans et al. 2005, 2006). These studies have mostly found what has already been shown with metazoan SR proteins. However, some of these proteins exhibit drastically different behavior in their mobility, implying important differences in the pre-mRNA splicing between plants and metazoans.

Several of the initial studies, conducted with plant cells and antibodies raised against human core splicing proteins, showed that the distribution pattern of splicing proteins changes with the cell cycle and the metabolic state of cells (Beven et al. 1995; Glyn and Leitch 1995; Cui and Moreno Diaz de la Espina 2003). Analyses during cell cycle progression showed that their overall expression levels, and the number, size, and location of their speckles, changed in a cell cycle-dependent manner, disappearing at metaphase and reappearing after the appearance of daughter nuclei (Glyn and Leitch 1995). Similar observations were also noted for SR splicing factors (Fang et al. 2004; Tillemans et al. 2005). Heat shock is known to inhibit RNA Pol II-mediated transcription and disrupt pre-mRNA splicing both in plants and animals and is a reflection of the metabolic activities of cells. Several investigations have shown that heat shock dramatically rearranges the distribution pattern of both core snRNP and non-snRNP spliceosomal proteins (Beven et al. 1995; Glyn and Leitch 1995; Ali et al. 2003). Immunocytochemistry of wheat root tips and cells with antibody against a core spliceosomal protein at elevated temperatures also displayed a similar pattern (Glyn and Leitch 1995). These studies show that heat shock generally leads to an increased accumulation of proteins in enlarged areas, implying their increased storage during low metabolic activity. This property does not appear to be limited to core spliceosomal proteins as SR proteins were also shown to accumulate in enlarged speckles after heat shock. Increasing the temperature to 37°C or 42°C resulted in the accumulation of SR45, RSZp22, and RSp31 in enlarged speckles, probably because of their increased storage and reduced demand due to lower transcription/splicing activities (Ali et al. 2003; Tillemans et al. 2005) (Fig. 3). The heat-dependent enlargement was inhibited by a phosphatase inhibitor, implying that heat shock-induced relocalization involves dephosphorylation of SR proteins (Ali et al. 2003). Interestingly, heat and cold also changed the alternative splicing pattern of several SR proteins (Palusa et al. 2007). It is possible that regulation of the alternative splicing pattern of these genes is related to a change in the subnuclear reorganization of SR splicing factors affected by heat and cold.

As pre-mRNA splicing occurs cotranscriptionally, several groups have studied the relationship between transcription and the dynamics of plant SR proteins representing the plant-specific SR family (SR45), SF2 family (SR1, SR30), and SC35-family (SR33). Inhibition of transcription by drugs or heat shock led to (a) increased accumulation of splicing factors in speckles eventually leading to increased speckle area and (b) cessation of all kinds of speckle movements including budding off and peripheral movements (Ali et al. 2003; Fang et al. 2004). In some cases, very different patterns such as appearance of thousands of "microspeckles" in trichomes was also observed (Fang et al. 2004). All SR proteins are phosphoproteins, whose subcellular localization and interaction with other components is regulated by their reversible phosphorylation (Graveley 2000). Consistent with these facts, the subcellular localization of plant SR proteins was shown to be dependent on the phosphorylation state of the cells. Inhibition of phosphorylation with the general protein kinase inhibitor staurosporine led to enlarged speckles, suggesting that the release of SR proteins from speckles requires their phosphorylation (Ali et al. 2003; Tillemans et al. 2005) (Fig. 3).

Time-lapse microscopy of FP-tagged SR proteins has allowed detailed examination of their mobility behavior in speckles in different plant cells and under different physiological conditions. These studies show that SR speckles in plant nuclei

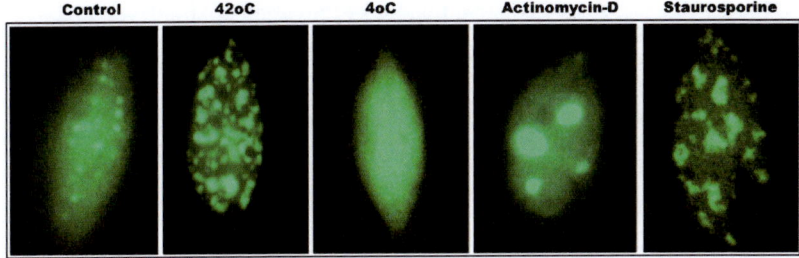

Fig. 3 Subcellular distribution of a plant-specific SR splicing factor. Transgenic Arabidopsis plants expressing GFP-SR45 display a characteristic speckle and diffuse nucleoplasmic distribution, which is changed to enlarged speckles by inhibition of transcription by heat (42°C) or actinomycin-D and by inhibition of protein kinase by staurosporine. In contrast, cold treatment (4°C), relocalized SR45 to a diffused nucleoplasmic pool (Ali et al. 2003)

display limited movements in a constrained area, suggesting that they are loosely anchored to less mobile nuclear components, or, alternatively, they may be restrained from mobility by physical barriers such as chromatin. The later explanation seems unlikely as Cajal bodies, which have approximately similar size, can move from one location to another (Boudonck et al. 1999). Whether this movement has any functional relevance is unknown, but confinement to a specific area may reflect an area of high demand such as a highly expressed gene(s), consistent with observations in SF2/ASF, which was recruited efficiently to a nearby activated gene (Misteli et al. 1997). In addition to these restricted movements, plant SR speckles also fuse with each other and bud off from speckles (Ali et al. 2003; Fang et al. 2004).

Mobility Analyses of Proteins Associated with Pre-mRNA Splicing at the Molecular Level

FRAP analyses with Arabidopsis SR splicing factors under control conditions suggest that, similar to their metazoan counterparts, plant SR proteins are also dynamically distributed between speckles and nucleoplasm and that their mobility is significantly lower than a freely diffusing protein of similar size. A major fraction of plant SR proteins also display rapid mobility both in speckles and nucleoplasm. Because of the use of different measures of mobility (diffusion vs. half-time of recovery), it is difficult to compare the mobility properties of different SR proteins reported in different articles (Fang et al. 2004; Ali and Reddy 2006). In any case, it is clear from these analyses that plant SR proteins roam the nuclear space with diffusion coefficients ranging from 0.38 to 1.0 $\mu m^2\ s^{-1}$ (Fang et al. 2004; Ali and Reddy 2006; Tillemans et al. 2006). Fluorescence in a bleached speckle of several SR proteins recovered within seconds, suggesting that SR proteins exchange continuously between speckles and nucleoplasm very rapidly (Fang et al. 2004).

Photobleaching analysis also allows the investigation of mechanisms that regulate SR protein mobility. As mentioned above, the subcellular localization of most SR proteins appears to be regulated by phosphorylation and transcription status of the cell. The metazoan SR splicing factors SF2/ASF and SC35 were shown to move in the nucleus independent of ATP and protein phosphorylation (Kruhlak et al. 2000; Phair and Misteli 2000). Furthermore, inhibition of transcription was shown to slightly accelerate the association of SF2/ASF splicing factor with speckles. In contrast, a FRAP analysis of the mobility of GFP-tagged SR45, a plant-specific splicing factor, and SR1/SRp34, a plant homolog of SF2/ASF, showed that ATP-depletion resulted in dramatic reduction in the mobile fraction and diffusion coefficients of both proteins, suggesting that the movement of plant SR proteins is dependent upon ATP (Ali and Reddy 2006). Whether this dependence on ATP is a general mechanism in plants remains to be seen. Under control conditions, the retarded movement of plant SR proteins, as compared to neutral proteins of similar size, is in agreement with a model that explains reduced diffusion resulting from interaction and assembly into a higher-order complex as was proposed for metazoan SF2/ASF and several other nuclear proteins (Phair and Misteli 2000; Misteli 2001). Under ATP-depleted conditions, results with plant SR proteins favor a scenario in which ATP plays an active role in their mobility. The apparent lack of any effect of ATP depletion on the mobility of SF2/ASF was interpreted as passive movement. The recovery in a bleached region results from the exchange of bleached molecules with fluorescent molecules from the surrounding areas. This could be interpreted as repeated dissociation/reassociation of SR proteins to their binding sites. The slowed recovery in the ATP-depleted cells indicates that the binding/unbinding of plant SR proteins is dependent upon ATP. Since ATP is required for the rearrangements of spliceosomal complexes during splicing, in the absence of ATP these proteins probably do not readily dissociate from larger complexes, resulting in retarded mobility of SR proteins. Alternatively, ATP depletion may result in hypophosphorylation of SR proteins, which leads to their reduced mobilities. The retarded mobility of SR45 and SR1 in ATP-depleted cells may also result from a general rearrangement of the nuclear environment due to depletion of energy as was observed for mRNAs and mRNP particles (Shav-Tal et al. 2004; Vargas et al. 2005). Several observations indicate that ATP depletion has a specific negative effect on the mobility of plant SR proteins. A global nuclear rearrangement would be expected to restrict the mobility of other nuclear proteins also. On the contrary, ATP depletion had no substantial effects on the mobility of GFP alone and of the nuclear localized NLS-GFP-GUS fusion that is twice the size of SR45. However, being neutral, GFP or NLS-GFP-GUS may not be in a complex large enough to be impeded by the smaller sizes of "pores" presumably generated by the rearrangement of chromatin by ATP depletion (Shav-Tal et al. 2004). This issue was resolved by conducting FRAP analysis with the deletion mutants of SR45. SR45 consists of an N-terminal serine/arginine-rich domain (RS1), a middle RNA recognition motif (RRM), and a C-terminal RS2 domain. If the reduced mobility is merely due to a nonspecific effect, a proportionally similar effect of ATP depletion on the mobilities of these domain deletion mutants was expected. On the contrary, however, individual domains of SR45 exhibited very different localization and kinetic properties in response to ATP depletion. First, ATP depletion did not change the distribution of RS1 of SR45; it remained in a diffused pattern as in the control. In contrast, a substantial fraction of RRM of SR45, which has a molecular weight comparable to RS1 of SR45, concentrated in numerous foci in the cytoplasm. In addition, a substantial fraction (82%) of RRM in these foci was rendered immobile. Second, despite being similar in size and with similar diffusion properties under control conditions, in the ATP-depleted cells, RRM and RS1 had different mobile fractions (15% for RRM and 100% for RS1) and disproportionate reduction in effective diffusion (1.8× for RS1 and 27× for RRM). Third, the RS1+RRM domain, although smaller than RS2 (46 kDa vs 56 kDa), had a significantly higher immobile fraction than the latter. These observations suggest a scenario in which ATP plays an active role in the mobility of SR45 and, therefore, the observed reduced mobility of plant SR proteins in ATP-depleted cells does not result from nonspecific general perturbation of the cellular environment.

Similarly, FRAP analyses showed that the mobility of other plant SR proteins is dependent upon transcription and phosphorylation (Fang et al. 2004; Tillemans et al. 2005; Ali and Reddy 2006). The reduced mobility of SR proteins in speckles after kinase inhibition probably results from their failure to release from speckles in a hypophosphorylated state. This explanation is supported by the observation with SR proteins in metazoans, where they are recruited to splicing sites in a hyperphosphorylated state and disrupting the phosphorylation of SF2/ASF by mutating the RS dipeptides to RG dipeptides abolished its recruitment to active transcription sites (Misteli et al. 1998). A possible explanation for the plant SR proteins is that in the absence of transcription, and hence splicing, splicing factors are not recruited to active transcription sites. Instead they remain tightly bound to their binding sites in the speckles, which would account for their retarded mobilities (Fig. 4). Consistent with the cytoplasm-localized activities of SR proteins in RNA metabolism, several metazoan splicing factors are known to shuttle between the cytoplasm and nucleus (Hunag and Steitz 2005). These analyses have primarily employed a heterokaryon-based approach. As expected, and consistent with observations in animal cells, with FRAP and FLIP analyses, a plant SR protein, RSZp22, was also shown to exhibit nucleo-cytoplasmic shuttling, suggesting that like their animal counterparts some plant SR proteins also shuttle and play roles in other aspects of RNA metabolism in the cytoplasm (Tillemans et al. 2006). Tillemans et al (2006) also showed that this shuttling was sensitive to inhibition of phosphorylation and ATP. Whether other plant SR splicing factors also display shuttling properties remains unknown. In any case, it is critical to establish a relationship between the various mobility properties of SR and other splicing-related proteins and their physiological and molecular functions.

Model for the Distribution and Kinetics of Plant SR Proteins

Based upon the FRAP kinetic analyses, we conclude that plant SR proteins constantly exchange between speckles and nucleoplasm under a steady-state equilibrium state (Fig. 4a). Hence, a balance of influx and efflux of SR proteins in and out of the speckles determines the morphology and size of speckles. An increased influx and/or decreased efflux would increase the size of the SR speckles and a decrease in the influx and/or increased efflux would decrease the size or completely dismantle the speckles. How ATP, inhibitors of transcription and protein phosphorylation would affect the movement of SR proteins and speckle size and shape is illustrated in Fig. 4. Inhibition of transcription would lead to a decline of pre-mRNAs, and hence splicing (Fig. 4b). At the completion of performing their task in splicing, most of the SR protein molecules are recycled for the next round of splicing and thus would accumulate in the speckles, a proposed site for the processing and recycling of splicing factors. In the absence of any pre-mRNA substrate, SR proteins would tend to stay in the speckles, eventually leading to an increased speckle size (Fig. 4b). When phosphorylated, SR proteins leave the speckle and participate in splicing,

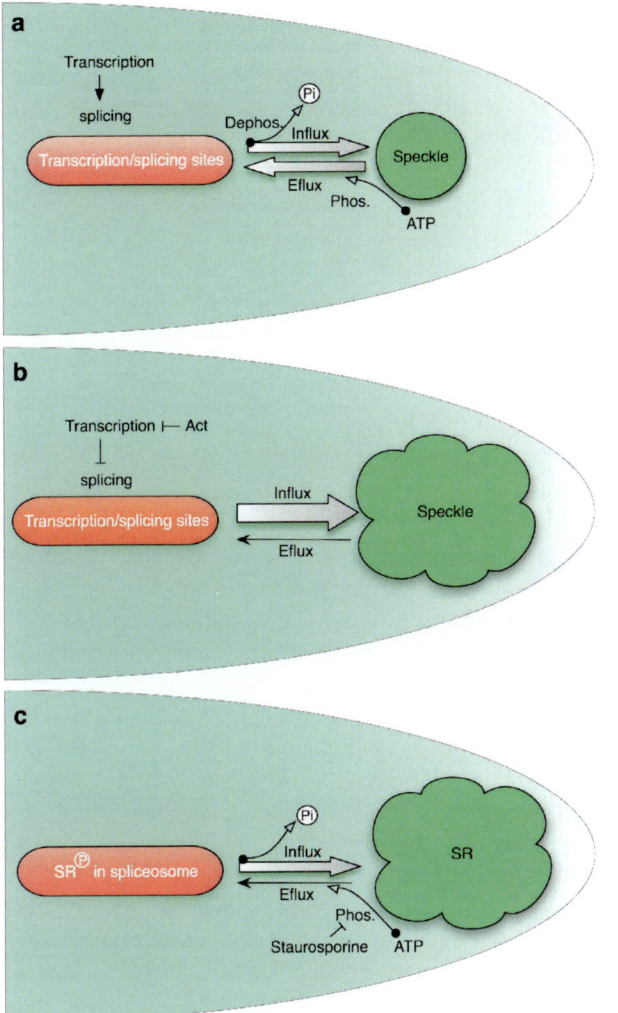

Fig. 4 A model for the dynamic exchange of plant SR splicing factors between speckles and nucleoplasm. **a** Under normal physiological conditions, speckles are maintained by a steady-state exchange of SR proteins between nucleoplasm and speckles. *Phos*, phosphorylation; *Dephos*, dephosphorylation. **b** Inhibition of transcription would diminish the population of pre-mRNA substrate of SR proteins, and therefore, in the absence of a substrate, SR45 remains in the speckle, resulting in its enlargement and consequently in reduced mobility of SR45. **c** SR proteins are phosphoproteins, and their phosphorylation by a LAMMER-type kinase, AFC2, likely regulates their interaction with other splicing proteins. Inhibition of phosphorylation by either protein kinase inhibitor or depletion of ATP results in alteration in phosphorylation status of SR45 and its sequestration into the speckle

and when dephosphorylated they enter speckles, probably for recycling/processing (Fig. 4c). In the absence of phosphorylation, they are not competent for recruitment to the splicing sites and remain in the speckles. A phosphoproteomic analysis of

Arabidopsis proteins has identified specific sites in several SR proteins (de la Fuente van Bentem et al. 2006). It would be interesting to show the role of these phosphorylation sites, by the point mutation approach, in regulating the subcellular distribution of splicing-related proteins. Additionally, these proteins were shown to be phosphorylated by Arabidopsis SR protein-specific kinase 4 (SRPK4). Similarly other protein kinases (LAMMER type) such as PK12 and AFC2 are also involved in phosphorylating SR proteins (Golovkin and Reddy 1999; Savaldi-Goldstein et al. 2000, 2003). Interestingly, the protein kinase AFC2 also displays a nuclear localization pattern very similar to SR proteins (unpublished data), suggesting that the subnuclear localization of signaling components likely plays a regulatory role in affecting the activity of splicing-related proteins.

Conclusions and Future Directions

Over the past several years, with advanced fluorescence microscopy techniques, our view about the organization of plant nuclei has changed from that of a static one to a highly dynamic one in which components of splicing are in a constant flux between various subnuclear domains. These studies suggest that during their life cycle plant splicing factors exist in different domains, which in turn likely impose spatial regulation on their activity. These studies in plants have so far focused on the regulation of the subnuclear organization in a tissue- and metabolic state-dependent manner, but with the introduction of new techniques such BiFC, combined with high-throughput analyses, we expect a rapid advancement in our understanding of how plant pre-mRNA splicing is spatially and temporally organized and how interactions of various splicing components are regulated. Such future studies in plants will not only be essential for understanding the functional organization of plant nuclei but will also uncover evolutionary differences between humans and plants, which will eventually lead to a better understanding of splicing in general. In summary, compartmentalization of splicing machinery in the nucleus may be providing an important spatial regulation mechanism analogous to the controlled regulation of trafficking of proteins across nuclear membranes, although a lipid bilayer membrane does not enclose SR-containing subnuclear domains. So far results indicate that splicing-related proteins rapidly and randomly roam the 3-D intranuclear space and interact with other proteins and RNA targets and hence provide an efficient means of supplying splicing proteins wherever they are needed. Regulation of the intranuclear localization and the mobility of SR proteins by ATP, transcription, and protein phosphorylation suggest that their activity is regulated spatially and temporally by a complex interaction of *trans-* and *cis*-components. The differences in the regulation of mobility of plant and metazoan SR proteins suggest that not all SR proteins exhibit similar dynamics and that such differences are likely to have important implications in pre-mRNA splicing.

Acknowledgements This work was supported by a grant from NSF.

References

Ali GS, Reddy AS (2006) ATP, phosphorylation and transcription regulate the mobility of plant splicing factors. J Cell Sci 119:3527-3538

Ali GS, Golovkin M, Reddy AS (2003) Nuclear localization and in vivo dynamics of a plant-specific serine/arginine-rich protein. Plant J 36:883-893

Beven AF, Simpson GG, Brown JW, Shaw PJ (1995) The organization of spliceosomal components in the nuclei of higher plants. J Cell Sci 108:509-518

Boudonck K, Dolan L, Shaw PJ (1998) Coiled body numbers in the Arabidopsis root epidermis are regulated by cell type, developmental stage and cell cycle parameters. J Cell Sci 111:3687-3694

Boudonck K, Dolan L, Shaw PJ (1999) The movement of coiled bodies visualized in living plant cells by the green fluorescent protein. Mol Biol Cell 10:2297-2307

Cui P, Moreno Diaz de la Espina S (2003) Sm and U2B proteins redistribute to different nuclear domains in dormant and proliferating onion cells. Planta 217:21-31

de la Fuente van Bentem S, Anrather D, Roitinger E, Djamei A, Hufnagl T et al. (2006) Phosphoproteomics reveals extensive in vivo phosphorylation of Arabidopsis proteins involved in RNA metabolism. Nucleic Acids Res 34:3267-3278

Docquier S, Tillemans V, Deltour R, Motte P (2004) Nuclear bodies and compartmentalization of pre-mRNA splicing factors in higher plants. Chromosoma 112:255-266

Fang Y, Hearn S, Spector DL (2004) Tissue-specific expression and dynamic organization of SR splicing factors in Arabidopsis. Mol Biol Cell 15:2664-2673

Fisher CE, Brown DM, Shaw J, Beswick PH, Donaldson K (1998) Respirable fibres: surfactant coated fibres release more Fe^{3+} than native fibres at both pH 4.5 and 7.2. Ann Occup Hyg 42:337-345

Glyn MC, Leitch AR (1995) The distribution of a spliceosome protein in cereal (Triticeae) interphase nuclei from cells with different metabolic activities and through the cell cycle. Plant J 8:531-540

Golovkin M, Reddy ASN (1999) An SC35-like protein and a novel serine/arginine-rich protein interact with Arabidopsis U1-70K protein. J Biol Chem 274:36428-36438

Graveley BR (2000) Sorting out the complexity of SR protein functions. RNA 6:1197-1211

Huang Y, Steitz JA (2005) SRprises along a messenger's journey. Mol Cell 17:613-615

Kalyna M, Barta A (2004) A plethora of plant serine/arginine-rich proteins: redundancy or evolution of novel gene functions? Biochem Soc Trans 32:561-564

Kruhlak MJ, Lever MA, Fischle W, Verdin E, Bazett-Jones DP et al. (2000) Reduced mobility of the alternate splicing factor (ASF) through the nucleoplasm and steady state speckle compartments. J Cell Biol 150:41-51

Lamond AI, Spector DL (2003) Nuclear speckles: a model for nuclear organelles. Nat Rev Mol Cell Biol 4:605-612

Li X, Manley JL (2005) New talents for an old acquaintance: the SR protein splicing factor ASF/SF2 functions in the maintenance of genome stability. Cell Cycle 4:1706-1708

Li X, Manley JL (2006) Cotranscriptional processes and their influence on genome stability. Genes Dev 20:1838-1847

Lorković ZJ, Barta A (2002) Genome analysis: RNA recognition motif (RRM) and K homology (KH) domain RNA-binding proteins from the flowering plant *Arabidopsis thaliana*. Nucleic Acids Res 30:623-635

Lorković ZJ, Barta A (2004) Compartmentalization of the splicing machinery in plant cell nuclei. Trends Plant Sci 9:565-568

Lorković ZJ, Hilscher J, Barta A (2004) Use of fluorescent protein tags to study nuclear organization of the spliceosomal machinery in transiently transformed living plant cells. Mol Biol Cell 15:3233-3243

Misteli T (2001) Protein dynamics: implications for nuclear architecture and gene expression. Science 291:843-847

Misteli T, Caceres JF, Spector DL (1997) The dynamics of a pre-mRNA splicing factor in living cells. Nature 387:523-527

Misteli T, Caceres JF, Clement JQ, Krainer AR, Wilkinson MF et al. (1998) Serine phosphorylation of SR proteins is required for their recruitment to sites of transcription in vivo. J Cell Biol 143:297-307

Palusa SG, Ali GS, Reddy ASN (2007) Alternative splicing of pre-mRNA of Arabidopsis serine/arginine-rich proteins:regulation by hormones and stresses. Plant J 49:1091-1107

Pendle AF, Clark GP, Boon R, Lewandowska D, Lam YW et al. (2005) Proteomic analysis of the Arabidopsis nucleolus suggests novel nucleolar functions. Mol Biol Cell 16:260-269

Phair RD, Misteli T (2000) High mobility of proteins in the mammalian cell nucleus. Nature 404:604-609

Reddy AS (2004) Plant serine/arginine-rich proteins and their role in pre-mRNA splicing. Trends Plant Sci 9:541-547

Reddy ASN (2007) Alternative splicing of pre-messenger RNAs in plants in the Genomic Era. Annu Rev Plant Biol 58:267-294

Reits EA, Neefjes JJ (2001) From fixed to FRAP: measuring protein mobility and activity in living cells. Nat Cell Biol 3: E145-147

Riera M, Redko Y, Leung J (2006) Arabidopsis RNA-binding protein UBA2a relocalizes into nuclear speckles in response to abscisic acid. FEBS Lett 580:4160-4165

Sanford JR, Longman D, Caceres JF (2003) Multiple roles of the SR protein family in splicing regulation. In: Jeanteur P, editor. Regulation of alternative splicing. New York: Springer. pp. 33-58

Saudan PJ, Shaw L, Brown MA (1998) Urinary calcium/creatinine ratio as a predictor of preeclampsia. Am J Hypertens 11:839-843

Savaldi-Goldstein S, Sessa G, Fluhr R (2000) The ethylene-inducible PK12 kinase mediates the phosphorylation of SR splicing factors. Plant J 21:91-96

Savaldi-Goldstein S, Aviv D, Davydov O, Fluhr R (2003) Alternative splicing modulation by a LAMMER kinase impinges on developmental and transcriptome expression. Plant Cell 15:926-938

Schmiedeberg L, Weisshart K, Diekmann S, Meyer Zu Hoerste G, Hemmerich P (2004) High- and low-mobility populations of HP1 in heterochromatin of mammalian cells. Mol Biol Cell 15:2819-2833

Shav-Tal Y, Darzacq X, Shenoy SM, Fusco D, Janicki SM et al. (2004) Dynamics of single mRNPs in nuclei of living cells. Science 304:1797-1800

Shaw PJ, Brown JW (2004) Plant nuclear bodies. Curr Opin Plant Biol 7:614-620

Shaw PJ, Beven AF, Leader DJ, Brown JW (1998) Localization and processing from a polycistronic precursor of novel snoRNAs in maize. J Cell Sci 111:2121-2128

Sprague BL, McNally JG (2005) FRAP analysis of binding: proper and fitting. Trends Cell Biol 15:84-91

Testillano PS, Sanchez-Pina MA, Olmedilla A, Fuchs JP, Risueno MC (1993) Characterization of the interchromatin region as the nuclear domain containing snRNPs in plant cells. A cytochemical and immunoelectron microscopy study. Eur J Cell Biol 61:349-361

Tillemans V, Dispa L, Remacle C, Collinge M, Motte P (2005) Functional distribution and dynamics of Arabidopsis SR splicing factors in living plant cells. Plant J 41:567-582

Tillemans V, Leponce I, Rausin G, Dispa L, Motte P (2006) Insights into nuclear organization in plants as revealed by the dynamic distribution of Arabidopsis SR splicing factors. Plant Cell 18:3218-3234

Vargas DY, Raj A, Marras SA, Kramer FR, Tyagi S (2005) Mechanism of mRNA transport in the nucleus. Proc Natl Acad Sci USA 102:17008-17013

Walter M, Chaban C, Schutze K, Batistic O, Weckermann K et al. (2004) Visualization of protein interactions in living plant cells using bimolecular fluorescence complementation. Plant J 40:428-438

Wang BB, Brendel V (2004) The ASRG database: identification and survey of *Arabidopsis thaliana* genes involved in pre-mRNA splicing. Genome Biol 5: R102

Wang BB, Brendel V (2006) Genomewide comparative analysis of alternative splicing in plants. Proc Natl Acad Sci USA 103:7175-7180

Regulation of Splicing by Protein Phosphorylation

R. Fluhr

Contents

Introduction 120
Transcriptional Processes That Impact on Splicing 121
    The Pace of Transcription Can Determine the Outcome of Splicing 121
    CTD of Pol II Interacts with a Variety of Pre-mRNA Factors 123
    Chromatin Structure Impacts on Transcription Rates and Splicing 124
Splicing Factors 125
    Role of Serine/Arginine-Rich Protein Phosphorylation in Splicing 125
    Phosphorylation-Dependent Localization of SR Proteins 127
    Phosphorylation in Postsplicing Functions of SR Proteins 128
Kinases That Phosphorylate Splicing Factors 130
    LAMMER Kinases Modulate Splicing 130
    Cellular Partitioning of SR Proteins by SRPK Activity 132
    Topoisomerase I 133
    Concluding Remarks 133
References 134

Abstract Most eukaryotic messenger RNAs are transcribed as precursors that necessitate specific and exact processing of intron boundaries. Furthermore, the choice of these boundaries appears to be fluid and adaptive to the rate of transcription and the developmental and physiological state of the cell. A central regulator of splicing reactions and choice are kinases that work through phosphorylation of specific factors like RNA polymerase II, which influences the pace of transcription and of SR splicing factors. While very different in their mechanisms both regulatory pathways will impact on splicing site choice. This chapter summarizes the biology of splicing-related phosphorylation activity, emphasizing plant-specific aspects in relation to the metazoan counterpart.

R. Fluhr
Plant Sciences, Weizmann Institute of Science, 76100, Rehovot, Israel
e-mail: robert.fluhr@weizmann.ac.il

Introduction

Pre-mRNA splicing is tightly coupled to transcription, polyadenylation, surveillance, and transport, which operate together with the spliceosomal system to produce a mature transcript. The maintenance of splicing throughout evolution has been attributed to its contribution to the enhancement of genomic complexity in which reshuffling of key domains can build new protein entities. If the outcome of splicing events were solely directed to the production of constitutive splicing events there would probably be no need to regulate the splicing machinery. Conceivably, the spliceosome pre-mRNA accessory proteins and *cis*-regulatory RNA sequences would adapt themselves to splice away unwanted sequence in a predicted manner. In this respect, if alternatively spliced genes were synthesized in a fixed ratio of transcript, once again the need for their regulation would be mute. However, this is not the case, as during different developmental stages or under different environmental conditions a complex, dynamic, and varying ratio of cellular transcripts is produced in humans and plants. The outcome of transcript structure is controlled at many levels; prominent among them is reversible phosphorylation of splicing-related proteins. This chapter summarizes the types of mechanisms by which phosphorylation can affect pre-mRNA splicing.

Recent evidence indicates that a high incidence (32%-60%) of alternative splicing is present in the human genome (reviewed in Modrek and Lee 2002). Plants are thought to exhibit less alternative splicing (10%-22%), and, unexpectedly, analysis in *Arabidopsis* as well as in rice showed that intron retention is the most common type of alternative splicing, comprising more than 50% of the alternative splicing types, followed by differential 5' and 3' choice (~30%) and exon skip (~10%) (Iida et al. 2004; Ner-Gaon et al. 2004; Ner-Gaon and Fluhr 2006; Wang and Brendel 2006) (see the chapter by B. J. Haas, this volume). The interest in counting and classifying splicing types, as well as the current research efforts in elucidating the molecular mechanism that drives differential splice choices, is inspired by the idea that the different transcripts will have biological ramifications. This paradigm has been documented in plants, and a few cases are worth mentioning. For example, Rubisco activase regulates the activity of Rubisco, the key enzyme for carbon assimilation in plants. Only the larger spliced isoform of the activase includes a thioredoxin motif that adds the capability to sense, when needed, light-induced changes in the cellular redox potential (Zhang and Portis 1999). The *Arabidopsis* gene FCA encodes an RNA binding protein that functions to promote floral transition. Of the multiple FLC transcripts only one is functional, and alternative processing of the FCA transcript limits, both spatially and temporally, the amount of functional FCA protein. (Macknight et al. 2002) (see the chapter by L. C. Terzi and G. G. Simpson, this volume). An exquisite example of the control that alternative splicing plays involves the efficacy of plant resistance genes (see the chapter by W. Gassmann, this volume). The gene that is responsible for hypersensitive response to the tobacco mosaic virus shows dynamic alternative splicing patterns that are essential in an unknown manner for the resistance function (Dinesh-Kumar and Baker 2000). The examples above illustrate the need for the control of alternative splicing as either a part of rapid and dynamic adaptation to the environment or for regulation during the slower pace of differentiation. Pre-mRNA processing that yields precisely spliced transcripts involves a continuous cotranscriptional process in which each stage can impact on the end result. The process is dictated from the start by the pace of transcription and interacting spliceosomal factors that are all intimately controlled by reversible phosphorylation mechanisms.

Transcriptional Processes That Impact on Splicing

The Pace of Transcription Can Determine the Outcome of Splicing

Phosphorylation processes play a critical role in determining the rate of transcription, and a kinetic coupling exists between transcript splicing choice and the pace of transcript elongation. This was the direct conclusion drawn from experiments in which RNA polymerase II (Pol II) sites were introduced into the DNA template that caused transcriptional pausing during Pol II-dependent transcription. The use of these templates resulted in differential inclusion of a skipped intron (Roberts et al. 1998). Thus, by delaying synthesis of an essential downstream RNA *cis*-element, a decision to splice or repress an exon occurred. Hence, in the simplest sense windows of splicing opportunity are determined by the particular transcription rate of the gene. Consistent with a "first come, first served" model, exon skipping in vivo was inhibited when transcription was slowed in yeast cells by polymerase mutants or when cells were treated with inhibitors of elongation (Howe et al. 2003; Kornblihtt et al. 2004). A simple competition-type model that can explain this result is shown in Fig. 1. The model hypothesizes that when the strengths of the 3′ splice sites differ a sputtering polymerase will facilitate the use of weaker sites because of the lack of competition from stronger, more distal sites. Thus examining the phosphorylation control mechanisms that dictate RNA polymerase II elongation rate is important.

A critical element in regulating the pace of Pol II transcription is the phosphorylation status of the terminal region. The Pol II C-terminal domain (CTD) is inherently unstructured yet evolutionarily conserved, and in fungi, plants, and animals it is composed of 25 to 52 tandem copies of the consensus repeat heptad $Y_1S_2P_3T_4S_5P_6S_7$ (Phatnani and Greenleaf 2006). A general sense of how the phosphorylation status of the CTD influences Pol II activity was obtained by examining global CTD phosphorylation patterns in vivo. This was carried out by the whole cell ChIP procedure using active transcription in prostate-derived LNCaP cells (Morris et al. 2005). In this procedure, small sub-gene-size chromosomal fragments and their accompanying proteins were immune-precipitated with phospho-specific antibodies (i.e., that bind to pS_5 and pS_2). The precipitated DNA was then amplified

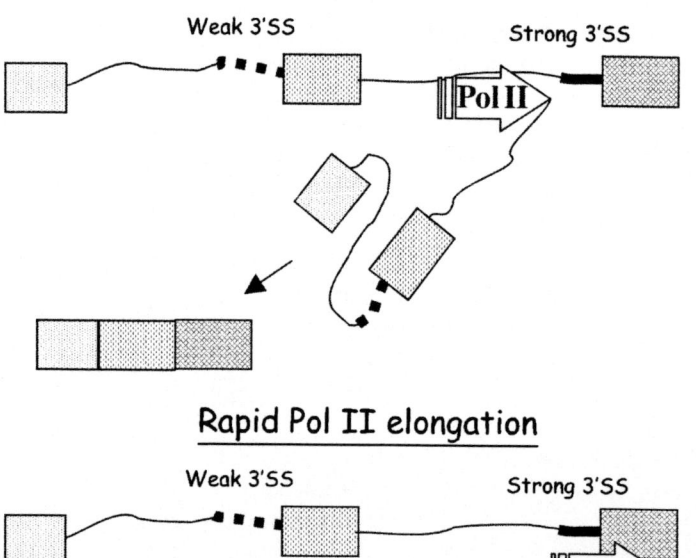

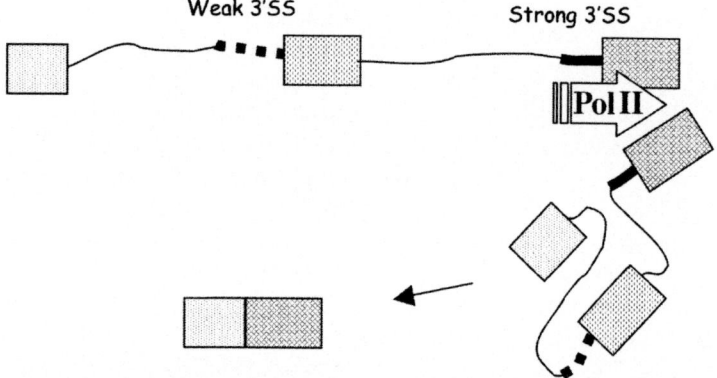

Fig. 1 Alternative splicing and Pol II elongation rate. When an exon is preceded by a weak 3′ splice site (*SS*) the rate of elongation can determine whether the exon will be included. In the top image, when the rate of Pol II elongation is slow the weak 3′ splice site has no competition and is used as a splice site. In the bottom image, the rapid elongation rate by Pol II exposes a stronger 3′ splice site that competes out the weaker site and results in "exon skipping." Thus, low transcriptional elongation rates favor exon inclusion, whereas high elongation rates favor skipping. In the case of the availability of 2 strong 3′ splicing sites the exon would be included constitutively, in a manner that is independent of the elongation rate

and sequenced to determine which particular part of the gene was enriched for a particular phospho moiety. In this way, both the phosphorylation status and transcriptional status could be monitored. The results showed that on most actively transcribed genes, the Pol II that was located near promoters contained higher levels of pS_5 and very little pS_2, whereas elongating polymerases appeared to be phosphorylated at both S_2 and S_5. These changes are thought to orchestrate the association of different factors that impact on transcription rates as well as splicing. The basal transcription factor TFIIH shows S_5 preference, while phosphorylation of S_2 by P-TEFb kinase complex is associated with elongation. The extent of exon skipping correlated with the efficiency with which Pol II transcripts reach the 3' end of the gene, supporting a kinetic coupling model in which the rate of transcript elongation determines the outcome of two competing splicing reactions that occur cotranscriptionally. Hence, as summarized in Fig. 1, rapid, highly processive transcription favors exon skipping, whereas slower, less processive transcription favors inclusion (Nogues et al. 2002).

A plethora of regulatory elements are involved in CTD phosphorylation and are candidates for understanding Pol II regulation of splicing. The hypophosphorylated form of RNA Pol II was found to be enriched at promoter-pause sites in *Drosophila* cells. In that case, P-TEFb, a heterdimer of the kinase Cdk9 and cyclin T could be shown to stimulate the promoter-paused Pol II to enter into productive elongation via its phosphorylation (Lis et al. 2000). The candidate homolog in *Arabidopsis*, CDKC;2, binds to RNPs and a novel (unknown) protein that has structural features shared with SR-like transcription factors (see below). This may provide a functional coupling between transcription control and RNA processing (Barroco et al. 2003). Extensive BLAST searches carried out in many species retrieved a large group of CDK9-like sequences that displayed strong phylogenetic affinity to CDK9 (14 from *Arabidopsis* and 8 from *Oryza*). The presence of a CDK9-like gene complex that appears to be specific to plants as well as conservation of CDK 7 and 8 suggests coevolution of the Pol II CTD and these CTD-directed CDKs (Guo and Stiller 2004). The plant-specific amplification of CDK9-like genes may implicate new and complex modes of regulation for Pol II-dependent elongation and hence impart control of splicing choice, although their functions have yet to be determined experimentally in plants.

CTD of Pol II Interacts with a Variety of Pre-mRNA Factors

Proteins that interact with the CTD also interact with SR splicing factors. For example, AtCyp59 belongs to a family of modular proteins in *Arabidopsis* consisting of a peptidyl-prolyl *cis-trans* isomerase (PPIase) domain, followed by an RNA recognition motif (RRM), and a C-terminal domain enriched in charged amino acids (see the chapter by A. Barta et al., this volume). AtCyp59 was shown to interact with the *Arabidopsis* spliceosomal SR protein SCL33/SR33. However, in the nucleus it localized in a punctuate pattern resembling transcription sites rather than colocalizing with SR proteins in nuclear speckles. Interestingly, in immunoprecipitation assays, AtCyp59 interacts with Pol II CTD, and in transgenic plants the phosphorylation status of CTD was negatively controlled by the level of AtCyp59 expression (Gullerova et al. 2006). These results may explain the diverse localization patterns of these proteins. Cyclophilins possess PPIase activity; for example, they catalyze *cis-trans* isomerization of peptide bonds preceding a proline. It was suggested that the chaperon-like activity associated with PPIase domains may contribute to keeping pre-mRNA in a more accessible conformation for binding of SR proteins and other factors in splice site recognition. Thus, in addition to phosphorylation/ dephosphorylation of splicing factors mediated by kinases and phosphatases that are discussed below, the PPIase activity may be involved in the regulation of protein-protein interactions during spliceosome assembly.

PPIase domains are also found in *Arabidopsis* nuclear proteins, CypRS64 and CypRS92, that contain a C-terminal domain followed by a charged domain with many SR/SP dipeptides. C-terminal domains rich in arginine/serine (RS) dipeptides are characteristic for splicing factors called SR proteins (Graveley 2000; Sanford et al. 2003). These proteins are localized to distinct nuclear bodies. CypRS64 interacts with the SR proteins SRp30 and SRp34/SR1, which play a role in 5′ splice site recognition. These findings, together with the observation that binding of SRp34/ SR1 to CypRS64 is phosphorylation dependent and that coexpression of CypRS64 with its binding partners resulted in relocalization of CypRS64 from the nuclear bodies to nuclear speckles, indicate functional interactions (Lorković et al. 2004). Hence, CypRS64 may regulate phosphorylation/dephosphorylation of SR proteins and other spliceosomal components or be involved in the dynamics of spliceosome assembly. Interestingly, by yeast two-hybrid analysis, CypRS64 was shown to interact with the SR protein-specific kinase SRPK4 (Lorković et al. 2004).

Arabidopsis CypRS64 and CypRS92 are nuclear proteins that are highly related to the human Srcyp (Dubourg et al. 2004). The mitosis-specific phosphorylation of Srcyp suggests regulation by the Cdc2-cyclin B kinase complex. Furthermore, in humans it has been identified as an interacting partner of the RNA polymerase II C-terminal domain (CTD) (Bourquin et al. 1997) and the SR protein-specific LAMMER kinases Clk/Sty that are discussed below (Nestel et al. 1996). Thus the exact direct function of cyclophilin-containing domains in splicing regulation is unknown. RS domains of splicing factors are not structured because of their repetitive sequence, and the PPIase activity of AtCyp59, CypRS64, and CypRS92 may be necessary for putting them into a correct conformation for protein-protein interactions or, alternatively, for being receptive to phosphorylation/ dephosphorylation cycles.

Chromatin Structure Impacts on Transcription Rates and Splicing

As the pace of transcription elongation can dictate the resultant splicing patterns, one might expect that the compactness of chromatin structure would impact on splicing. Indeed, the application of the histone deacetylation inhibitor trichostatin A favors exon skipping. It is possible that the state of hyperacylation favors rapid transcription through the more loosely bound core histones (Nogues et al. 2002; Kornblihtt 2006). Consistent with this scenario, the SWI/SNF chromatin-remodeling factor was shown to interact with Pol II via its ATPase subunit called Brahma (Brm)

(Batsche et al. 2006). Overexpression of Brm in humans promotes alternative exon inclusion by multiple interactions with small nuclear RNAs and other factors. For example, Sam68, an RNA binding protein, is found to be associated with splicing regulatory elements that are present in the gene CD44, which is made up of a set of constitutive and alternative exons. Sam68 mediates the inclusion of variable exons when it is stimulated by the ERK MAP kinases. The distribution of Brm and Pol II and their interaction were examined along the CD44 gene by the application of ChIP technology. The results implied that the presence of SWI/SNF complex induced stalling of the polymerase. When the ChIP analysis was extended to the detection of CTD phosphorylation status, a dramatic dynamic distribution in the phosphorylation pattern was noted. In the splicing variable region Brm was shown to associate with pS_5 CTD Pol II species, while in the constant region pS_2 was dominant. While it remains to be seen whether the CTD phosphorylation status is a cause or effect of this process, the consequence of chromatin remodeling complexes can be seen as follows. In MAP kinase-stimulated cells the association of Sam68 at the alternatively spliced region may cause Brm association that brings about a pS_5 increase in Pol II CTD and a slowdown of Pol II activity, which then impacts on splicing.

Arabidopsis thaliana encodes 42 putative SNF2-like ATPases (http://chromdb.org), and at least four belong to the canonical SWI2/SNF2 subfamily (Verbsky and Richards 2001). Silencing of *Arabidopsis BRAHMA* (At2g46020) results in reduced fertility, curly leaves, homeotic transformations during flower development, and photoperiod-independent early flowering by derepression of *CONSTANS* (*CO*), *FLOWERING LOCUS T* (*FT*), and *SUPPRESSION OF OVEREXPRESSION OF CONSTANS1* (*SOC1*) (Farrona et al. 2004). It will be of interest to examine the bearing these mutations have on splicing.

Splicing Factors

Role of Serine/Arginine-Rich Protein Phosphorylation in Splicing

Pre-mRNA splicing takes place on the supraspliceosome that is composed of active spliceosomes connected to each other by the pre-mRNA (Azubel et al. 2006). The major spliceosome contains small nuclear ribonucleoprotein particles (snRNPs) and a U2/U6 snRNA structure that has been proposed to form the active site for catalysis of the transesterifications. The snRNPs recognize splice sites and branchpoint sequences in pre-mRNA and aid in splicing. The relatively abundant cellular proteins that contribute to ribonucleoprotein particles are SR (serine/arginine rich) proteins that in the interphase nucleus reside in speckles from which they are recruited by phosphorylation to participate in pre-mRNA synthesis. The importance of phosphorylation events in the assembly of this structure was first established in seminal experiments showing that treatment of HeLa nuclear splicing extracts with protein phosphatase 1 (PP1) prevents prespliceosome formation and stable binding of snRNPs to the pre-mRNA. PP1 does not inhibit splicing if added to preassembled spliceosome structures. However, the addition of purified SR protein splicing factors restores spliceosome formation and splicing to PP1-treated extracts (Mermoud et al. 1994). These observations are consistent with SR proteins being primary targets of the spliceosome and their action being regulated by phosphorylation.

The family of SR proteins in *Arabidopsis* comprises at least 19 proteins that range in size from 21 to 45 kDa and serve similar functions (Lopato et al. 1999; Reddy 2004) (see the chapters by A. Barta et al. and G. S. Ali and A. S. N. Reddy, this volume). The SR proteins have modular domain structure with one or two RNA recognition motifs (RRMs) at the N-terminus and an arginine/serine-rich (RS) domain at the C-terminus. Plant SR proteins can appear as conserved orthologs of animal SR proteins, for example, SF2/ASF (splicing factor 2/alternative splicing factor), SC35 (spliceosomal component 35), SCLs (SC35-like), and 9G8. However, there are many plant-specific factors, for example, atRSZ33, which interacts with SR proteins, that also must be considered (Lopato et al. 2002). Because of their implication in the early recognition of splice sites, SR proteins represent ideal factors to regulate alternative splicing. Thus the RNA-binding specificity and the ability of SR proteins to interact with other proteins justify in-depth analysis of phosphorylation properties of SR proteins in order to understand the molecular basis of splicing and alternative splicing.

SR proteins are likely targeted at multiple levels to ensure that the appropriate factors are present at the right time and at the right place to accomplish their functions. One type of regulation determines the steady-state level of SR proteins through specific transcriptional and posttranscriptional processes of protein turnover and is not discussed here. In this respect, it is of interest that kinase components that regulate the phosphorylation status of SR proteins may also be involved in their turnover. For example, the stability of the SR protein SRp55 is determined by the proteasome-mediated pathway that is controlled by the expression of the SR protein LAMMER kinase (Clk/Sty kinase). Thus, for this SR protein, a specific phosphorylation-mediated degradation pathway exists that does not seem to affect other SR proteins such as SC35 or ASF/SF2 (Lai et al. 2003).

The SR proteins are phosphoproteins characterized by a high content of arginine and serine, which often form a regular repetition of RS dipeptides that can represent 76% to 80% of the domain. One characteristic of RS domains is their ability to become heavily phosphorylated on Ser residues (Bourgeois et al. 2004). This phosphorylation, for example in the case of SF2/ASF, is required for its specific interaction with U1-70K and U1 snRNP (Xiao and Manley 1998). In one model the RS domain of RNA enhancer-bound SR protein interacts with other splicing factors that also contain an RS domain, and in this way, facilitate recruitment of the spliceosome (Hertel and Graveley 2005). For example, in animals the SR proteins SF2/ASF and SC35 have been shown to interact with the U2AF subunit in a RS domain-dependent manner (Wu and Maniatis 1993). Other experiments implicate that RS domains of SR proteins can engage directly in protein-RNA interactions (Shen et al. 2004). Taking charge modifications into consideration, a likely model for SR action is that hypophosphorylated RS domains interact with RNA, perhaps through electrostatic interactions between positively charged arginines, while phosphorylated RS domains are involved in protein interactions with other RS domains.

Plant factors have been shown to have complex levels of interaction as well. The plant-specific RS-rich protein atRSZ33 interacts with a variety of splicing factors and has a RNA recognition motif (RRM) and two zinc knuckles embedded in a basic RS region. The atRSZ33 phosphoprotein was shown to concentrate in nuclear speckles and is enriched in roots and flowers. These interactions suggest that its main activity is in spliceosome assembly. Deletion mapping of the polypeptide showed that both zinc knuckles, together with a small part of the RS and the RRM domain, are required for efficient interactions (Lopato et al. 2002). Zinc interacting motifs are known for their sensitivity to the cellular redox status mediated by the presence of cysteine residues (Wu et al. 1996). The presence of these motifs may imply that atRSZ33-like factors integrate both redox and phosphorylation cellular status during spliceosomal assembly.

An intriguing type of regulation of SR proteins includes modification by phosphorylation of RS domains by signal transduction-dependent pathways. The main question arising is how SR proteins regulate splicing in the context of a specific cell type or in response to extracellular signals.

Phosphorylation-Dependent Localization of SR Proteins

In addition to the modulation of SR proteins turnover or spliceosomal interactions, the activity of an existing pool of SR proteins is regulated by changing their intracellular or subnuclear localization. This can be achieved through control of phosphorylation or dephosphorylation of the RS domain that affects the nucleocytoplasmic shuttling of some SR proteins-typically for recruitment to active transcription sites (Caceres et al. 1998). A subset of SR proteins shuttles continuously between the nucleus and the cytoplasm, suggesting that the role of SR proteins in gene expression may not be limited to nuclear pre-mRNA splicing but also includes cytoplasmic functions. Importantly, the modulation of the intracellular localization of specific SR proteins can result in changes in alternative splicing patterns. For example, during poststroke ischemia the intracellular Ca^{2+} concentration in brain tissue rises. When ischemia is induced the splicing regulatory protein tra$2^{\beta}1$ as well as other SR proteins accumulate in the cytoplasm in a hyperphosphorylated state. This was detected with the use of mAb104 monoclonal antibody that reacts to phospho motifs on SR proteins (Roth et al. 1991). Significantly, a concomitant increase was observed in the inclusion in the ICH-1S alternative exon of the *ICH-1* gene. The result is consistent with the observation that the presence of excess of SR protein factors ASF/SF2 and SC35 can promote the complementary effect, namely, the skipping of this exon. Hence, as part of ischemia stress signaling the pre-mRNA splicing pathways are regulated by causing relocalization of proteins that are involved in splice site selection (Daoud et al. 2002).

In *Arabidopsis*, distinct tissue-specific expression patterns were observed for SR1/atSRp34 and atSRp30 (Fang et al. 2004). These factors localized in a cell type-dependent speckled pattern that moved within a constrained nuclear space in a cell cycle-dependent fashion. With the technique of photobleaching followed by fluorescence recovery a rapid exchange rate of splicing factors could be observed. Furthermore, dynamic organization of plant speckles was closely related to the transcriptional activity of the cells (see the chapter by G. S. Ali and A. S. N. Reddy, this volume). The rapidity of phosphorylation effects on splicing factor organization was illustrated in analysis of the wild-type SR protein RSZp22 that displays a predominant nuclear speckled organization and some nucleolus localization. Deletion of the RS domain results in predominant nucleolar accumulation. In addition, modulation of the phosphorylation/dephosphorylation cycle of the RS domain by the addition of phosphorylation inhibitors impacts on this subnuclear distribution. Thus the RS domain controls atRSZp22 nucleoplasm and nucleolus shuttling through its phosphorylation level (Tillemans et al. 2005, 2006). In another example, SR protein RSp31 can transiently associate with the nucleolus in a cell type-dependent manner that is modified by the phosphatase inhibitors okadaic acid and staurosporine (Docquier et al. 2004). In that case, inhibition of kinase activity led to a perinucleolar redistribution of RSp31 (Tillemans et al. 2006). The fact that plants contain twice as many splicing factors as animals and that there is a distinct nucleolar localization for some of these factors points to differences in their regulation. For example, SR45 contains two RS domains separated by an RRM and has no animal homolog (Golovkin and Reddy 1999). With the technique of recovery after fluorescent bleaching (FRAP) of fluorescently tagged SR45 and SR1/Srp34, the movement of these SR proteins was found to be very sensitive to ATP levels, transcription, and protein kinase inhibitors (Ali and Reddy 2006). This is in contrast to the results obtained in animals and suggests that in plants, in the absence of active splicing, the factors remain more tightly bound to the speckles.

Phosphorylation in Postsplicing Functions of SR Proteins

The role of SR proteins in mRNA processing is not restricted to their function as splicing factors. They are initially recruited for splicing in their hyperphosphorylated form, gradually changing to a partially dephosphorylated stage that may mark the ribonucleoprotein complex for export as shown in Fig. 2. In the dephosphorylated state, they can recruit the export adapter NXF1 and cooperate with the postsplicing exon-junction complex (EJC) (Huang et al. 2004). The NXF1 or TAP sequence has not been detected in the *Arabidopsis* genome. However, other factors containing nuclear transport-like domains thought to be involved in nuclear trafficking have been identified and may replace this function (Lorković and Barta

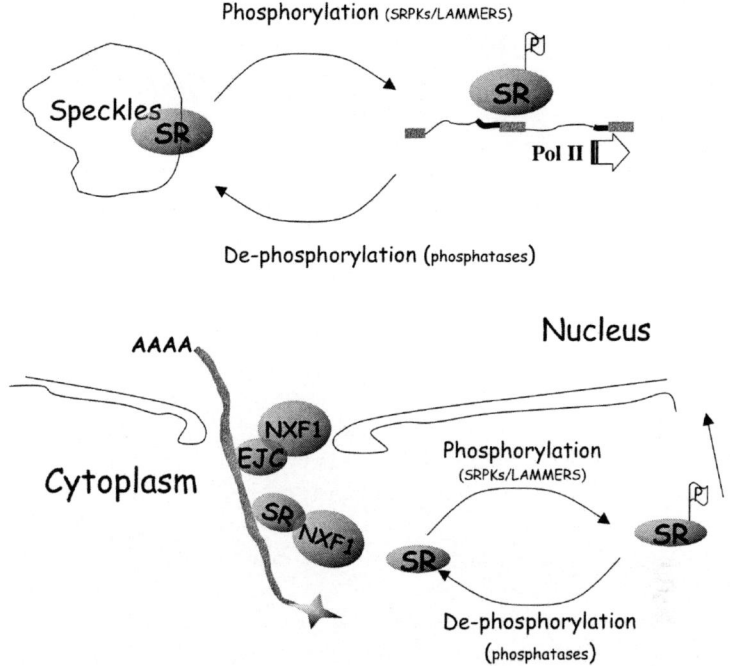

Fig. 2 Cycles of phosphorylation that control SR protein movement and splicing. Splicing speckles are dynamic structures that respond to the levels of Pol II transcription, fluctuating in shape in response to transcription as the splicing factors diffuse in and out. When transcription is halted, splicing speckles enlarge and the splicing factors redistribute from sites of transcription back into the splicing speckles. Conversely, when transcription rates are very high, splicing factors redistribute out of the speckles and into the nucleoplasm. The shuttling of SR proteins is controlled by the kinase activity of a few kinases. Hyperphosphorylation of SR proteins leads to disassembly of splicing speckles, and phosphorylated SR proteins are targeted to nascent transcripts via interactions with the CTD of RNA pol II and exonic enhancer sequences. Once the spliceosome has assembled the dephosphorylation of SR proteins by phosphatases is necessary for catalysis and the movement of SR proteins back to the speckles. Splicing results in the dephosphorylation of bound SR proteins and in the deposition of exon junction complexes (*EJC*) upstream of exon-exon boundaries in the spliced RNA. In animals, hypophosphorylated forms of the SR proteins help to recruit the export adapter NXF1, which orchestrates passage of the mRNP through the nuclear pore. A subset of SR proteins shuttle between the nucleus and the cytoplasm and are active in mRNA escort and during translation. Translation is accompanied by RNP remodeling, leading to the release and reimport of rephosphorylated SR proteins. Kinases that have been identified as orchestrating the controlled shuttling of SR proteins are LAMMER and SRPK kinases

2002). The EJC is a set of proteins deposited on maturing pre mRNA and plays a role in mRNA quality control. During the elimination of introns, EJCs are deposited upstream of the exon-exon junctions. The EJCs remain associated with spliced mRNAs until after nucleocytoplasmic export, when the majority of EJCs are removed by the first or "pioneering" round of translation. If it is not removed, for example, because of the presence of an in-frame premature termination codon, that mRNA is recognized as defective and targeted for degradation by the process of nonsense-mediated decay (NMD) (Stutz and Izaurralde 2003). Thus an important role for the EJC is the communication of positional information regarding the previous positions of introns and hence mRNA quality.

In plants, many components of the postsplicing EJC involved in mRNA export and NMD/mRNA surveillance are conserved except for NXF1/TAP (Pendle et al. 2005) (see the chapters by D. A. Belostotsky and J. W. S. Brown and P. J. Shaw, this volume). Surprisingly, these were localized to the nucleoli and may indicate a function for the plant nucleolus in storage or assembly of EJC subcomplexes, mRNA export, or NMD. In *Arabidopsis*, the *Agrobacterium tumefaciens*-mediated transient expression assay was applied to RSZp22, a factor homologous with human SRp20 and 9G8. The fluorescence loss in the nucleus brought about by photobleaching of cytoplasmic regions indicated nuclear-cytoplasmic shuttling. This was further supported by the use of leptomycin B (LMB), a potent nuclear export inhibitor. Proteins exported from the nucleus can possess a distinct nuclear export signal that forms a ternary complex with an export receptor, termed CRM1/exportin1 or XPO1. This interaction is inhibited by LMB. When LMB was applied and the cytoplasm was photobleached, the fluorescence loss was much diminished, indicating that RSZp22 shuttles between the nucleus and the cytoplasm (Tillemans et al. 2006).

After SR-assisted mRNA export, the SR proteins are rephosphorylated in the cytoplasm and recycled to the nucleus. As discussed above, differences in the phosphorylation status of SR proteins are correlated with their cellular positioning that can impact on functional splicing. However, this shuttling has even wider impact on cellular metabolism than was thought previously. For example, it was shown that SF2/ASF associates with translating ribosomes and could stimulate translation of reporter mRNA in *Xenopus* oocytes or stimulate translation in vitro with a HeLa cell-free translation system (Sanford et al. 2004).

The SR proteins are dispersed in the nucleoplasm and also localized to numerous nuclear speckles (about 30) that are called splicing factor compartments (SFCs) (Misteli 2000; Lamond and Spector 2003). Transcription and splicing likely proceed at the periphery of and outside of SFCs, with SR proteins shuttling between SFCs and active sites of transcription and splicing in a phosphorylation-dependent manner as discussed above. The modification of SR proteins by phosphorylation influences their recruitment to sites of transcription, their interaction with other proteins and RNAs, and their nucleo-cytoplasmic shuttling as summarized in Fig. 2.

Kinases That Phosphorylate Splicing Factors

LAMMER Kinases Modulate Splicing

The RS domains of SR proteins are the targets for specific phosphorylation by several protein kinases, including Clk/Sty1-4 of the LAMMER family of protein kinases and SRPK1 and -2 kinases. In the simplest sense, overexpression of

SR-specific kinases leads to their hyperphosphorylation and subsequent disassembly of splicing factor compartments (Misteli 2000).

Clk/Sty itself has a serine/arginine-rich noncatalytic N-terminal region that is important for its association with SR splicing factors. In vitro, Clk/Sty was shown to efficiently phosphorylate the SR family member ASF/SF2 on serine residues located within the RS domain and can colocalize with SR proteins into nuclear speckles. The overexpression of the active Clk/Sty kinase induced redistribution of SR proteins, suggesting that LAMMER kinases regulate SR protein activity and compartmentalization (Colwill et al. 1996a). SRPK1 and Clk/Sty phosphorylate the same SR proteins in vitro, but SRPK1 shows higher specific activity toward ASF/SF2. The Clk/Sty kinase phosphorylated Ser-Arg, Ser-Lys, or Ser-Pro sites, whereas SRPK1 had a strong preference for Ser-Arg sites (Colwill et al. 1996b). LAMMER kinases can autophosphorylate on both Ser/Thr and Tyr residues, and when autophosphorylation occurs both the pattern of phosphorylation of the SR protein substrate and the spectrum of SR protein substrates changes. For example, phosphorylation of ASF/SF2 is sensitive to changes in Tyr, but not Ser/Thr autophosphorylation, while that of SC35 displays the opposite pattern. In contrast, phosphorylation of a third SR protein, SRp40, is unaffected by autophosphorylation (Prasad and Manley 2003). In this respect, it is of interest that a specific inhibitor of LAMMER kinase, TG003, a benzothiazole compound, was shown to inhibit SF2/ASF-dependent splicing of β globin pre-mRNA in vitro and suppress serine/arginine-rich protein phosphorylation and dissociation of nuclear speckles, providing a new tool for exploring LAMMER kinase function.

Elegant evidence for the in vivo role of LAMMER kinases in development comes from experiments in *Drosophila*. The fly LAMMER kinase homolog called DOA was found to phosphorylate the Drosophila SR protein RBP1, as well as Tra and Tra2. DOA mutations disrupt the splicing of the doublesex pre-mRNA, a key regulator of sex determination, by controlling alternative splice site selection in this pre-mRNA (Du et al. 1998). The regulation of LAMMER kinases can be brought about not only by their upregulation or activation but also by alternative splicing. During erythroleukemia cell differentiation alternative splicing of LAMMER generates transcripts encoding a full-length kinase and a truncated catalytically inactive protein. In undifferentiated cells the full-length transcript was more abundant than the truncated form and the ratio switched at the latter stages of differentiation (Garcia-Sacristan et al. 2005). The interaction between active and inactive kinase isoforms likely impacts on splicing. Similarly, in the mouse, LAMMER kinases come in two forms: The 55-kDa isoform is primarily localized to the nucleus, and the 105-kDa isoform localizes to the cytoplasm (Yun et al. 2000).

A recombinant LAMMER-class protein kinase (AFC-2) from *Arabidopsis* was expressed and showed autophosphorylation and phosphorylation activity. AFC-2 phosphorylated four SR proteins (SR33, SR45, SRZ21, and SRZ22) (Golovkin and Reddy 1999); in contrast, AFC1 was inactive. The tobacco LAMMER kinase (PK12) is closely related to the *Arabidopsis* AFC-2 and autophosphorylates in vitro on serine, threonine, and tyrosine residues, thereby making it a member of the dual-specificity protein kinases. Interestingly, the kinase activity, as well as transcript level, was rapidly and transiently increased when plants were treated with the plant hormone ethylene (Sessa et al. 1996). The recombinant PK12 protein was localized to the nucleus and phosphorylated both plant and animal SR proteins (Savaldi-Goldstein et al. 2000). Plant LAMMER kinases differ from animal Clk/sty kinases in that they do not contain an RS-rich region in the N-terminus, suggesting that structural features other than an RS domain are also important in the interaction of LAMMER kinases with other proteins. Interestingly, in an attempt to screen for signal transduction factors from plants, AFC1 but not AFC2 from *Arabidopsis* was shown to suppress a mating defect in the fus3 and kss1 signal transduction mutant of *Saccharomyces cerevisiae* (Bender and Fink 1994). Whether this is the result of serendipitous complementation or reveals novel biological significance is not clear. Mutation of the tobacco LAMMER motif resulted in the detection of aberrantly localized kinase protein to a ring structure in the nucleus (Savaldi-Goldstein et al. 2003). Transgenic lines that showed high levels of tobacco LAMMER polypeptide expression and kinase activity were diminutive and had prolonged life cycles. Significantly, plants with elevated kinase activity displayed a shift in the alternatively spliced mRNAs of splicing factors atSRp30, atSRp34/SR1, and U1-70K, indicating that the activity of the plant LAMMER kinase modulates splicing.

Cellular Partitioning of SR Proteins by SRPK Activity

Once exported to the cytoplasm, SR proteins are reimported to the nucleus by interacting with the import receptor hMtr10/Transportin-SR in a phosphorylation-dependent manner (Yun et al. 2003). The SRPK family of kinases specific for SR proteins are also localized in the cytoplasm and are crucial for initiating nuclear import of SR proteins in a phosphorylation-dependent manner. Structurally, the bipartite kinase catalytic core of SRPK is separated by a unique spacer sequence that controls its cytoplasmic localization. Deletion of the spacer had little effect on kinase activity, but caused a quantitative translocation of the kinase to the nucleus and aberrant aggregation of splicing factors in the nucleus. The subcellular localization of SRPK depends on the cell cycle, and these kinases are translocated to the nucleus at the G_2/M boundary. (Ding et al. 2006) (see the chapter by G. S. Ali and A. S. N. Reddy, this volume).

The exquisite interplay between the kinase classes LAMMER and SRPK was illustrated by identifying a docking motif through the analysis of crystal structure of an active fragment of human SRPK1 bound to an SR protein peptide. The motif directs phosphorylation of ASF/SF2 by SRPK1 to the N-terminal component of the RS domain. As discussed above, this will cause SR proteins to associate within nuclear speckles. In contrast, the LAMMER kinase shows preference for phosphorylation of the C-terminal component of the ASF/SF2 RS domain that stimulates release of SR proteins from speckles. Thus the docking motif facilitates sequential phosphorylation of SR proteins by SRPK1 and LAMMER kinases (Ngo et al.

2005). Furthermore, it was shown that while SRPK1 efficiently phosphorylated select stretches of amino acids in the N-terminal portion of the RS domain in ASF/SF2, the LAMMER kinase was able to phosphorylate all available RS serine residues. This observation is consistent with the idea that LAMMER and SRPK kinases can work sequentially (Velazquez-Dones et al. 2005).

The importance of the phosphorylated status of nuclear protein was noted in a proteomic survey of phospholabeled protein. Of 79 phosphorylation sites recovered, 50 were on SR proteins. The preferred sites of phosphorylation were serines that make up a conserved RSP motif (van Bentem et al. 2006). *Arabidopsis* SRPK 4 showed strong activity using the SR protein RSp31 as a substrate that also resulted in RpSP site phosphorylation. Proteome analysis by mass spectra of isolated plant nuclear proteins was carried out on the peptides from the nuclear extracts. They showed that the phospho-peptide signatures in vitro and in vivo were very similar and contained a prominent RpSP site that is conserved among many members of other RNA metabolism proteins (van Bentem et al. 2006). The results implicate SRPK 4 as one of the in vivo kinases that target these sites. As shown by Genevestigator analysis (https://www.genevestigator.ethz.ch/; Zimmermann et al. 2005) the transcript level of this kinase is regulated 1.5-fold by a variety of environmental stress factors including senescence, heat, and osmotic stress, but the relevance of SRPK 4 transcript abundance to splicing remains to be seen.

Topoisomerase I

Topoisomerase I is primarily known as a DNA nicking/closing enzyme that has implications for chromatin condensation and decondensation during the cell cycle. It displays a phosphodiesterase and ligase activity (Wang 2002). It interacts with SR proteins via a distinct activity domain in which Topo I acts as a kinase (Rossi et al. 1996). Topo I depletion results in the hypophosphorylation of SR proteins and impairs exonic splicing enhancer (ESE)-dependent but not constitutive splicing in murine B lymphoma-derived cell lines. Ectopic expression of Topo I restores both wild-type phosphorylation of SR proteins and ESE-dependent splicing (Soret et al. 2003). Similar to LAMMER kinase, specific inhibitors exist, and diospyrin derivatives have been shown to inhibit SR phosphorylation but not topoisomerase activity and promise to add a new tool in research in structure function relationships (Tazi et al. 2005). Information on plant-related Topo I activities in relation to splicing is lacking.

Concluding Remarks

The RS domain has been shown to engage in both protein-protein interactions and protein-RNA interactions that are germane to splicing. The extensive phosphorylation of SR proteins by SRPK, LAMMER kinases, and topoisomerase I lead to profound changes in the protein- and RNA-interaction properties of the RS domains. It is a challenge to identify signal transduction pathways that will engage in specific splice-dependent changes. One example is regulation of fibronectin expression. The fibronectin transcript contains an exon called EDA that is included at low levels in the adult but at high levels during embryogenesis or during wound healing. Protein kinase B/Akt is a highly conserved metazoan kinase present in the cytoplasm that responds to a large variety of signals that involve cell survival. It is a downstream component in phosphoinositide (PI) 3-kinase signaling (Fayard et al. 2005). It was shown to impact on the inclusion of the EDA exon by its ability to modify 9G8 and SF2/ASF SR protein activity by direct phosphorylation. This activity enhances the reimport of free cytoplasmic SR proteins into the nucleus. This potentiates SR protein binding to transcribed mRNA and increases the inclusion of the EDA exon and also enhances the translation of mRNAs containing the EDA exon. In this case high activities of LAMMER kinases and SRPK have effects opposite to that of AKT (Blaustein et al. 2005). This remarkable signal transduction loop results in more translated EDA-containing special transcript, leading to a rapid and concerted response to the cellular physiological input. If this regulatory scenario is paradigmatic, then specificity of the response is achieved by a novel kinase input via SR proteins that is used by the cell for a particular event. The function of LAMMER and SRPK kinase would then be to serve as basal splicing regulators that maintain the phosphorylation of the SR proteins in a state that is receptive for perturbations that will initiate specific responses. The integration of our understanding of the impact that phosphorylation has on splicing into specific cellular processes is part of the future challenges in pre-mRNA splicing biology.

References

Ali GS, Reddy ASN (2006) ATP, phosphorylation and transcription regulate the mobility of plant splicing factors. J Cell Sci 119:3527–3538
Azubel M, Habib N, Sperling R, Sperling J (2006) Native spliceosomes assemble with pre-mRNA to form supraspliceosomes. J Mol Biol 356:955–966
Barroco RM, De Veylder L, Magyar Z, Engler G, Inze D, Mironov V (2003) Novel complexes of cyclin–dependent kinases and a cyclin–like protein from *Arabidopsis thaliana* with a function unrelated to cell division. Cell Mol Life Sci 60:401–412
Batsche E, Yaniv M, Muchardt C (2006) The human SWI/SNF subunit Brm is a regulator of alternative splicing. Nat Struct Mol Biol 13:22–29
Bender J, Fink GR (1994) Afc1, a Lammer kinase from *Arabidopsis thaliana*, activates Ste12-dependent processes in yeast. Proc Natl Acad Sci USA 91:12105–12109
Blaustein M, Pelisch F, Tanos T, Munoz MJ, Wengier D, Quadrana L, Sanford JR, Muschietti JP, Kornblihtt AR, Caceres JF, Coso OA, Srebrow A (2005) Concerted regulation of nuclear and cytoplasmic activities of SR proteins by AKT. Nat Struct Mol Biol 12:1037–1044
Bourgeois CF, Lejeune F, Stevenin J (2004) Broad specificity of SR (serine/arginine) proteins in the regulation of alternative splicing of pre-messenger RNA. Prog Nucl Acid Res Mol Biol 78:37–88

Bourquin JP, Stagljar I, Meier P, Moosman P, Silke J, Baechi T, Georgiev O, Schaffner W (1997) A serine/arginine–rich nuclear matrix cyclophilin interacts with the C–terminal domain of RNA polymerase II. Nucl Acids Res 25:2055–2061

Caceres JF, Screaton GR, Krainer AR (1998) A specific subset of SR proteins shuttles continuously between the nucleus and the cytoplasm. Gene Dev 12:55–66

Colwill K, Pawson T, Andrews B, Prasad J, Manley JL, Bell JC, Duncan PI (1996a) The Clk/Sty protein kinase phosphorylates SR splicing factors and regulates their intranuclear distribution. EMBO J 15:265–275

Colwill K, Feng LL, Yeakley JM, Gish GD, Caceres JF, Pawson T, Fu XD (1996b) SRPK1 and Clk/Sty protein kinases show distinct substrate specificities for serine/arginine–rich splicing factors. J Biol Chem 271:24569–24575

Daoud R, Mies G, Smialowska A, Olah L, Hossmann KA, Stamm S (2002) Ischemia induces a translocation of the splicing factor tra2–1 and changes alternative splicing patterns in the brain. J Neurosci 22:5889–5899

Dinesh-Kumar SP, Baker BJ (2000) Alternatively spliced N resistance gene transcripts: Their possible role in tobacco mosaic virus resistance. Proc Natl Acad Sci USA 97:1908–1913

Ding JH, Zhong XY, Hagopian JC, Cruz MM, Ghosh G, Feramisco J, Adams JA, Fu XD (2006) Regulated cellular partitioning of SR protein–specific kinases in mammalian cells. Mol Biol Cell 17:876–885

Docquier S, Tillemans V, Deltour R, Motte P (2004) Nuclear bodies and compartmentalization of pre–mRNA splicing factors in higher plants. Chromosoma 112:255–266

Du C, McGuffin ME, Dauwalder B, Rabinow L, Mattox W (1998) Protein phosphorylation plays an essential role in the regulation of alternative splicing and sex determination in *Drosophila*. Mol Cell 2:741–750

Dubourg B, Kamphausen T, Weiwad M, Jahreis G, Feunteun J, Fischer G, Modjtahedi N (2004) The human nuclear SRcyp is a cell cycle–regulated cyclophilin. J Biol Chem 279:22322–22330

Fang YD, Hearn S, Spector DL (2004) Tissue-specific expression and dynamic organization of SR splicing factors in *Arabidopsis*. Mol Biol Cell 15:2664–2673

Farrona S, Hurtado L, Bowman JL, Reyes JC (2004) The *Arabidopsis thaliana* SNF2 homolog AtBRM controls shoot development and flowering. Development 131:4965–4975

Fayard E, Tintignac LA, Baudry A, Hemmings BA (2005) Protein kinase B/Akt at a glance. J Cell Sci 118:5675–5678

Garcia-Sacristan A, Fernandez-Nestosa MJ, Hernandez P, Schvartzman JB, Krimer DB (2005) Protein kinase clk/STY is differentially regulated during erythroleukemia cell differentiation: a bias toward the skipped splice variant characterizes postcommitment stages. Cell Res 15:495–503

Golovkin M, Reddy ASN (1999) An SC35–like protein and a novel serine/arginine–rich protein interact with Arabidopsis U1-70K protein. J Biol Chem 274:36428–36438

Gullerova M, Barta A, Lorković ZJ (2006) AtCyp59 is a multidomain cyclophilin from *Arabidopsis thaliana* that interacts with SR proteins and the C–terminal domain of the RNA polymerase II. RNA 12:631–643

Guo Z, Stiller JW (2004) Comparative genomics of cyclin–dependent kinases suggest co–evolution of the RNAP IIC–terminal domain and CTD–directed CDKs. BMC Genomics 5:69

Hertel KJ, Graveley BR (2005) RS domains contact the pre–mRNA throughout spliceosome assembly. Trends Biochem Sci 30:115–118

Howe KJ, Kane CM, Ares M (2003) Perturbation of transcription elongation influences the fidelity of internal exon inclusion in *Saccharomyces cerevisiae*. RNA 9:993–1006

Huang, YQ, Yario TA, Steitz JA (2004) A molecular link between SR protein dephosphorylation and mRNA export. Proc Natl Acad Sci USA 101:9666–9670

Iida K, Seki M, Sakurai T, Satou M, Akiyama K, Toyoda T, Konagaya A, Shinozaki K (2004) Genome–wide analysis of alternative pre–mRNA splicing in *Arabidopsis thaliana* based on full–length cDNA sequences. Nucl Acids Res 32:5096–5103

Kornblihtt AR (2006) Chromatin, transcript elongation and alternative splicing. Nat Struct Mol Biol 13:5–7

Kornblihtt AR, De la Mata M, Fededa JP, Munoz MJ, Nogues G (2004) Multiple links between transcription and splicing. RNA 10:1489–1498

Lai MC, Lin RI, Tarn WY (2003) Differential effects of hyperphosphorylation on splicing factor SRp55. Biochem J 371:937–945

Lamond AI, Spector DL (2003) Nuclear speckles: A model for nuclear organelles. Nat Rev Mol Cell Biol 4:605–612

Lis JT, Mason P, Peng J, Price DH, Werner J (2000) P–TEFb kinase recruitment and function at heat shock loci. Gene Dev 14:792–803

Lopato S, Kalyna M, Dorner S, Kobayashi R, Krainer AR, and Barta A (1999) atSRp30, one of two SF2/ASF–like proteins from *Arabidopsis thaliana*, regulates splicing of specific plant genes. Gene Dev 13:987–1001

Lopato S, Forstner C, Kalyna M, Hilscher J, Langhammer U, Indrapichate K, Lorković ZJ, Barta A (2002) Network of interactions of a novel plant–specific Arg/Ser–rich protein, atRSZ33, with atSC35–like splicing factors. J Biol Chem 277:39989–39998

Lorković ZJ, Barta A (2002) Genome analysis: RNA recognition motif (RRM) and K homology (KH) domain RNA–binding proteins from the flowering plant *Arabidopsis thaliana*. Nucl Acids Res 30:623–635

Lorković ZJ, Lopato S, Pexa M, Lehner R, Barta A (2004) Interactions of *Arabidopsis* RS domain containing cyclophilins with SR proteins and U1 and U11 small nuclear ribonucleoprotein–specific proteins suggest their involvement in pre–mRNA splicing. J Biol Chem 279:33890–33898

Macknight R, Duroux M, Laurie R, Dijkwel P, Simpson G, Dean C (2002) Functional significance of the alternative transcript processing of the Arabidopsis floral promoter *FCA*. Plant Cell 14:877–888

Mermoud JE, Cohen PTW, Lamond AI (1994) Regulation of mammalian spliceosome assembly by a protein–phosphorylation mechanism. EMBO J 13:5679–5688

Misteli T (2000) Cell biology of transcription and pre–mRNA splicing: nuclear architecture meets nuclear function. J Cell Sci 113:1841–1849

Modrek B, Lee C (2002) A genomic view of alternative splicing. Nat Genet 30:13–19

Morris DP, Michelotti GA, Schwinn DA (2005) Evidence that phosphorylation of the RNA polymerase II carboxyl–terminal repeats is similar in yeast and humans. J Biol Chem 280:31368–31377

Ner–Gaon H, Fluhr R (2006) Whole–genome microarray in *Arabidopsis* facilitates global analysis of retained introns. DNA Res 13:111–121

Ner–Gaon H, Halachmi R, Savaldi–Goldstein S, Rubin E, Ophir R, Fluhr R (2004) Intron retention is a major phenomenon in alternative splicing in *Arabidopsis*. Plant J 39:877–885

Nestel FP, Colwill K, Harper S, Pawson T, Anderson SK (1996) RS cyclophilins: Identification of an NK–TR(1)–related cyclophilin. Gene 180:151–155

Ngo JCK, Chakrabarti S, Ding JH, Velazquez–Dones A, Nolen B, Aubol BE, Adams JA, Fu XD, Ghosh G (2005) Interplay between SRPK and Clk/Sty kinases in phosphorylation of the splicing factor ASF/SF2 is regulated by a docking motif in ASF/SF2. Mol Cell 20:77–89

Nogues G, Kadener S, Cramer P, Bentley D, Kornblihtt AR (2002) Transcriptional activators differ in their abilities to control alternative splicing. J Biol Chem 277:43110–43114

Pendle AF, Clark GP, Boon R, Lewandowska D, Lam YW, Andersen J, Mann M, Lamond AI, Brown JWS, Shaw PJ (2005) Proteomic analysis of the *Arabidopsis* nucleolus suggests novel nucleolar functions. Mol Biol Cell 16:260–269

Phatnani HP, Greenleaf AL (2006) Phosphorylation and functions of the RNA polymerase IICTD. Gene Dev 20:2922–2936

Prasad J, Manley JL (2003) Regulation and substrate specificity of the SR protein kinase Clk/Sty. Mol Cell Biol 23:4139–4149

Reddy ASN (2004) Plant serine/arginine–rich proteins and their role in pre–mRNA splicing. Trends Plant Sci 9:541–547

Roberts GC, Gooding C, Mak H.Y, Proudfoot NJ, Smith CWJ (1998) Co–transcriptional commitment to alternative splice site selection. Nucl Acids Res 26:5568–5572

Rossi F, Labourier E, Forne T, Divita G, Derancourt J, Riou JF, Antoine E, Cathala G, Brunel C, Tazi J (1996) Specific phosphorylation of SR proteins by mammalian DNA topoisomerase I. Nature 381:80–82

Roth MB, Zahler AM, Stolk JA (1991) A conserved family of nuclear phosphoproteins localized to sites of polymerase–II transcription. J Cell Biol 115:587–596

Sanford JR, Gray NK, Beckmann K, Caceres JF (2004) A novel role for shuttling SR proteins in mRNA translation. Gene Dev 18:755–768

Savaldi–Goldstein S, Sessa G, Fluhr R (2000) The ethylene–inducible PK12 kinase mediates the phosphorylation of SR splicing factors. Plant J 21:91–96

Savaldi–Goldstein S, Aviv D, Davydov O, Fluhr R (2003) Alternative splicing modulation by a LAMMER kinase impinges on developmental and transcriptome expression. Plant Cell 15:926–938

Sessa G, Raz V, Savaldi S, Fluhr R (1996) PK12, a plant dual–specificity protein kinase of the LAMMER family, is regulated by the hormone ethylene. Plant Cell 8:2223–2234

Shen HH, Kan JLC, Green MR (2004) Arginine–serine–rich domains bound at splicing enhancers contact the branchpoint to promote prespliceosome assembly. Mol Cell 13:367–376

Soret J, Gabut M, Dupon C, Kohlhagen G, Stevenin J, Pommier Y, Tazi J (2003) Altered serine/arginine–rich protein phosphorylation and exonic enhancer–dependent splicing in mammalian cells lacking topoisomerase 1. Cancer Res 63:8203–8211

Stutz F, Izaurralde E (2003) The interplay of nuclear mRNP assembly, mRNA surveillance and export. Trends Cell Biol 13:319–327

Tazi, J, Bakkour N, Soret J, Zekri L, Hazra B, Laine W, Baldeyrou B, Lansiaux A, Bailly C (2005) Selective inhibition of topoisomerase I and various steps of spliceosome assembly by diospyrin derivatives. Mol Pharmacol 67:1186–1194

Tillemans V, Dispa L, Remacle C, Collinge M, Motte P (2005) Functional distribution and dynamics of *Arabidopsis* SR splicing factors in living plant cells. Plant J 41:567–582

Tillemans V, Leponce I, Rausin G, Dispa L, Motte P (2006) Insights into nuclear organization in plants as revealed by the dynamic distribution of *Arabidopsis* SR splicing factors. Plant Cell 18:3218–3234

van Bentem SDF, Anrather D, Roitinger E, Djamei A, Hufnagl T, Barta A, Csaszar E, Dohnal I, Lecourieux D, Hirt H (2006) Phosphoproteomics reveals extensive in vivo phosphorylation of Arabidopsis proteins involved in RNA metabolism. Nucl Acids Res 34:3267–3278

Velazquez–Dones A, Hagopian JC, Ma CT, Zhong XY, Zhou HL, Ghosh G, Fu XD, Adams JA (2005) Mass spectrometric and kinetic analysis of ASF/SF2 phosphorylation by SRPK1 and Clk/Sty. J Biol Chem 280:41761–41768

Verbsky ML, Richards EJ (2001) Chromatin remodeling in plants. Curr Opin Plant Biol 4:494–500

Wang BB, Brendel V (2006) Genomewide comparative analysis of alternative splicing in plants. Proc Natl Acad Sci USA 103:7175–7180

Wang JC (2002) Cellular roles of DNA topoisomerases: A molecular perspective. Nat Rev Mol Cell Biol 3:430–440

Wu JY, Maniatis T (1993) Specific interactions between proteins implicated in splice-site selection and regulated alternative splicing. Cell 75:1061–1070

Wu XS, Bishopric NH, Discher DJ, Murphy BJ, Webster KA (1996) Physical and functional sensitivity of zinc finger transcription factors to redox change. Mol Cell Biol 16:1035–1046

Xiao SH, Manley JL (1998) Phosphorylation–dephosphorylation differentially affects activities of splicing factor ASF/SF2. EMBO J 17:6359–6367

Yun B, Lee K, Farkas R, Hitte C, Rabinow L (2000) The LAMMER protein kinase encoded by the Doa locus of Drosophila is required in both somatic and germline cells and is expressed as both nuclear and cytoplasmic isoforms throughout development. Genetics 156:749–761

Yun CY, Velazquez–Dones AL, Lyman SK, Fu XD (2003) Phosphorylation-dependent and -independent nuclear import of RS domain–containing splicing factors and regulators. J Biol Chem 278:18050–18055

Zhang N, Portis AR (1999) Mechanism of light regulation of Rubisco: A specific role for the larger Rubisco activase isoform involving reductive activation by thioredoxin–f. Proc Natl Acad Sci USA 96:9438–9443

Zimmermann P, Hennig L, Gruissem W (2005) Gene–expression analysis and network discovery using Genevestigator. Trends Plant Sci 10:407–409

mRNA Cap Binding Proteins: Effects on Abscisic Acid Signal Transduction, mRNA Processing, and Microarray Analyses

J.M. Kuhn, V. Hugouvieux, and J.I. Schroeder(✉)

Contents

Introduction 140
mRNA Binding Protein Mutations That Affect ABA Signaling 140
A Mutation in the mRNA Cap Binding Protein, *abh1*, Causes ABA Hypersensitivity..... 141
abh1 Reveals Events in the mRNA Metabolism of Flowering Time Regulators 144
Disruption of *CBP20* Also Confers ABA Hypersensitivity and Drought Tolerance........ 145
References 147

Abstract The plant hormone abscisic acid (ABA) intricately regulates a multitude of processes during plant growth and development. Recent studies have established a connection between genes participating in various steps of cellular RNA metabolism and the ABA signal transduction machinery. In this chapter we focus on the plant nuclear mRNA cap binding proteins, CBP20 and CBP80. We summarize and report recent findings on their effects on cellular signal transduction networks and mRNA processing events. *ABA hypersensitive 1* (*abh1*) harbors a gene disruption in the *Arabidopsis CBP80* gene. Loss-of-function mutation of *ABH1* can also result in an early flowering phenotype in the *Arabidopsis* accession C24. *abh1* revealed noncoding *cis*-natural antisense transcripts (cis-NATs) at the *CONSTANS* locus in wild-type plants with elevated *cis*-NAT expression in the mutant. *abh1* also revealed an influence on the splicing of the MADS box transcription factor *Flowering Locus C* pre-mRNA, which may result in the regulation of flowering time. Furthermore, new experiments analyzing complementation of *cpb20* with site-directed *cpb20* mutants provide evidence that the CAP binding activity of CBP20 is essential for the observed *cbp*-associated phenotypes.

J.I. Schroeder
Division of Biological Sciences, Cell and Developmental Biology Section,
and Center for Molecular Genetics, University of California San Diego,
9500 Gilman Drive, La Jolla, CA 92093-0116, USA
e-mail: julian@biomail.ucsd.edu

In conclusion, mutants in genes participating in RNA processing provide excellent tools to uncover novel molecular mechanisms for the regulation of RNA metabolism and of signal transduction networks in wild-type plants.

Introduction

The phytohormone abscisic acid (ABA) participates in the regulation of several physiologically important stress and developmental responses throughout the life cycle of plants. During seed maturation, ABA is associated with the acquisition of nutritive reserves, but also coregulates desiccation tolerance and seed dormancy (Koornneef et al. 1998; Finkelstein et al. 2002; Zhu 2002). In vegetative growth stages, ABA has been implicated in the regulation of adaptive responses to various environmental conditions such as drought, salt, and cold stress (Koornneef et al. 1998; Leung and Giraudat 1998; Verslues and Zhu 2005). Upon experiencing drought conditions, ABA exerts tight control of the closure of stomatal pores to limit water loss by transpiration (Schroeder et al. 2001; Hetherington and Woodward 2003; Fan et al. 2004; Israelsson et al. 2006).

Several genes and second messenger molecules have been identified that take part in ABA signal transduction (MacRobbie 1998; Schroeder et al. 2001; Fedoroff 2002; Verslues and Zhu 2005; Xie et al. 2005; Christmann et al., 2006). Much of this work has been achieved through a combination of molecular genetic, biophysical, and cell biological characterizations of ABA signaling components. Among the ABA signaling components, genes encoding for phosphorylation events, protein phosphatases, protein kinases, and transcription factors have been identified as crucial elements in ABA signal transduction (Meyer et al. 1994; Leung et al. 1994; Schmidt et al. 1995; Li et al. 2000; Mustilli et al. 2002; Xie et al. 2005; Christmann et al. 2006; Mori et al. 2006).

mRNA Binding Protein Mutations That Affect ABA Signaling

Recent studies point to a novel direction for influencing transcript abundance of ABA-modulated genes via posttranscriptional mRNA processing. Several genes that function in plant RNA metabolism have been implicated in the regulation of ABA signaling: HYL1, ABH1, SAD1, AKIP1, CBP20, LOS4, AHG2/PARN, STABILIZED1, and SAD2 (Lu and Fedoroff 2000a; Hugouvieux et al. 2001; Xiong et al., 2001a; Li et al. 2002b; Kuhn and Schroeder 2003; Papp et al. 2004; Nishimura et al. 2005a; Gong et al. 2005a; Lee et al. 2006a; Verslues et al. 2006a; Riera et al. 2006a) (see the chapter by V. Chinnusamy et al., this volume). The proteins encoded by mutant genes that affect ABA responses are implicated in participating in several aspects of mRNA metabolism. In addition to ABA hypersensitivity, *hyl1* displays pleiotropic phenotypes (Lu and Fedoroff 2000b). The HYL1 protein harbors a double-stranded

RNA binding domain and has been implicated in microRNA biogenesis (Han et al. 2004; Vazquez et al. 2004; Kurihara et al. 2006). Apart from HYL1, our present knowledge about the targets and the detailed mechanisms by which these proteins modulate ABA signaling is fragmentary. SAD1 exhibits high homology to Sm-like small nuclear ribonucleoproteins of human and mouse, which are known to function in splicing, export, and degradation (Xiong et al. 2001b). The STABILIZED1 protein, a pre-mRNA splicing factor, also plays a role in pre-mRNA splicing (Lee et al. 2006b). Two proteins, the DEAD box RNA helicase LOS4 and the importin β protein SAD2 were found to encode proteins participating in nuclear RNA export (Gong et al. 2005b; Verslues et al. 2006b) (see the chapter by Chinnusamy et al., this volume). The ABA hypersensitive mutant *ahg2* was shown to have a mutation in the gene encoding a poly(A)-specific ribonuclease (PARN), which is known from animal systems to function in mRNA degradation (Nishimura et al. 2005b). Interestingly, the RNA binding protein AKIP1, whose function is yet unknown, relocalizes to nuclear speckles upon ABA treatment (Li et al. 2002a; Riera et al. 2006b). The mechanisms of AKIP regulation in *Vicia faba* (Li et al. 2002a) and its closest relative, UBA2, among three *Arabidopsis* homologs appear to differ (Riera et al. 2006b). How these mutants and the mRNA cap binding protein mutants discussed below precisely affect ABA signaling is not yet known. Two hypotheses have been discussed (Hugouvieux et al. 2001; Xiong et al. 2001b; Fedoroff 2002; Nishimura et al. 2005b), which do not strictly exclude one another. In a hypothetical model these mutants could alter expression levels of transcripts that encode important rate-limiting ABA signaling proteins. An alternative model could be, that by way of these proteins, ABA actually modulates aspects of RNA metabolism, thus affecting ABA responses. However, the precise functions of the majority these proteins remain unknown. In this chapter we summarize and report recent findings on mRNA cap binding proteins and their effects on cellular signal transduction networks and mRNA processing in plants.

A Mutation in the mRNA Cap Binding Protein, *abh1*, Causes ABA Hypersensitivity

The *ABA hypersensitive 1* (*abh1*) mutant was initially isolated in a genetic screen for abscisic acid (ABA) response mutants in the Columbia-0 ecotype (Hugouvieux et al. 2001). The *abh1* mutant confers increased sensitivity to ABA during seed germination and stomatal closure and reduces wilting during drought exposure (Hugouvieux et al. 2001). *abh1* mutant plants also exhibit a slightly slowed growth and a serrated leaf phenotype (Hugouvieux et al. 2001, 2002). *abh1* did not show significant changes in responses to other known plant hormones except for a slight gibberellic acid insensitivity (Hugouvieux et al. 2001), which correlates with ABA hypersensitivity. *abh1* mutation causes ABA-hypersensitive increases in cytoplasmic calcium transients in guard cells, enhances slow anion channel activity, and causes a decrease in inward-rectifying potassium channel currents (Hugouvieux et al.

2001, 2002). The modulation of these mechanisms in *abh1* plants is consistent with the amplification of early ABA signal transduction events in *abh1* guard cells (Hugouvieux et al. 2001, 2002). The *ABH1* gene encodes the large subunit of the dimeric *Arabidopsis* nuclear cap-binding complex (CBC), which additionally consists of a 20-kDa subunit (AtCBP20) (Hugouvieux et al. 2001; Kmieciak et al. 2002). In yeast and human HeLa cells the CBC has been shown to participate in the splicing of introns from pre-mRNA, in 3' mRNA processing and in nuclear mRNA export (Izaurralde et al. 1994; Gorlich et al. 1996; Lewis et al. 1996a; Lewis et al. 1996b; Flaherty et al. 1997; Lewis and Izaurralde 1997; Makarov et al. 2002; Zhou et al. 2002; Balatsos et al. 2006; Cheng et al. 2006). Other studies showed that the nuclear CBC can mediate the initial round of translation (Fortes et al. 2000; Ishigaki et al. 2001). More recently, the CBC was shown to be associated with a pioneer round of translation and nonsense-mediated RNA decay (NMD) as a surveillance mechanism for proper mRNA integrity in mammalian cells (Ishigaki et al. 2001; Maquat 2004; Hosoda et al. 2005; Gao et al. 2005). Thus the nuclear CBC appears to function in a multitude of pre-mRNA processing steps.

Despite the diverse functions of the CBC in nonplant systems, only few genes display differential expression levels compared to wild-type controls in both leaves of *abh1* (Hugouvieux et al. 2001) and in the analogous *Saccharomyces cerevisiae CBP80* mutant, *gcr3* (Uemura and Jigami 1992). The CBP80 protein is encoded by single genes in both *Saccharomyces cerevisiae* and *Arabidopsis* and is not essential for survival of both organisms (Uemura and Jigami 1992; Hugouvieux et al. 2001), indicating that CBP80 may play a facilitating function in its multiple mRNA processing functions. Furthermore, the nonessential nature of CBP80 and CBP20 (discussed below) in plants and yeast raises the question of whether an additional different class of nuclear cap binding protein may exist. Microarray experiments show that in non-ABA-treated plants mRNA expression levels of only a few genes are affected in *abh1* mutant leaves compared to wild-type plants (Hugouvieux et al. 2001) (Hugouvieux, Kuhn, and Schroeder, unpublished). One of the transcripts with reduced expression levels in *abh1* has been shown to function in ABA signaling: a negative regulator of ABA signaling, the *AtPP2C* protein phosphatase 2C (Sheen 1998; Kuhn et al. 2006; Yoshida et al. 2006). In *abh1*, *PP2CA* mRNA is reduced compared to wild type (Hugouvieux et al. 2001; Kuhn et al. 2006). This correlates with *pp2ca* mutants that were isolated in forward genetic screens, which demonstrated that loss-of-function alleles in *PP2CA* show ABA hypersensitivity (Kuhn et al. 2006; Yoshida et al. 2006). However, no direct impact of *abh1* on AtPP2CA pre-mRNA splicing could be identified (Kuhn et al. 2006), implicating a more complex mechanism for the downregulation of this transcript. The RD20 protein, which also shows reduced transcript levels in *abh1*, may also have a function in early ABA signaling as a Ca^{2+} binding protein (Takahashi et al. 2000). In addition, several genes with reduced transcript levels in *abh1* are predicted to participate in the regulation of oxidative stress (Hugouvieux et al. 2001). This is consistent with reports in which reactive oxygen species have been linked to early ABA signal transduction (Pei et al. 2000; Guan et al. 2000; Murata et al. 2001). Based on several independent microarray experiments and Northern blot analyses, the lectin-related gene *AtPP2-A1* is strongly downregulated in *abh1* (Fig. 1a) (Hugouvieux and Schroeder, unpublished). Interestingly, the lectin-related *AtPP2-A1* transcript levels can be restored by overexpression of the lectin-related *AtPP2-A1* cDNA in *abh1*. However, *abh1* plant lines overexpressing *AtPP2-A1* do not show altered ABA sensitivity or complementation of *abh1* (Fig. 1b) (Hugouvieux, Kuhn, and Schroeder, unpublished).

Interestingly, although microarray experiments in the *absence* of ABA show only few transcripts that are differentially expressed among *abh1* and wild-type plants (Hugouvieux et al. 2001 and data not shown), this picture changes dramatically when

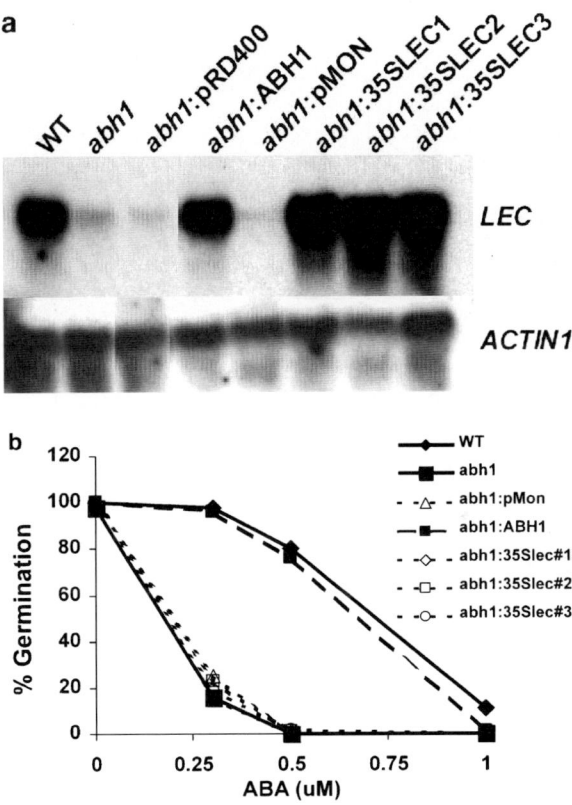

Fig. 1 *abh1* causes strong downregulation of a lectin-related transcript *AtPP2-A1* (At4g19840). **a** Northern blot analysis of (from left) of the lectin-like *AtPP2-A1* in wild-type (*WT*) and *abh1*, *abh1* plants expressing an empty control vector (*abh1*:pRD400), the *ABH1* gene (*abh1*:*ABH1*), or *AtPP2-A1* cDNA (abh1:35SLEC1–3). Overexpression of *AtPP2-A1* cDNA restores *AtPP2-A1* transcript levels, whereas vector (pRD400 and pMON)-transformed *abh1* lines did not. Hybridization signals with *ACTIN1* cDNA (*ACT1*) are shown as loading controls. **b** Comparison of seed germination rates of plant lines introduced in **a** shows that overexpression of *AtPP2-A1* does not complement the ABA hypersensitive phenotype of *abh1* with seeds being exposed to 0, 0.3, 0.5, and 1 μM ABA at 6 days.

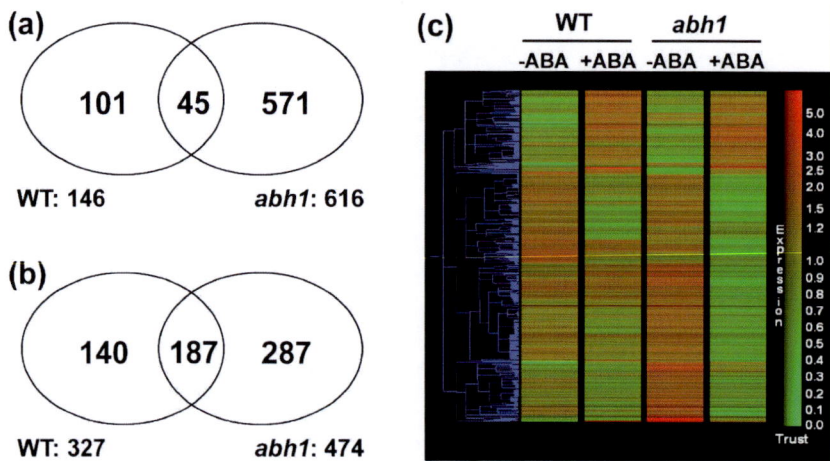

Fig. 2 *abh1* dramatically changes gene expression in response to abscisic acid. *Numbers* represent transcripts with significant and at least 1.5-fold changes in expression levels in wild-type (*WT*) and *abh1* plants in response to treatment with ABA (Affymetrix MAS5, Microsoft Excel and Access). **a** Transcripts with decreased expression in response to ABA in wild-type (*WT*) and *abh1* plants. **b** Transcripts that respond with increased expression levels in wild-type (*WT*) and *abh1* after ABA treatment. **c** Cluster analysis of abscisic acid-regulated transcript levels from microarray data (ATH1, Affymetrix; analysis in GeneSpring (TM) 4.2 software) in wildtype and *abh1* leaves with more than 3 fold changes in transcript abundance. Note the large number of genes that are abscisic acid down-regulated in *abh1* compared to wildtype. Note also that in the absence of ABA *abh1* and wildtype clusters show much less differences than in the presence of ABA. Green indicates decreased expression, while red indicates elevated expression levels.

transcript expression in *abh1* and wild type is compared in ABA-treated plant tissues. When global expression levels are compared in ABA-treated wild-type and *abh1* plants, 571 transcripts show significant and specific downregulation in *abh1*, while only 146 genes exhibit decreased expression levels in wild type with the use of identical criteria during the analysis (Fig. 2a,c). A similar phenomenon can be observed for genes, which respond with increased expression to treatment with ABA (Fig. 2b). (See microarray data accompanying this chapter at http://www.ncbi.nlm.nih.gov/projects/geo/, GSE7112.) These data indicate that ABA-induced mRNA levels are critically linked to the proper functioning of ABH1.

abh1 Reveals Events in the mRNA Metabolism of Flowering Time Regulators

Apart from affecting ABA signal transduction, *ABH1* was also shown to function in the regulation of flowering time in a very late flowering *Arabidopsis* background (Bezerra et al. 2004). *abh1* can act as a suppressor of the very late flowering Sf2 FRI allele in the *Arabidopsis* Columbia ecotype. Loss-of-function mutations in *ABH1* result in decreased expression of the MADS box transcription factor *FLC* and early flowering in this late flowering background (Bezerra et al. 2004). An early flowering phenotype was also found in an *abh1* allele in the C24 ecotype (Kuhn et al. 2007). Extensive knowledge exists of genes participating in flowering time regulation and the resulting signaling network. The early flowering time phenotype of *abh1* thus provides an approach to analyze effects of *abh1* loss of function on mRNA processing in plants and in a multicellular eukaryote, as such analyses have thus far not been pursued in a multicellular organism. By using a combination of RNA population profiling of 19 flowering related genes in the *abh1* mutant, previously unreported mRNA processing mechanisms associated with three major flowering time regulators were identified (Kuhn et al. 2007). The study revealed a regulatory function of *FLC* intron 1 during pre-mRNA splicing in wild type, a process, which is strongly affected in *abh1* as evident by the accumulation of intron 1 containing splice intermediates (Kuhn et al. 2007). These data may point to a possible novel mechanism of how flowering time regulation may be affected by *FLC* pre-mRNA processing. *abh1*, moreover, facilitated the identification of premature intronic polyadenylation of transcripts encoding the *FLM* MADS box transcription factor (Kuhn et al. 2007). In both *abh1* and wild-type plants, premature polyadenylation of *FLM* transcripts within the long first intron could be shown, a molecular event that may add an additional layer of control to fine-tune *FLM* mature transcript abundance (Kuhn et al. 2007). The study also revealed noncoding *cis*-natural antisense transcripts (*cis*-NATs) at the *CONSTANS* (*CO*) locus in wild-type plants, which are highly expressed in the *abh1* background (Kuhn et al. 2007). The mRNA processing events that were revealed via analysis of the early flowering *abh1* mutant were found to also occur in wild-type plants, but to a lesser degree. Thus, by enhancing steps in RNA metabolism present in wild type, *abh1* proved useful as a tool for uncovering novel mechanisms that also occur in wild-type plants during the regulation of flowering time genes (Kuhn et al. 2007).

Interestingly, the elevated expression of polyadenylated *cis*-NATs at the *CONSTANS* locus had no impact on diurnal transcript regulation of the mature *CO* mRNA (Kuhn et al. 2007). These observations are consistent with other reports in which no effects of *cis*-NATs on native mRNA expression have been identified (Wang et al. 2005; Faghihi and Wahlestedt 2006; Wang et al., 2006). These findings imply a yet to be identified function of *CO cis*-NATs in *Arabidopsis* plants such as quality control mechanisms, guidance during nuclear mRNA export, or translational control (Makalowska et al. 2005). Interestingly, analyses of splicing of mRNAs of the lectin-related gene *AtPP2-A1* also point to the presence of such *cis*-NATs at this locus (Kuhn and Schroeder, unpublished). For the *CONSTANS* gene, by enhancing steps in the RNA metabolism present in wild type, *abh1* has become a useful tool for discovering novel mechanisms in plant mRNA processing.

Disruption of *CBP20* Also Confers ABA Hypersensitivity and Drought Tolerance

A mutation in the gene encoding the small subunit of the dimeric cap binding complex, *CBP20*, was recently identified in a genetic screen for identifying mutants with pleiotropic alterations in vegetative development and stress responses (Papp et al. 2004). Loss of

CBP20 confers ABA hypersensitivity during seed germination, reduced stomatal conductance, and enhanced tolerance to drought conditions (Papp et al. 2004). Similarly to *abh1*, *cbp20* also displays a serrated leaf phenotype (Papp et al. 2004), further underscoring the necessity that ABH1 and CBP20 need to operate in an intact complex for full functionality. Crystal structure analyses of the human CBC showed that both subunits of the complex share tight interactions (Mazza et al. 2001; Calero et al. 2002). The CBC structure further revealed that the 5′ RNA cap structure binds to the CBP20 subunit (Mazza et al. 2001; Calero et al. 2002). Binding of CBP20 to the 5′ cap structure is mainly mediated by two stacked tyrosine residues in the binding pocket. Previous mutagenesis studies of CBP20 showed a greatly reduced affinity for the 5′ cap structure upon mutation of one tyrosine in vitro (Mazza et al. 2001). Since in both *abh1* and *cbp20* the respective proteins are not expressed in the analyzed mutants (Hugouvieux et al. 2001; Papp et al. 2004), structure function analyses of the cap binding complex can be pursued in the *cbp20* background. Complementation of *cbp20* with a *CBP20* cDNA was achieved based on the serrated leaf phenotype (Fig. 3). However, mutation of tyrosine 37 (Y37A) in AtCBP20 does not restore the wild-type phenotype (Fig. 3). These data provide evidence that the 5′ mRNA cap binding function of the CBC is required for complementation of cap binding mutant phenotypes (Fig. 3). Moreover, these findings further suggest that the conserved tyrosine residue also confers specific binding of the CBC to the 5′ cap structure in vivo. Moreover, the data indicate that the presence of the CBC alone is not sufficient for CBC function but requires the binding of the complex to mRNA.

In contrast to the cytosolic cap binding activities performed by the elongation initiation factor 4E (eIF4E), for which five putative genes are encoded in the *Arabidopsis* genome, both *ABH1* and *CBP20* encode single genes (Hugouvieux

Fig. 3 Mutation of the 5′ cap structure binding pocket in CBP20 abolishes nuclear cap binding complex function. Representative leaves of wild-type (*WT*) and *cbp20* plants transformed with a control plasmid (pGreen), *CBP20* cDNA and *CBP20* cDNA harboring a mutation in the putative cap structure binding pocket (Y37A, *CBP20m*) are shown. While *CBP20* transcripts can complement the serrated leaf phenotype of *cbp20*, the mutated *CBP20* transcript (*CBP20m*), for which mRNA binding is reduced in in vitro experiments (Mazza et al. 2001), cannot confer complementation of this phenotype.

et al. 2001; Kmieciak et al. 2002). The mechanisms by which *abh1* and *cbp20* affect transcript abundance in such a selective way without resulting in more severe phenotypes and how ABA hypersensitivity is conferred in *abh1*, *cbp20* and other mRNA binding proteins will be interesting for further studies. Similar to its functions in yeast and mammalian cells, ABH1 has been shown to affect several RNA processing events. Therefore the *Arabidopsis* nuclear cap binding complex may have impacts on some transcripts via mRNA processing mechanisms, which are rate limiting for early ABA signaling events, as has been shown in *Arabidopsis* accessions and backgrounds for which *abh1* shows early flowering (Bezerra et al. 2004; Kuhn et al. 2007). Another plausible mechanism of regulation could be that the plant CBC activity depends on the activation by cell cycle or stress-induced proteins (Wilson et al. 1999; Wilson and Cerione 2000). This could also explain why *abh1* and *cbp20* loss-of-function mutations exhibit relatively normal growth and gene expression until exposed to stressful or environmental conditions as, for example, mediated by ABA (Fig. 2).

The identification of gene disruption mutants of the CBP80 and CBP20 cap binding proteins in plants provides many possibilities for further elucidation of CBC function in a multicellular organism and can complement research on this machinery in nonplant systems, given that analyses have mainly been focused on loss-of-function mutants in yeast and gene silencing in tissue culture lines. Together with newly emerging technologies for investigating, for instance, splicing of transcripts on a genomewide level (Clark et al. 2002; Johnson et al. 2003), mutants like *abh1* and *cbp20* will accelerate research for gaining knowledge of fundamental processes for fine-tuning gene expression at defined stages of a plant's life cycle.

Acknowledgements We thank Csaba Koncz for kindly providing *cbp20* seeds. This research was supported by the National Institutes of Health (R01-GM-060396), by National Science Foundation (MCB0417118) grants to J.I.S., and by a European Molecular Biology Organization fellowship to J.M.K.

References

Balatsos NAA, Nilsson P, Mazza C, Cusack S, Virtanen A (2006) Inhibition of mRNA deadenylation by the nuclear cap binding complex (CBC). J Biol Chem 281:4517–4522
Bezerra IC, Michaels SD, Schomburg FM, Amasino RM (2004) Lesions in the mRNA cap-binding gene *ABA HYPERSENSITIVE 1* suppress *FRIGIDA*-mediated delayed flowering in *Arabidopsis*. Plant J 40:112–119
Calero G, Wilson KF, Ly T, Rios-Steiner JL, Clardy JC, Cerione RA (2002) Structural basis of m7GpppG binding to the nuclear cap-binding protein complex. Nat Struct Biol 9:912–917
Cheng H, Dufu K, Lee CS, Hsu JL, Dias A, Reed R (2006) Human mRNA export machinery recruited to the 5' end of mRNA. Cell 127:1389–1400
Christmann A, Moes D, Himmelbach A, Yang Y, Tang Y, Grill E (2006) Integration of abscisic acid signalling into plant responses. Plant Biol (Stuttg) 8:314–325
Clark, TA, Sugnet, CW, Ares, M. Jr. (2002) Genomewide analysis of mRNA processing in yeast using splicing-specific microarrays. Science 296:907–910.

Faghihi M, Wahlestedt C (2006) RNA interference is not involved in natural antisense mediated regulation of gene expression in mammals. Genome Biol 7:R38

Fan LM, Zhao Z, Assmann SM (2004) Guard cells: a dynamic signaling model. Curr Opin Plant Biol 7:537–546

Fedoroff NV (2002) Cross-talk in abscisic acid signaling. Sci STKE 2002:RE10

Finkelstein RR, Gampala SSL, Rock CD (2002) Abscisic acid signaling in seeds and seedlings. Plant Cell 14:S15-S45

Flaherty SM, Fortes P, Izaurralde E, Mattaj IW, Gilmartin GM (1997) Participation of the nuclear cap binding complex in pre-mRNA 3' processing. Proc Natl Acad Sci USA 94:11893–11898

Fortes P, Inada T, Preiss T, Hentze MW, Mattaj IW, Sachs AB (2000) The yeast nuclear cap binding complex can interact with translation factor eIF4G and mediate translation initiation. Mol Cell 6:191–196

Gao Q, Das B, Sherman F, Maquat LE (2005) Cap-binding protein 1-mediated and eukaryotic translation initiation factor 4E-mediated pioneer rounds of translation in yeast. Proc Natl Acad Sci USA 102:4258–4263

Gong Z, Dong CH, Lee H, Zhu J, Xiong L, Gong D, Stevenson B, Zhu JK (2005) A DEAD Box RNA helicase is essential for mRNA export and important for development and stress responses in *Arabidopsis*. Plant Cell 17:256–267

Gorlich D, Kraft R, Kostka S, Vogel F, Hartmann E, Laskey RA, Mattaj IW, Izaurraide E (1996) Importin provides a link between nuclear protein import and U snRNA export. Cell 87:21–32

Guan LM, Zhao J, Scandalios JG (2000) *Cis*-elements and *trans*-factors that regulate expression of the maize Cat1 antioxidant gene in response to ABA and osmotic stress: H_2O_2 is the likely intermediary signaling molecule for the response. Plant J 22:87–95

Han MH, Goud S, Song L, Fedoroff N (2004) The *Arabidopsis* double-stranded RNA-binding protein HYL1 plays a role in microRNA-mediated gene regulation. Proc Natl Acad Sci USA 101:1093–1098

Hetherington AM, Woodward FI (2003) The role of stomata in sensing and driving environmental change. Nature 424:901–908

Hosoda N, Kim YK, Lejeune F, Maquat LE (2005) CBP80 promotes interaction of Upf1 with Upf2 during nonsense-mediated mRNA decay in mammalian cells. Nat Struct Mol Biol 12:893–901

Hugouvieux V, Kwak JM, Schroeder JI (2001) An mRNA cap binding protein, ABH1, modulates early abscisic acid signal transduction in *Arabidopsis*. Cell 106:477–487

Hugouvieux V, Murata Y, Young JJ, Kwak JM, Mackesy DZ, Schroeder JI (2002) Localization, ion channel regulation, and genetic interactions during abscisic acid signaling of the nuclear mRNA cap-binding protein, ABH1. Plant Physiol 130:1276–1287

Ishigaki Y, Li X, Serin G, Maquat LE (2001) Evidence for a pioneer round of mRNA translation: mRNAs subject to nonsense-mediated decay in mammalian cells are bound by CBP80 and CBP20. Cell 106:607–617

Israelsson M, Siegel RS, Young J, Hashimoto M, Iba K, Schroeder JI (2006) Guard cell ABA and CO_2 signaling network updates and Ca^{2+} sensor priming hypothesis. Curr Opin Plant Biol 9:654–663.

Izaurralde E, Lewis J, McGuigan C, Jankowska M, Darzynkiewicz E, Mattaj IW (1994) A nuclear cap binding protein complex involved in pre-mRNA splicing. Cell 78:657–668

Johnson JM et al. (2003) Genome-wide survey of human alternative pre-mRNA splicing with exon junction microarrays. Science 302:2141–2144

Kmieciak M, Simpson CG, Lewandowska D, Brown JWS, Jarmolowski A (2002) Cloning and characterization of two subunits of *Arabidopsis thaliana* nuclear cap-binding complex. Gene 283:171–183

Koornneef M, Leon-Kloosterziel KM, Schwartz SH, Zeevaart JAD (1998) The genetic and molecular dissection of abscisic acid biosynthesis and signal transduction in *Arabidopsis*. Plant Physiol Biochem 36:83–89

Kuhn JM, Boisson-Dernier A, Dizon MB, Maktabi MH, Schroeder JI (2006) The protein phosphatase *AtPP2CA* negatively regulates abscisic acid signal transduction in *Arabidopsis*, and effects of *abh1* on *AtPP2CA* mRNA. Plant Physiol 140:127–139

Kuhn JM, Breton G, Schroeder JI (2007) mRNA metabolism of flowering time regulators in wild type *Arabidopsis* revealed by a nuclear cap binding protein mutant, *abh1*. Plant J 50:1049-1062

Kuhn JM, Schroeder JI (2003) Impacts of altered RNA metabolism on abscisic acid signaling. Curr Opin Plant Biol 6:463-469

Kurihara Y, Takashi Y, Watanabe Y (2006) The interaction between DCL1 and HYL1 is important for efficient and precise processing of pri-miRNA in plant microRNA biogenesis. RNA 12:206-212

Lee BH, Kapoor A, Zhu J, Zhu JK (2006) STABILIZED1, a stress-upregulated nuclear protein, is required for pre-mRNA splicing, mRNA turnover, and stress tolerance in *Arabidopsis*. Plant Cell 18:1736-1749

Leung J, Bouvier-Durand M, Morris PC, Guerrier D, Chefdor F, Giraudat J (1994) *Arabidopsis* ABA response gene ABI1: features of a calcium-modulated protein phosphatase. Science 264:1448-1452

Leung J, Giraudat J (1998) Abscisic acid signal transduction. Annu Rev Plant Biol 49:199-222

Lewis JD, Gorlich D, Mattaj IW (1996a) A yeast cap binding protein complex (yCBC) acts at an early step in pre-mRNA splicing. Nucl Acids Res 24:3332-3336

Lewis JD, Izaurralde E (1997) The role of the cap structure in RNA processing and nuclear export. Eur J Biochem 247:461-469

Lewis JD, Izaurralde E, Jarmolowski A, McGuigan C, Mattaj IW (1996b) A nuclear cap-binding complex facilitates association of U1 snRNP with the cap-proximal 5' splice site. Genes Dev 10:1683-1698

Li J, Kinoshita T, Pandey S, Ng CK, Gygi SP, Shimazaki K, Assmann SM (2002) Modulation of an RNA-binding protein by abscisic-acid-activated protein kinase. Nature 418:793-797

Li J, Wang XQ, Watson MB, Assmann SM (2000) Regulation of abscisic acid-induced stomatal closure and anion channels by guard cell AAPK kinase. Science 287:300-303

Lu C, Fedoroff N (2000) A mutation in the *Arabidopsis HYL1* gene encoding a dsRNA binding protein affects responses to abscisic acid, auxin, and cytokinin. Plant Cell 12:2351-2366

MacRobbie EA (1998) Signal transduction and ion channels in guard cells. Philos Trans R Soc Lond B Biol Sci 353:1475-1488

Makalowska I, Lin CF, Makalowski W (2005) Overlapping genes in vertebrate genomes. Comput Biol Chem 29:1-12

Makarov EM, Makarova OV, Urlaub H, Gentzel M, Will CL, Wilm M, Luhrmann R (2002) Small nuclear ribonucleoprotein remodeling during catalytic activation of the spliceosome. Science 298:2205-2208

Maquat LE (2004) Nonsense-mediated mRNA decay: splicing, translation and mRNP dynamics. Nat Rev Mol Cell Biol 5:89-99

Mazza C, Ohno M, Segref A, Mattaj IW, Cusack S (2001) Crystal structure of the human nuclear cap binding complex. Mol Cell 8:383-396

Meyer K, Leube MP, Grill E (1994) A protein phosphatase 2C involved in ABA signal transduction in *Arabidopsis thaliana*. Science 264:1452-1455

Mori IC et al. (2006) CDPKs CPK6 and CPK3 function in ABA regulation of guard cell S-type anion- and Ca^{2+}-permeable channels and stomatal closure. PLoS Biol 4:e327

Murata Y, Pei ZM, Mori IC, Schroeder J (2001) Abscisic acid activation of plasma membrane Ca^{2+} channels in guard cells requires cytosolic NAD(P)H and is differently disrupted upstream and downstream of reactive oxygen species production of *abi1-1* and *abi2-1* protein phosphatase 2C mutants. Plant Cell 13:2513-2523

Mustilli AC, Merlot S, Vavasseur A, Fenzi F, Giraudat J (2002) *Arabidopsis* OST1 protein kinase mediates the regulation of stomatal aperture by abscisic acid and acts upstream of reactive oxygen species production. Plant Cell 14:3089-3099

Nishimura N, Kitahata N, Seki M, Narusaka Y, Narusaka M, Kuromori T, Asami T, Shinozaki K, Hirayama T (2005) Analysis of *ABA hypersensitive germination2* revealed the pivotal functions of PARN in stress response in *Arabidopsis*. Plant J 44:972-984

Papp I, Mur LA, Dalmadi A, Dulai S, Koncz C (2004) A mutation in the *Cap Binding Protein 20* gene confers drought tolerance to *Arabidopsis*. Plant Mol Biol 55:679–686

Pei ZM, Murata Y, Benning G, Thomine S, Klusener B, Allen GJ, Grill E, Schroeder JI (2000) Calcium channels activated by hydrogen peroxide mediate abscisic acid signalling in guard cells. Nature 406:731–734

Riera M, Redko Y, Leung J (2006) *Arabidopsis* RNA-binding protein UBA2a relocalizes into nuclear speckles in response to abscisic acid. FEBS Lett 580:4160–4165

Schmidt C, Schelle I, Liao YJ, Schroeder JI (1995) Strong regulation of slow anion channels and abscisic acid signaling in guard cells by phosphorylation and dephosphorylation events. Proc Natl Acad Sci USA 92:9535–9539

Schroeder JI, Allen GJ, Hugouvieux V, Kwak JM, Waner D (2001) Guard cell signal transduction. Annu Rev Plant Physiol Plant Mol Biol 52:627–658

Sheen J (1998) Mutational analysis of protein phosphatase 2C involved in abscisic acid signal transduction in higher plants. Proc Natl Acad Sci USA 95:975–980

Takahashi S, Katagiri T, Yamaguchi-Shinozaki K, Shinozaki K (2000) An *Arabidopsis* gene encoding a Ca^{2+}-binding protein is induced by abscisic acid during dehydration. Plant Cell Physiol 41:898–903

Uemura H, Jigami Y (1992) GCR3 encodes an acidic protein that is required for expression of glycolytic genes in *Saccharomyces cerevisiae*. J Bacteriol 174:5526–5532

Vazquez F, Gasciolli V, Crete P, Vaucheret H (2004) The nuclear dsRNA binding protein HYL1 is required for microRNA accumulation and plant development, but not posttranscriptional transgene silencing. Curr Biol 14:346–351

Verslues PE, Zhu JK (2005) Before and beyond ABA: upstream sensing and internal signals that determine ABA accumulation and response under abiotic stress. Biochem Soc Trans 33:375–379

Verslues PE, Guo Y, Dong CH, Ma W, Zhu JK (2006) Mutation of SAD2, an importin beta-domain protein in *Arabidopsis*, alters abscisic acid sensitivity. Plant J 47:776–787

Wang H, Chua NH, Wang XJ (2006) Prediction of *trans*-antisense transcripts in *Arabidopsis thaliana*. Genome Biol 7:R92

Wang XJ, Gaasterland T, Chua NH (2005) Genome-wide prediction and identification of *cis*-natural antisense transcripts in *Arabidopsis thaliana*. Genome Biol 6:R30

Wilson KF, Cerione RA (2000) Signal transduction and post-transcriptional gene expression. Biol Chem 381:357–365

Wilson KF, Fortes P, Singh US, Ohno M, Mattaj IW, Cerione RA (1999) The nuclear cap-binding complex is a novel target of growth factor receptor-coupled signal transduction. J Biol Chem 274:4166–4173

Xie Z, Ruas P, Shen QJ (2005) Regulatory networks of the phytohormone abscisic acid. Vitam Horm 72:235–269

Xiong L, Gong Z, Rock CD, Subramanian S, Guo Y, Xu W, Galbraith D, Zhu JK (2001) Modulation of abscisic acid signal transduction and biosynthesis by an Sm-like protein in *Arabidopsis*. Dev Cell 1:771–781

Yoshida T, Nishimura N, Kitahata N, Kuromori T, Ito T, Asami T, Shinozaki K, Hirayama T (2006) ABA-hypersensitive germination3 encodes a protein phosphatase 2C (AtPP2CA) that strongly regulates abscisic acid signaling during germination among Arabidopsis protein phosphatase 2Cs. Plant Physiol 140:115–126

Zhou Z, Licklider LJ, Gygi SP, Reed R (2002) Comprehensive proteomic analysis of the human spliceosome. Nature 419:182–185

Zhu JK (2002) Salt and drought stress signal transduction in plants. Annu Rev Plant Biol 53:247–273

Messenger RNA 3′ End Formation in Plants

A.G. Hunt

Contents

Introduction 152
Messenger RNA 3′ Ends in Plants. 155
   Plant Polyadenylation Signals 155
   3′ End Microheterogeneity in Plant Transcription Units. 157
   Alternative Polyadenylation in Plants 158
The Polyadenylation Apparatus in Plants 160
   The Complement of *Arabidopsis* Plant Polyadenylation Factor Subunits. 160
   Characteristics of Individual Subunits of the Plant Polyadenylation Complex 162
   CPSF Subunits. 162
   RNA Binding Proteins in the Polyadenylation Complex. 167
Cellular Signaling and 3′ End Formation. 168
Summary and Prospects. 169
References 170

Abstract Messenger RNA 3′ end formation is an integral step in the process that gives rise to mature, translated messenger RNAs in eukaryotes. With this step, a pre-messenger RNA is processed and polyadenylated, giving rise to a mature mRNA bearing the characteristic poly(A) tract. The poly(A) tract is a fundamental feature of mRNAs, participating in the process of translation initiation and being the focus of control mechanisms that define the lifetime of mRNAs. Thus messenger RNA 3′ end formation impacts two steps in mRNA biogenesis and function. Moreover, mRNA 3′ end formation is something of a bridge that integrates numerous other steps in mRNA biogenesis and function. While the process is essential for the expression of most genes, it is also one that is subject to various forms of regulation, such that both quantitative and qualitative aspects of gene expression may be modulated via the polyadenylation complex. In this review, the current status of understanding of

A.G. Hunt
Department of Plant and Soil Sciences, University of Kentucky, 301A Plant Science Building
Lexington, KY 40546-0312, USA
e-mail: aghunt00@uky.edu

mRNA 3' end formation in plants is discussed. In particular, the nature of mRNA 3' ends in plants is reviewed, as are recent studies that are beginning to yield insight into the functioning and regulation of plant polyadenylation factor subunits.

Abbreviations CPSF: Cleavage and polyadenylation specificity factor; CstF: Cleavage stimulatory factor; CFIm: Mammalian cleavage factor I; CFIy: Yeast cleavage factor; CPF: Yeast cleavage and polyadenylation factor; PAP: Poly(A) polymerase; UTR: Untranslated region; nts: Nucleotides; FUE: Far upstream element; NUE: Near upstream element; CS: Cleavage site; CE: Cleavage element; ETR1: Ethylene receptor 1; TIR: Toll/interleukin-1 receptor homology domain; Fip: Factor interacting with poly(A) polymerase; FLAG: So-called FLAG epitope

Introduction

The polyadenylation of messenger RNAs is a central feature of gene expression in eukaryotes and is mediated by an evolutionarily conserved protein complex of more than 15 subunits (Wahle and Ruegsegger 1999; Zhao et al. 1999; Proudfoot 2004). This complex recognizes sequence elements in the 3' regions of the primary transcript (the polyadenylation signal) and subsequently mediates the processing and polyadenylation of the RNA. In mammals, the polyadenylation signal (usually AAUAAA) is recognized by the so-called cleavage and polyadenylation specificity factor (or CPSF; Murthy and Manley 1995), the U-rich downstream element by the cleavage stimulatory factor (CstF; MacDonald et al. 1994), and sequences 5' of AAUAAA by the mammalian cleavage factor I (CFIm; Venkataraman et al. 2005). In yeast, subunits of cleavage factor I (CFIy) bind the so-called A-rich positioning element (Gross and Moore 2001) and the efficiency element situated 5' of the positioning element (Valentini et al. 1999). Various subunits of the cleavage and polyadenylation factor (CPF) bind to sequences near or at the cleavage site (Barabino et al. 2000; Dichtl and Keller 2001; Dichtl et al. 2002). The pre-mRNA is thought to be processed by the 73-kDa subunit of CPSF (Ryan et al. 2004; Mandel et al. 2006) and subsequently polyadenylated. The poly(A) tail is added by a characteristic, conserved nucleotidyltransferase, poly(A) polymerase. Distinctive poly(A) binding proteins control poly(A) length and mediate nucleocytoplasmic transport. Other subunits of the complex (listed in Table 1) participate in various ways; recent reviews (Wahle and Ruegsegger 1999; Zhao et al. 1999; Proudfoot 2004) may be consulted for details regarding these details. (Note that Table 1 includes a summary of the subunit terminologies that are used in this review.)

Messenger RNA 3' end formation is not an isolated process, proceeding in a spatially and temporally uncoupled manner with respect to other steps in mRNA biogenesis and monitoring. Rather, it is coupled to the many other processes that combine to produce a mature mRNA. Some polyadenylation complex subunits interact with the distinctive C-terminal domain of RNA polymerase II as well as

Table 1 Eukaryotic polyadenylation factor subunits

Subunit (mammalian terminology)[1]	Yeast counterpart[2]	Arabidopsis counterpart(s)[3]
PAP	Pap1	At1g17980
		At2g25850
		At3g06560
		At4g32850
CPSF160	Yhh1	At5g51660
CPSF100	Ydh1	At5g23880
CPSF73a	Ysh1	At1g61010 [AtCPSF73(I)]
CPSF73b	[4]	At2g01730 [AtCPSF73(II)]
CPSF30	Yth1	At1g30460
hFip1	Fip1	At5g58040
		At3g66652
CstF77	Rna14	At1g17760
CstF64	Rna15	At1g71800
CstF50	[4]	At5g60940
hPfs2	Pfs2	At5g13480 (FY)
CFIm68/59	[4]	-
CFIm25	[4]	At4g29820
		At4g25550
hClp1	Clp1	At3g04680
		At5g39930
hPcf11p	Pcf11	At1g66500
		At4g04885
		At5g43620
		At2g36480
Symplekin	Pta1	At5g01400
		At1g27590/At1g27595[5]
[1]	Hrp1	-
PabN	Pab1, Nab2	At5g10350
		At5g51120
		At5g65260

The *Arabidopsis* proteins and their genes were identified by BLASTP (Altschul et al. 1997), using the respective mammalian or yeast protein sequences.
[1]Mammalian polyadenylation factor subunit
[2]Yeast counterpart of the corresponding mammalian subunit
[3]*Arabidopsis* genes (Arabidopsis Gene Identification designation provided) that encode orthologs of the corresponding mammalian subunit
[4]No corresponding protein
[5]These *Arabidopsis* genes each encode proteins with symplekin-related domains; as noted by Herr et al. (2006), these two domains are present in a single rice gene: no obvious corresponding *Arabidopsis* gene is apparent; see text for discussion

with components of the basal transcription initiation machinery (Calvo and Manley, 2003; Bentley 2005). Functional interactions of polyadenylation factor subunits with other transcription factors have been reported as well (Xing et al. 2004; Calvo and Manley 2005; Kavanagh et al. 2006). Other polyadenylation factor subunits interact with various splicing factors (Niwa et al. 1990; Gunderson et al. 1994, 1997; Millevoi et al. 2002, 2006). Still others interact with transcription elongation and termination factors (Hammell et al. 2002; Buratowski 2005), and elongation and termination are necessarily coupled to mRNA 3' end formation. Among the manifestations of these interactions are that various polyadenylation subcomplexes associate with the transcribed gene at different stages during transcription; some can be found associated with the promoters of expressed genes in chromatin immunoprecipitation experiments, while others display a dynamic similar to that of phosphorylated RNA polymerase II, present at low levels at the promoter and at higher levels downstream from the transcription initiation site (Licatalosi et al. 2002; Ahn et al. 2004; Kim et al. 2004; Calvo and Manley 2005; Venkataraman et al. 2005).

A number of the subunits of the eukaryotic polyadenylation complex are associated with other processes in the cell. CPSF and CstF subunits function, in concert with cytoplasmic RNA-binding proteins, in the masking and polyadenylation of cytoplasmic RNAs during oogenesis and early embryonic development (Bilger et al. 1994; Dickson et al. 1999, 2001; Mendez et al. 2000; Hofmann et al. 2002; Barnard et al. 2004; Rouget et al. 2006). CPSF and symplekin act in concert with U7 in the formation of 3' ends of cell cycle-regulated histone mRNAs (Dominski et al. 2005a; Kolev and Steitz 2005), and several proteins that function in polyadenylation also play roles in the maturation of other small nuclear RNAs (Morlando et al. 2002; Nedea et al. 2003; Dichtl et al. 2004; Morlando et al. 2004; Baillat et al. 2005). In *C. elegans*, CstF serves as a link between mRNA 3' end formation and *trans*-splicing (Evans et al. 2001). Thus many aspects of the formation and modification of RNA 3' ends in the cell, apart from mRNA polyadenylation, seem to involve subsets of the canonical polyadenylation complex.

By most accounts, the core components of the polyadenylation apparatus are present in all cell types, at all stages of development. However, there are some interesting deviations from this general rule, differences that may be relevant to the functions of the polyadenylation complex in plants. In particular, mammals possess testis-specific isoforms of CstF64 (Wallace et al. 1999; Dass et al. 2001a, 2001b; Wallace et al. 2004) and poly(A) polymerase (PAP; Kashiwabara et al. 2000, 2002; Lee et al. 2000; Le et al. 2001; Zhuang et al. 2004). These isoforms have been proposed to play specialized roles in RNA metabolism during male gametogenesis, and they serve notice that different tissues may possess different forms of the polyadenylation apparatus.

Messenger RNA 3' end formation is guided by distinctive sets of sequences present in the precursor RNA. Historically, the prototypical polyadenylation signal was recognized as AAUAAA, a motif present in a majority of mammalian mRNAs, usually within 30 nt (5', or upstream) of the cleavage/polyadenylation site. Also historically, it was found that poly(A) signals in yeast and plants were different from those in mammals, most notably in that AAUAAA is much less prevalent in 3'-UTRs in the latter groups of organisms (Hunt 1994; Rothnie 1996; Li and Hunt 1997; Zhao et al. 1999). However, recent bioinformatics studies have raised the possibility that the tripartite arrangement of sequence signal that guides mRNA

3' end formation in yeasts and plants is also a feature of mammalian polyadenylation signals (Legendre and Gautheret 2003; Hu et al. 2005; Tian et al. 2005). Thus, as is the case with the subunits of the complex, there would seem to be a common theme to polyadenylation signals in eukaryotes. The nature of this tripartite arrangement in plants is discussed below.

Genomewide surveys indicate that a large number of transcription units in mammals are subject to alternative polyadenylation-the production of mRNAs that differ in the location of their 3' ends (Beaudoing and Gautheret, 2001; Sorek et al. 2004; Tian et al. 2005, 2007; Yan and Marr, 2005). This phenomenon has the potential to greatly expand the coding capacity of genomes, as it allows for the creation of different terminal exons as well as for exon skipping, both of which leads to an alteration of the protein-coding capacity of the mRNA. Alternative polyadenylation is likely accomplished by a plethora of mechanisms, but a common theme involves interplay between splicing and polyadenylation activities (Tian et al. 2007); this may be one rationalization for the numerous direct interactions between polyadenylation and splicing factors that have been reported.

Messenger RNA 3' Ends in Plants

Plant Polyadenylation Signals

The best-characterized aspect of mRNA 3' end formation in plants is the nature of the polyadenylation signal, the *cis* element(s) that guide mRNA polyadenylation. Early work on the nature of polyadenylation signals in plants led to a tripartite model for the structure of a plant polyadenylation signal (Fig. 1; Hunt

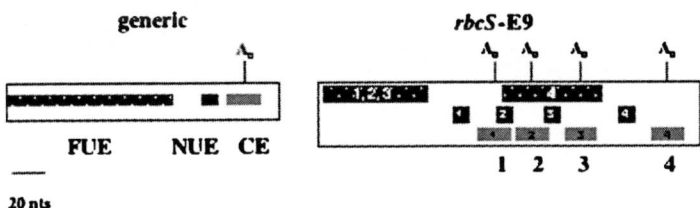

Fig. 1 Structure of plant polyadenylation signals. On the left is illustrated a generic polyadenylation signal, indicating the relative positions of the three *cis* elements that constitute a plant polyadenylation signal. On the right is shown the structure of the polyadenylation signal from the pea *rbcS*-E9 gene, deduced from deletion (Hunt and MacDonald 1989) and linker substitution (Mogen et al. 1992) analyses. The different elements (FUE, NUE, CE) are shaded as indicated in the representation of a generic site. *FUE*, far-upstream element; *NUE*, near upstream element; *CE*, cleavage element; A_n, cleavage/polyadenylation site. These elements are discussed in the text. *Numbers* beneath the representation of the *rbcS*-E9 signal denote the four poly(A) sites in this 3'-UTR; *numbers* within the illustrated *cis* elements designate the sites that are regulated by the respective element. Note that sites 1, 2, and 3 are all controlled by the same FUE.

1994; Rothnie 1996). Thus three elements were identified-one situated 50 or more nt 5′ (or upstream) of the cleavage/polyadenylation site (the so-called far-upstream element, or FUE), one situated between 10 and 30 nt 5′ of the cleavage/polyadenylation site (the near-upstream element, or NUE), and the cleavage/polyadenylation site itself (the cleavage site, or CS). The FUE was poorly defined in terms of sequence and was suggested to be a combination of U+G-rich motifs, perhaps related to the sequence UUUGA (based on results obtained with the cauliflower mosaic virus FUE; Sanfacon et al. 1991; Rothnie et al. 1994). The NUE was typified by the motif AAUAAA, but was also found to extend for more than 6 nt, and could be little more than an extended A-rich region (Li and Hunt 1995). The CS itself was found to reside within tracts of elevated U content.

This model was arrived at by focusing on a handful of plant polyadenylation signals (Hunt and MacDonald 1989; MacDonald et al. 1991; Sanfacon et al. 1991; Mogen et al. 1992; Wu et al. 1993, 1994; Ohtsubo and Iwabuchi 1994; Rothnie et al. 1994; Li and Hunt 1995). More recent bioinformatics analyses have established that this model is an accurate general representation of polyadenylation signals in plants. Graber et al. (1999) noted distinctive biases in overall sequence composition in regions of *Arabidopsis* and rice 3′-UTRs that corresponded to the FUE, NUE, and CS regions defined by functional analysis. In particular, these authors documented an extended U-rich region between 30 and 150 nt upstream of the CS in both rice and *Arabidopsis*, a more limited A-rich region centered around 20 nt upstream from the CS, and a decided bias toward U immediately 5′ of the CS. Using a more extensive data set of *Arabidopsis* sequences, and including sequences downstream from the CS itself, Loke et al. (2005) presented similar results. These authors found a strong bias toward U between 50 and 150 nt upstream from the CS, an equally strong bias toward A between 10 and 30 nt upstream from the CS, and an highly U-rich region that extended from 10 nt upstream to 15 nt downstream from the CS. These authors termed the latter the cleavage element, or CE, and proposed that the CE itself is determined by the occurrence of a YA dinucleotide embedded within a U-rich region. These features-U-richness that extends for more than 100 nt upstream from the cleavage/polyadenylation site, a distinct A-rich putative NUE, and very high U content near the conserved YA dinucleotide that defines the poly(A) site itself-have also been noted in large-scale analyses of *Arabidopsis* and rice cDNAs (Alexandrov et al. 2006; Dong et al. 2007).

RNA structure has been suggested to influence mRNA 3′ end formation in mammals (Brown et al. 1991; Graveley et al. 1996; Zarudnaya et al. 2003; Wu and Alwine 2004). Loke et al. (2005) presented observations suggestive of a similar influence in plants. These authors noted a correlation between computer-predicted secondary structures and the effects of mutations on 3′ end processing in vivo. The matter of the influence of structure of RNA 3′ end processing is still murky in all cases in which such effects have been described. However, correlations such as described by Loke et al. are intriguing and need to be explained in the context of a model for the plant polyadenylation apparatus.

The nature of the different *cis* elements that constitute plant polyadenylation signals raises interesting questions as to specificity and efficiency. None of the three elements (including the NUE, which may be analogous to AAUAAA) is particularly conserved at the level of nucleotide sequence, and it is probable that any of these elements might occur many times in a pre-mRNA. This is especially true in introns that, like 3'-UTRs, are A+U rich (Simpson and Filipowicz, 1996; Brown and Simpson, 1998; Lorković et al. 2000). Indeed, a multiplicity of small, redundant U-rich tracts appears to be one determinant of intron specification (Baynton et al. 1996; Ko et al. 1998; Simpson et al. 2004). The involvement of U-richness in introns, FUEs, and CS/CEs leads to questions as to the recognition and delineation of these U-rich regions. For example, is a U-rich region in an intron differentiated from the FUE in a poly(A) signal, and how are these in turn distinguished from the CS/CE? [The nature of this problem is conveyed by considering the arrangements of these elements in the pea *rbcS*-E9 gene (Fig. 1)]. Are there different sets of factors that recognize intron-localized and poly(A)-related motifs, or is a common set of factors involved, one that may recruit different processing complexes, depending on the nature of the affiliated *cis* elements? These questions have no answer at the present; however, the suggestive similarities between introns and polyadenylation signals speak to possible commonalities in these processes and recall the interplay between splicing and polyadenylation that is apparent in mammals (Tian et al. 2007).

3' End Microheterogeneity in Plant Transcription Units

The mRNAs that arise from any particular plant transcription unit are typically microheterogeneous at their 3' ends; that is to say, the 3' ends of such mRNAs may occur at any of a number of specific sites that are spread over tens of nucleotides along the transcription unit. The mechanisms that underlie this microheterogeneity have been explored in the case of two different poly(A) sites, from the pea *rbcS*-E9 and T-DNA-encoded octopine synthase genes (MacDonald et al. 1991; Mogen et al. 1992). The results of these studies indicate that closely spaced poly(A) sites are each controlled by their own NUE/CE combination, and that more than one NUE/CS combination may function with the same FUE. However, as seen in the case of the *rbcS*-E9 gene (Mogen et al. 1992), there may exist more than one FUE as well as NUE and CS. There is considerable potential and observed overlap between these functional units (see the *rbcS*-E9 example in Fig. 1).

That there exists extensive 3' end microheterogeneity in plants raises questions as to possible functional or regulatory roles for the phenomenon. In principle, it is possible that mRNAs with different 3' ends might possess different inherent stabilities (or translatabilities), since features that govern stability and translation often are found in 3'-UTRs. Careful studies of the stabilities of sets of mRNAs that are microheterogeneous at their 3' ends have not been conducted in plant systems, but precedence of a sort for such a mode of regulation exists in the form of studies in mammalian systems (Hansen et al. 1996; Miyamoto et al. 1996; Touriol et al. 1999; Caballero et al. 2004). There has been a report of differential loading of mRNAs displaying 3' end microheterogeneity onto ribosomes in plants (Skadsen and Knauer 1995), lending credence to the possibility of regulation in this manner. However, the contributions of 3' end microheterogeneity to gene expression have not been adequately studied, and the above considerations must be considered as hypothetical.

One curious feature of mRNA 3' end formation in plants is that plant mRNAs may possess nontemplated bases, apart from adenosine, immediately preceding the poly(A) tract (Jin and Bian 2004). This phenomenon seems widespread, and apparently involves the addition (or insertion) of one-three bases before the poly(A) tract. The mechanism by which such events occur is not known, but it is reasonable to infer that the phenomenon reflects some aspect of the biochemistry of the polyadenylation apparatus. Whether the phenomenon has any impact on gene expression is not known. However, the occurrence of nontemplated bases (other than adenosine) before the poly(A) tract has other ramifications. For example, the occurrence of nontemplated bases in mature mRNAs will impact the computational approaches taken to match tags (such as 3'-SAGE tags) with genomic DNA sequence, and thus catalogue polyadenylation sites genomewide.

Alternative Polyadenylation in Plants

A primary means by which gene expression might be affected at the level of mRNA 3' end formation is through alternative polyadenylation, or the choice of alternative 3' ends within a transcription unit. (A distinction must be drawn between the 3' end microheterogeneity discussed in the previous section and alternative 3' end processing that involves widely spaced polyadenylation sites; it is the latter phenomenon, which adds or subtracts exons or sizeable tracts of RNA sequence, that is the focus of this section.) This mode of regulation can impact gene expression through effects on mRNA stability or translatability, as well as by altering the protein product encoded by the gene. The likely scope of the contributions of alternative polyadenylation to gene expression is revealed by the observation that a large proportion (half or more) of all mammalian genes may be subject to alternative polyadenylation (Tian et al. 2005; Yan and Marr 2005). Several examples of this mode of regulation in plants have been reported; these include the alternative processing of S locus glycoprotein and S-receptor kinase mRNAs in *Brassica oleracae* (Tantikanjana et al. 1993; Giranton et al. 1995), transcripts encoding ETR1-related in *Prunus persica* (Bassett et al. 2002), mRNAs encoding a TIR motif-containing receptor-like protein in *Arabidopsis* (Meyers et al. 2002), mRNAs encoded by a cotton gene that encodes both lysine-ketoglutarate reductase and saccharopine dehydrogenase (Tang et al. 2002), spinach ascorbate peroxidase-encoding mRNAs (Ishikawa et al. 1997), transcripts encoded by the flowering-associated FCA gene (Quesada et al. 2003), and mRNAs encoding CPSF30-related proteins in *Arabidopsis* (Delaney et al. 2006).

The possible scope of alternative polyadenylation such as described in the preceding paragraph is not precisely known, but may be extensive. Analysis of large collections of cDNAs indicate that between 0.5% and 3% of *Arabidopsis* genes may be subject to this sort of alternative processing (Haas et al. 2003; Iida et al. 2004; Nagasaki et al. 2005; Xiao et al. 2005). These numbers may be massive underestimates, as more than 25% of all *Arabidopsis* genes possess so-called MPSS tags that lie within allegedly nonterminal exons or introns (Meyers et al. 2004). Analyses of cDNA collections from rice are in agreement with the results obtained with *Arabidopsis* (Kikuchi et al. 2003; Nagasaki et al. 2005). These various surveys are not exhaustive, and the differences in the two approaches (cDNA libraries vs. sequence tags) remain to be adequately explained. However, these studies indicate that alternative polyadenylation is widespread in plants. Taken along with the specific examples listed in this section, these considerations reveal a large possible scope of alternative polyadenylation in plants, a scope that can and does have sizeable impacts on plant growth and development.

In one instance, involving alternative processing of mRNAs encoded by the so-called FCA gene in *Arabidopsis*, a mechanism for alternative polyadenylation has been deduced (Quesada et al. 2003; Simpson et al. 2003). Briefly, FCA-derived transcripts may be polyadenylated to yield a "normally" processed mRNA encoding a full-sized FCA polypeptide, or at a site within the third intron of the FCA gene, yielding a truncated mRNA and potential polypeptide. Presumably, the poly(A) site associated with the full-length FCA transcript is a typical plant site. In contrast, the site within the third intron is one that dependent on FCA itself. FCA is an RNA-binding protein (Macknight et al. 1997) that also interacts physically and functionally with FY, a subunit of the plant polyadenylation apparatus. Apparently, FCA brings FY, and the rest of the polyadenylation complex, to the third intron, thereby promoting 3' end processing within the intron. Thus, by this mechanism, polyadenylation occurs at a site that is not recognized by any of the RNA-binding activities of the core complex. This mechanism has significant implications. It provides for a means by which RNA binding proteins distinct from the evolutionarily conserved 3' processing machinery may promote polyadenylation at noncanonical sites. Moreover, it renders as problematic the a priori prediction of alternative polyadenylation sites based solely on sequence characteristics. The plant polyadenylation signal is relatively nondescript, but it does lend itself to predictive algorithms that are modest in their success rates (Ji et al. 2007). However, if alternative polyadenylation occurs at sites that are not governed by the rules that define a canonical plant polyadenylation site, then predicting alternative polyadenylation becomes unrealistic. Admittedly, these caveats are based on but a single well-established example; however, they raise significant problems (and research opportunities) for the determination of the true scope of alternative polyadenylation in plant gene expression.

The Polyadenylation Apparatus in Plants

For the most part, the eukaryotic polyadenylation apparatus is widely conserved, with a similar set of proteins found in all of the eukaryotic lineages (Table 1). While many functions for these various subunits have been determined in mammals and yeast, the plant subunits are less well-studied. In this section, focus is placed on the complement of *Arabidopsis* polyadenylation factor subunits. This is done because of the availability of an extensively annotated complete genome sequence, and because virtually all of the experimentation on plant polyadenylation factor subunits has involved *Arabidopsis*. Other plant genomes possess similar complements of polyadenylation-related genes (Q. Q. Li, A. G. Hunt et al., unpublished observations), so the following considerations will likely apply to most plants. To facilitate discussion, *Arabidopsis* proteins will be preceded with the notation At (e.g., AtCPSF100 is the *Arabidopsis* ortholog of the 100-kDa subunit of mammalian cleavage and polyadenylation specificity factor).

The Complement of Arabidopsis Plant Polyadenylation Factor Subunits

The *Arabidopsis* genome possesses complements of probable polyadenylation factor-encoding genes that mirror the sets found in mammals and yeast (Table 1). As indicated in the following sections (and summarized in Fig. 2), most of these proteins can be conceptually linked with PAP and other established polyadenylation factor subunits (such as the members of the CPSF complex), thereby establishing their functionality as polyadenylation factor subunits. However, this remains to be established for some of these genes, such as the Clp1 and Pcf11 orthologs. The functions of the proteins identified solely by similarity searches should be taken as tentative. Many polyadenylation factor subunits are encoded by single genes, while others are encoded by modest gene families. The presence of possible duplicate genes for many polyadenylation factor subunits, including proteins that are highly conserved, raises the possibility of functional redundancy, or alternatively of functional specialization. The former is suggested by the report that mutations expected to inactivate one of the two *Arabidopsis* symplekin-related genes are not lethal (Herr et al. 2006). The latter may also apply to some of these gene families; thus T-DNA disruption of one of the two *Arabidopsis* Fip domain-containing proteins is lethal (Forbes et al., 2006), suggesting that the two *Arabidopsis* proteins play distinct roles in gene expression, roles that in the case of the AtFip1(V) gene may not be satisfied by the gene resident on chromosome III. Also, each of the two genes that encode *Arabidopsis* CPSF73-related proteins is essential (Xu et al. 2004, 2006), again suggestive of specialization . A clue as to this possible specialization may come from the observations that animals also possess two distinct CPSF73-related proteins (Dominski et al. 2005b), and that they have distinct roles in pre-mRNA

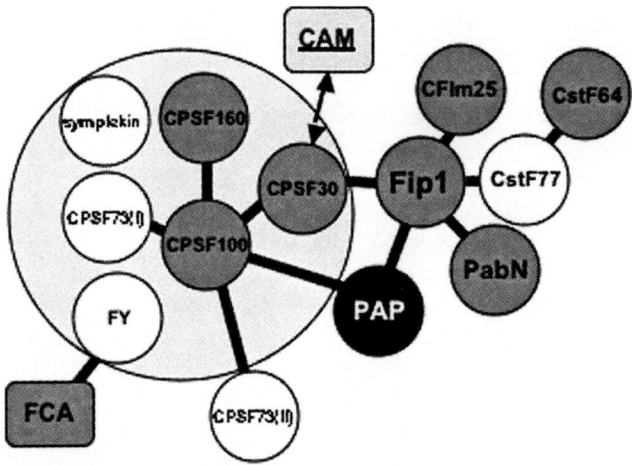

Fig. 2 Summary of protein-protein and protein-RNA interactions involving *Arabidopsis* polyadenylation factor subunits. Polyadenylation factor subunits are represented with *circles*, and factors that impinge on the complex are represented with *rounded rectangles*. RNA binding proteins are designated with *gray circles*. The *light gray circle* that encompasses the CPSF subunits is meant to indicate that the enclosed subunits are part of an identifiable complex [Delaney et al. 2006; Herr et al. 2006; note that CPSF73(II) is not part of this complex]. The *double arrow* linking calmodulin (*CAM*) with CPSF30 denotes a regulatory interaction that alters the activity of CPSF30. Other designations are as in Table 1 and the text.

processing. However, whether the characteristics of the animal proteins are reflected in plants is not known. Further exploration of this question-redundancy vs. specialization-is needed, and may be expected to yield new insight into the functioning of these various proteins.

Inspection of the set of plant polyadenylation-related genes that are readily identified by sequence searches reveals a possible or probable absence of proteins that have clear and important roles in other eukaryotes. For example, the *Arabidopsis* and rice genomes possess no obvious genes that might encode counterparts of the set of large subunits of CFIm (Table 1). These subunits consist of a set of related proteins and have a generic modular structure made up of an N-terminal RRM-type RNA-binding domain and a C-terminal serine+arginine-rich (RS) domain that is typical of splicing factors(Ruegsegger et al. 1996, 1998; Dettwiler et al. 2004). *Arabidopsis* has sizeable families of genes that encode RRM- and RS domain- containing proteins (Lorković and Barta 2002) and also a subset of genes that could encode proteins with both RRM and RS domains. However, sequence comparisons do not identify individual members, or groups, of genes in these sets that are obviously analogous in function to the 59- or 68-kDa subunits of CFIm. These different considerations render ambiguous the matter of the existence of CFIm68- or CFIm59-like proteins in plants. It is possible, even probable, that one or more of the 18 RRM-RS proteins in *Arabidopsis* function as does CFIm59/CFIm68 in RNA processing in mammals, but further research into the nature of these proteins, and their possible interactions with polyadenylation factor subunits such as CFIm25, is needed to resolve this issue.

Another example of a "missing" ortholog is that of Hrp1p-like proteins. Hrp1 is an integral subunit of the yeast CFI polyadenylation factor and binds to the so-called polyadenylation efficiency element (Kessler et al. 1997; Chen and Hyman 1998), a UA-rich motif located 5' or upstream from the A-rich positioning element and polyadenylation site. Hrp1 possesses two tandemly situated RRM-type RNA binding domains that are involved in the recognition of the UA-rich motif. Similarity searches of the *Arabidopsis* databases do not yield obvious counterparts to Hrp1. A number of proteins (10 or more) possess significant similarity to the second RRM motif of Hrp1 (unpublished observations), but the extent of similarity does not permit unambiguous identification of Hrp1-like *Arabidopsis* proteins. As is the case for CFIm59/CFIm68, it is probable that one or more of the RRM-containing *Arabidopsis* proteins (perhaps some of the 13 hnRNP-like proteins, or of the 130+ other RRM-containing proteins that are not poly(A) binding proteins or associated with splicing) play roles analogous to that fulfilled by Hrp1 in 3' end formation. However, this remains an open question.

Characteristics of Individual Subunits of the Plant Polyadenylation Complex

CPSF Subunits

As indicated in Table 1, plants possess genes that encode most of the complement of proteins that make up the core eukaryotic polyadenylation apparatus. Many of these genes have been studied to various extents, and observations that touch on interesting facets of mRNA 3' end formation have been reported. As might be expected, given the central roles of the factor in polyadenylation in mammalian systems and the essential nature of the proteins in yeast, plants appear to possess a complex analogous in many ways to CPSF. *Arabidopsis* possesses a single gene encoding CPSF100. This gene and its protein product are essential, as AtCPSF100-deficient mutants are embryo lethal (Tzafrir et al. 2004). Additionally, alleles of AtCPSF100 in which a conserved glycine at the N-terminus is altered displayed an interesting RNA silencing phenotype (Herr et al., 2006) This allele was isolated in a screen for mutants that displayed elevated silencing of a reporter transgene; the rationalization of this outcome was that the AtCPSF100 mutant is somewhat deficient in transcription termination and/or polyadenylation, such that increased quantities of read-through RNAs derived from the transgene are produced, triggering silencing.

AtCPSF100 resides in a complex that also includes AtCPSF73(I), AtCPSF160, FY, the *Arabidopsis* symplekin (Herr et al. 2006), and AtCPSF30 (Delaney et al.

2006). Consistent with this, AtCPSF160, AtCPSF100, AtCPSF73(I), AtCPSF73(II), and AtCPSF30 can all be found in the nucleus (Delaney et al. 2006; Xu et al. 2006). Curiously, different studies have yielded somewhat different, although not exclusive, results. Thus Herr et al. (2006) reported that FLAG-tagged AtCPSF100 copurified with AtCPSF160, AtCPSF73(I), *Arabidopsis* symplekin, and FY. However, AtCPSF73(II), AtCPSF30, and the various AtPAP isoforms were not detected in these studies, even though these proteins do interact with AtCPSF100 in two-hybrid and in vitro assays (Elliott et al. 2003; Xu et al. 2006). Other experiments involving immunoprecipitation of CPSF100 from nuclear extracts did reveal a copurification of AtCPSF30 with AtCPSF100 (Delaney et al. 2006), consistent with the two-hybrid and in vitro studies. Differences in the purification schemes [Xu et al. (2006) used concentrated nuclear extracts, where as Herr et al. (2006) used extracts from leaf material] might explain the differences in the two reports, especially the inconsistent results with AtCPSF30. Regardless, the totality of the three studies of this protein indicate the existence of a complex that includes AtCPSF100, AtCPSF160, AtCPSF73(I), FY, *Arabidopsis* symplekin, and probably AtCPSF30. They also leave open the possible association, with this complex, of other proteins such as AtCPSF73(II) and AtPAP. Additionally, the report of Herr et al. (2006) provides for a tangible link between mRNA 3' end formation and transcription termination in plants.

Arabidopsis possesses two genes [termed here AtCPSF73(I) and AtCPSF73(II), the Roman numerals representing the chromosome on which the respective genes reside] that encode CPSF73-like proteins. Both are essential genes, and misregulation of expression of these leads to distinctive developmental defects (Xu et al. 2004, 2006). In particular, plants hemizygous for the wild-type AtCPSF73(II) gene were defective in female gametogenesis (Xu et al. 2004), while plants overexpressing AtCPSF73(I) had altered male gametogenesis (Xu et al. 2006). These studies suggest that the two CPSF73-like proteins have different functions. Consistent with this, only AtCPSF73(I) is detected in complexes containing FLAG-tagged AtCPSF100 (Herr et al. 2006). However, both CPSF73-related proteins interact with the AtCPSF100 in two-hybrid and in vitro assays; these conflicting observations have yet to be resolved. Interestingly, in mammals there are also two CPSF73-related proteins. One [related to CPSF73(I)] has been implicated as the endonuclease involved in the cleavage step of mRNA 3' end formation (Ryan et al. 2004; Mandel et al. 2006), while the other [related to CPSF73(II)] has been reported to function in cell cycle progression but is otherwise not needed for growth (Dominski et al. 2005b). Whether analogous functions may be ascribed to the two *Arabidopsis* CPSF73 proteins is not clear.

In *Arabidopsis*, CPSF30 is encoded by a complex gene whose transcripts are alternatively processed; the result is an ability to encode AtCPSF30 or a larger protein in which all but the C-terminal 13 amino acids of CPSF30 have been fused (at their C-terminus) to additional sequences (Delaney et al. 2006). A similar arrangement is apparent in the rice genome, and ESTs that span the two parts of the protein have been deposited in databases (Delaney et al. 2006). The additional sequences possess a domain that is found in other *Arabidopsis* proteins that interact with calcineurin-interacting protein kinases (Ok et al. 2005), and as well in mammalian proteins that have been implicated in splicing (Imai et al. 1998; Stoilov et al. 2002). Virtually all of the AtCPSF30 protein is present in the larger variant, indicating that it is probable that both forms play some role in 3′ end processing, or conversely that the RNA binding activity of AtCPSF30 may have functions in other processes (such as pre-mRNA splicing). AtCPSF30 is also distinctive in that its RNA-binding activity can be inhibited in vitro by calmodulin, in a calcium-dependent fashion (Delaney et al. 2006). Thus calcium signaling has the potential to impact RNA processing through AtCPSF30 in at least two ways. The physiological significance of these interactions remains to be studied.

The yeast counterpart of CPSF30, Yth1p, is an essential protein (Barabino et al. 2000; Tacahashi et al. 2003). In contrast, AtCPSF30 is dispensable for growth and survival, even though the gene that encodes AtCPSF30, At1g30460, is the only *Arabidopsis* gene that encodes an obvious CPSF30/Yth1p homolog (Delaney et al. 2006). That this gene is nonessential is thus surprising. The reasons for the discrepancy between yeast and *Arabidopsis* are not known, but a number of interesting possibilities merit discussion. In plants, CPSF30 may not be essential because of the presence of other proteins that function as does CPSF30 in mRNA 3′end formation. While BLAST searches do not yield obvious candidates for such hypothetical proteins, *Arabidopsis* does possess a large family of CCCH zinc finger proteins (Delaney et al. 2006), one or more of which may be able to replace AtCPSF30 in mRNA 3′end formation. Moreover, there exists a possibility that other proteins, unrelated at the amino acid sequence (or even motif organization) level, may be able to provide the activity of CPSF30 in plants. These matters remain to be addressed experimentally.

FY

Plants possess proteins that are somewhat related to yeast Pfs2 (Table 1). This class of proteins is typified by its characteristic array of seven WD repeats, and it is within this array that the most extensive sequence similarity is seen. The plant Psf2-related proteins have been identified independent of studies on RNA processing as FY, a protein that is involved in the autonomous pathway of flowering time determination (Simpson et al., 2003) (see the chapter by L. C. Terzi and G. G. Simpson, this volume) As mentioned above, FY interacts with an RNA binding protein, FCA, to effect polyadenylation at a site within an "early" intron of the FCA transcript; the net result of this interaction is a reduction of production of full-length FCA-encoding mRNAs, and a concomitant alteration of expression of downstream FCA-dependent genes (such as FLC). It has been proposed that FCA binding to the third intron of transcripts encoded by its own gene serves to recruit FY, and concomitantly the 3′ processing apparatus, to the third intron. This mechanism is interesting, as it may represent a more general means by which alternative polyadenylation may occur. In addition to the WD repeat-containing part of the protein, FY possesses a C-terminal domain that consists of an array of repetitive proline- and glutamine-rich regions. Such regions are often associated with protein-protein interactions; indeed, the prior knowledge that so-called WW motifs have a propensity to interact with proline-rich domains was a factor in the studies that linked FY with FCA (which is a WW-containing protein). The possibility thus arises that other (hypothetical) interacting partners of FY may enable polyadenylation at sites bound by these proteins. (One intriguing possibility is that these hypothetical proteins may, as a group, constitute the recognition factors for one of the components of the plant polyadenylation signal. Such a model, while unprecedented and admittedly very speculative, would help to explain the heterogeneous nature of these *cis* elements; it may be that the nucleotide composition that has been determined in analyses of large collections of sequences may be the result of summation of many individual RNA-protein interactions, each with a decided specificity.)

Mutant alleles that are missing part or most of their C-termini, but retain the seven WD repeats, are viable but late flowering (Henderson et al. 2005). In contrast, a mutant allele that encodes a protein truncated within the first of the seven WD repeats is lethal (Henderson et al. 2005), suggestive of a more general and essential role for FY beyond one in flowering. Conditional RNA silencing of FY in tobacco also indicates an essential role for FY. FY is part of a complex that includes AtCPSF100, since it can be immunoprecipitated from cell-free extracts along with tagged AtCPSF100 (Herr et al. 2006). These observations, taken together, lend themselves to a model in which FY is an integral part of the core polyadenylation machinery and acts in alternative polyadenylation by bringing the core to noncanonical sites on a transcript.

CstF Subunits

Among the more conserved of polyadenylation factor subunits are the 77- and 64-kDa subunits of CstF. While not as extensively studied as other subunits, the *Arabidopsis* CstF64, like its yeast and mammalian counterparts, is an RNA-binding protein, and it interacts with AtCstF77 (Yao et al. 2002). The *Arabidopsis* CstF77 also interacts with AtFip1 (Forbes et al. 2006); given the interactions between AtFip1 and AtPAP, this observation provides a conceptual link between AtPAP and AtCstF, thereby buttressing the assumption that the plant CstF77 and CstF64 homologs are involved in mRNA 3' end formation. Curiously, the same may not, at this time, be said about the possible *Arabidopsis* CstF50 ortholog. In contrast to what is seen in mammals, the AtCstF50 apparently does not interact with AtCstF77 (Yao et al. 2002; A. G. Hunt et al., unpublished observations). The ramifications of this are not obvious; it is possible that plants, like yeast, actually have no counterpart to CstF50. Alternatively, plant CstF50-like proteins may associate with the complex in ways different from those that have been established in the mammalian complex.

Eukaryotic CstF64 proteins may be represented by a two-domain model: One domain is a canonical RRM-type RNA binding domain, while the other (at the C-terminus of the protein) is a conserved domain that mediates interactions with transcription termination factors in mammals and yeast (Qu et al. 2007). While similar studies with plant CstF64 proteins have not been conducted, the characteristics of the so-called ESP1 gene in *Arabidopsis* lend support to a similar link between 3' end formation and transcription termination in plants. ESP1, like AtCPSF100 (=ESP5) and symplekin (=ESP4) has a molecular transcriptional readthrough phenotype, suggestive of a defect in transcription termination. ESP1 encodes a polypeptide that has a domain that is related to the C-terminal domain of eukaryotic CstF64 proteins; however, ESP1 lacks a recognizable RNA-binding domain, and thus is not a canonical CstF64. Nonetheless, the association of the C-terminal motif with a deficit in transcription termination lends support to the idea that plant CstF64 proteins, like their eukaryotic counterparts, help to mediate between 3' end formation and termination of transcription.

Fip1

Plants possess small gene families (two members each in *Arabidopsis* and rice) whose predicted products contain a conserved motif that is also found in the yeast Fip1 and its mammalian counterpart (Forbes et al. 2006). Outside of the Fip1 motif (which is only 20 amino acids), one member of these families (At3g66652 in *Arabidopsis*) is apparently unrelated to Fip1 and its human counterpart, and these proteins have not been studied. The other member of the family in *Arabidopsis* (At5g58040), while possessing little amino acid sequence similarity with other eukaryotic Fip domain-containing proteins apart from the Fip motif, does have a more general domain structure that resembles its human counterpart (Forbes et al., 2006). This includes a novel acidic N-terminal domain and a C-terminal portion that includes an arginine-rich RNA-binding domain. The *Arabidopsis* protein [AtFip1(V)] interacts with AtPAP(IV), AtCstF77, AtCPSF30, and at least one AtPabN isoform, and the isolated N-terminus (responsible for the protein-protein interaction) by itself can stimulate nonspecific poly(A) polymerase activity by purified AtPAP(IV) (Forbes et al. 2006). These protein-protein interactions all involve the acidic N-terminal part of the protein; while it is conceivable that all of these interactions may exist in a single static complex, it is more likely that the various interactions, all determined in two-hybrid or in vitro studies, are reflective of a dynamic that involves reordering or remodeling around a small core that includes AtFip1(V).

Symplekin

Plants possess genes whose predicted products are similar to symplekin and its yeast counterpart Pta1p (Table 1). Mutations in one *Arabidopsis* symplekin gene were identified in the RNA silencing screen of Herr et al. (2006), and the symplekin isoform encoded by this gene can be found in a complex with AtCPSF100 [as well as FY, AtCPSF160, and AtCPSF73(I); Herr et al. 2006]. Thus it seems likely that symplekin is an authentic plant polyadenylation factor subunit. Curiously, it also seems to be part of (or perhaps one of) the connection between 3' end formation and termination of transcription. Whether symplekin and AtCPSF100 (and ESP1, for that matter) converge on a common component of the transcription machinery, or whether they act by different and somewhat independent means, is not known.

Poly(A) Polymerases

Arabidopsis possesses a modest (4 member) gene family that encodes enzymes similar to canonical nuclear PAPs (Table 1; Addepalli et al., 2004). (In this review, canonical nuclear PAPs are typified by the mammalian PAP-α or β, or yeast Pap1p, and are distinct from the growing set of noncanonical PAPs that include GLD2, Trf4, and Cid1/13). Within this family are interesting similarities and differences. One of the *Arabidopsis* PAPs [AtPAP(III); Addepalli et al., 2004] is much smaller than the other PAPs, and the truncation that leads to this difference involves loss or deletion of the C-terminal parts of the protein. This situation is somewhat analogous to that seen in mammals, where one of the three proteins clearly related to canonical nuclear PAPs is truncated at its C-terminus (Kashiwabara et al. 2000; Lee et al. 2000; Le et al. 2001). In mammals, the C-terminal extension seen in larger PAP isoforms has been associated with posttranslational modifications such as phosphorylation (Colgan et al. 1996; Bond et al. 2000); these modifications have been implicated in the cell cycle control of PAP activity. Whether similar control regulates plant PAP activity is not known, but the amino acid composition of the C-termini of the three larger *Arabidopsis* PAPs (high S+T content) would be consistent with analogous regulatory mechanisms. Regardless, AtPAP(III) lacks this domain, and thus would not be subject to the (admittedly hypothetical) controlling mechanisms. Interestingly, the truncated mammalian protein has been associated with spermatogenesis (Kashiwabara et al. 2002; Zhuang et al. 2004). While detailed studies of plant PAP genes have not been done, perusal of public domain microarray data indicates that the gene encoding the AtPAP(III) gene (At3g06560) is pollen specific (Meeks 2005; A. G. Hunt, unpublished observations). This suggests an interesting parallel between mammals and plants.

RNA Binding Proteins in the Polyadenylation Complex

The nature of the plant polyadenylation signal suggests an involvement of at least three distinct RNA-protein interactions in the course of mRNA 3' end formation, involving the FUE, NUE, and CE, respectively. This in turn implies an involvement of three (at a minimum) distinct RNA-binding proteins in the process. RNA-binding activity has been demonstrated for several of the *Arabidopsis* proteins listed in Table 1. As stated above, AtCstF64 binds RNA in vitro (Yao et al. 2002); this activity involves the N-terminal RRM-containing domain of the protein. However, the sequence preference of the *Arabidopsis* protein has not been determined, and it is not possible to associate the plant protein with a specific motif or element. It is not possible to make an inference based on studies in mammals and yeast, since the mammalian CstF64 associates with the UG-rich downstream element (MacDonald et al. 1994) whereas the yeast counterpart (Rna15) binds to the A-rich positioning element (Gross and Moore 2001). The AtFip1(V) protein also binds RNA (Forbes et al. 2006), much as does its human counterpart (Kaufmann et al. 2004). This activity maps to the C-terminal portion of the *Arabidopsis* protein and does not require the N-terminal portion that is implicated in the protein-protein interactions mentioned above. The plant protein shows a preference for poly(G) (Forbes et al. 2006), but a clear-cut affinity for one of the three polyadenylation-associated elements cannot be detected (Forbes 2005). AtCPSF30 binds RNA (Delaney et al., 2006), with a preference for poly(U) among the four homopolymers (K. J. Delaney and A. G. Hunt, unpublished observations). This suggests a possible affinity for the cleavage element, or perhaps the FUE. However, such an expectation has not been demonstrated experimentally. Based on other analogous systems, it is expected that AtCPSF160 (Murthy and Manley 1995), AtCPSF100 (Dichtl and Keller 2001), and the three AtPabN isoforms (Kuhn and Wahle 2004) will also bind RNA. Taken together, these considerations reveal a plethora of possible RNA-protein interactions. However, many questions still remain as to the actual factors that recognize the different parts of the plant polyadenylation signal, and to other possible roles that RNA binding may play in the process.

Cellular Signaling and 3′ End Formation

Given that mRNA 3′ end formation is central to gene expression, it might be expected that the process would not be impacted by regulatory mechanisms. However, the occurrence of alternative mRNA processing affords a level of regulation that extends beyond the matter of mere addition of the poly(A) tract to a mRNA. As has been discussed above, there exists interesting precedent for regulation of gene expression, and developmental process, at the level of alternative mRNA polyadenylation. These examples raise important questions regarding the mechanisms by which cellular signaling may impact mRNA 3′ end formation, and in particular poly(A) site choice.

As described above, alternative poly(A) site choice is an important determinant of flowering time, as it is one controlling element in the autonomous pathway (see the chapter by L. C. Terzi and G. G. Simpson, this volume). The alternative processing that yields different FCA mRNA isoforms is controlled in much the same manner as the timing of flowering, indicating that some step in the process is regulated in response to environmental cues. One possible focus for such regulatory mechanisms may be the RNA-binding protein FCA itself; in addition to bridging between FCA pre-mRNAs and FY, FCA also binds abscisic acid (Razem et al., 2006). Moreover, ABA inhibits the interaction of FCA with FY, providing one means by which hormones might impact the timing of flowering; thus, by preventing the binding of FY to FCA, ABA seems to delay flowering by increasing the production of full-length FCA, an negative regulator of FLC (which in turn is a negative regulator of flowering). This example offers a paradigm by which alternative RNA processing may be regulated-alternative (noncanonical) polyadenylation sites might be controlled by interactions with fairly specific RNA binding proteins, and the events that link the RNA-protein interaction with 3' end formation might in turn be regulated, perhaps directly (as in the case with ABA and FCA) by environmental cues.

Cellular signals may impinge less directly, via second messengers. For example, calmodulin inhibits the RNA-binding activity of the *Arabidopsis* CPSF30 protein in a calcium-dependent fashion (Delaney et al. 2006). This suggests that RNA processing that is mediated by AtCPSF30 may be reduced in cells in which Ca-dependent calmodulin signaling has been activated. The physiological consequences of this anticipated outcome have not been studied. However, as AtCPSF30 is not essential, it stands to reason that there may exist a set of AtCPSF30-dependent genes or poly(A) sites, and that inhibition of AtCPSF30 by calmodulin should alter poly(A) site choice within these genes.

These two examples provide two general mechanisms for linking polyadenylation with cellular signaling. In one mechanism, core polyadenylation factor subunits would be subject to direct control by second messengers or signaling cascades (via phosphorylation or ubiquitination). In the second, other RNA-binding factors would be brought to the core through interactions with subunits in the core. Via these interactions and modifications, mRNA 3' end formation would be regulated as far as poly(A) site choice is concerned, and perhaps also in terms of overall activity.

Summary and Prospects

As the understanding of mRNA 3' end formation in plants has grown over the past several years, a number of interesting issues have arisen. Clearly, alternative polyadenylation makes an important contribution to the sculpting of the plant proteome. However, both an understanding of the scope of the phenomenon and an ability to reliably predict occurrences of alternative polyadenylation, given just genome sequence data, are sorely lacking. Recent studies have begun to grapple with the problem of poly(A) site prediction in plants (Ji et al. 2007), but the possible involvement of relatively specific RNA-binding proteins in alternative polyadenylation raises difficult questions regarding the utility of such prediction methods for the identification of sites of alternative (perhaps regulated) polyadenylation. While a comprehensive picture of the plant polyadenylation machinery is still forthcoming, the beginnings of a working model may be assembled (Fig. 2). It is apparent that the *Arabidopsis* CPSF-related subunits associate in a discreet complex, and that many other subunits are associated with the process through one of the Fip1 orthologs. At the present, the relationship of the Pcf11 and Clp orthologs, or of

CstF50, to these complexes is not clear. (Indeed, it may be that not all of these proteins are authentic polyadenylation factor subunits, as their inclusion in this list is based solely on amino acid sequence similarity.) Subunits that interact with both CPSF and Fip1 include AtCPSF30 and PAP; whether this reflects bridging functions between two complexes, or alternatively the existence of remodeling attendant with progression through the reaction, is not clear. Two observations-the dispensability of AtCPSF30 for growth and the involvement of a very small part of AtFip1(V) in several interactions with other polyadenylation-related proteins-are suggestive of a dynamic nature of the complex, with a relatively modest core acting with different combinations of a larger array of factors, rather than in concert with the entire range of polyadenylation factor subunits. Most interesting, however, is the prospect that cellular signaling may impact directly and with important consequence on the polyadenylation machinery. The links from both hormone and calcium signaling to mRNA 3' end formation speak to numerous possible contributions of this RNA processing event to the regulation of gene expression.

While much has been learned about mRNA 3' end formation in plants, many important questions remain without answers. Specific subunits have yet to be associated with any of the three parts of the plant polyadenylation signal. At least two subunits of CPSF may be linked with transcription termination, but connections of mRNA 3' end formation with other steps in the process of gene expression have yet to be spelled out. Links between polyadenylation and splicing would seem to be especially important, given the growing realization of the impact of alternative polyadenylation on gene expression and the growing precedent that is apparent in other eukaryotes. Such links have yet to be established, or even explored to much of an extent. Finally, for this review, two very different mechanisms for regulating polyadenylation have been described. The scope of such mechanisms remains largely a matter of conjecture, but one that has great potential import for gene regulatory mechanisms.

References

Addepalli B, Meeks LR, Forbes KP, Hunt AG (2004) Novel alternative splicing of mRNAs encoding poly(A) polymerases in Arabidopsis. Biochim Biophys Acta 1679:117–128

Ahn SH, Kim M, Buratowski S (2004) Phosphorylation of serine 2 within the RNA polymerase II C-terminal domain couples transcription and 3' end processing. Mol Cell 13:67–76

Alexandrov NN, Troukhan ME, Brover VV, Tatarinova T, Flavell RB, Feldmann KA (2006) Features of Arabidopsis genes and genome discovered using full-length cDNAs. Plant Mol Biol 60:69–85

Altschul SF, Madden TL, Schaffer AA, Zhang J, Zhang Z, Miller W, Lipman DJ (1997) Gapped BLAST and PSI-BLAST: a new generation of protein database search programs. Nucleic Acids Res 25:3389–3402

Baillat D, Hakimi MA, Naar AM, Shilatifard A, Cooch N, Shiekhattar R (2005) Integrator, a multiprotein mediator of small nuclear RNA processing, associates with the C-terminal repeat of RNA polymerase II. Cell 123:265–276

Barabino SM, Ohnacker M, Keller W (2000) Distinct roles of two Yth1p domains in 3'-end cleavage and polyadenylation of yeast pre-mRNAs. EMBO J 19:3778–3787

Barnard DC, Ryan K, Manley JL, Richter JD (2004) Symplekin and xGLD-2 are required for CPEB-mediated cytoplasmic polyadenylation. Cell 119:641–651

Bassett CL, Artlip TS, Callahan AM (2002) Characterization of the peach homologue of the ethylene receptor, PpETR1, reveals some unusual features regarding transcript processing. Planta 215:679–688

Baynton CE, Potthoff SJ, McCullough AJ, Schuler MA (1996) U-rich tracts enhance 3' splice site recognition in plant nuclei. Plant J 10:703–711

Beaudoing E, Gautheret D (2001) Identification of alternate polyadenylation sites and analysis of their tissue distribution using EST data. Genome Res 11:1520–1526

Bentley DL. 2005. Rules of engagement: co-transcriptional recruitment of pre-mRNA processing factors. Curr Opin Cell Biol 17:251–256

Bilger A, Fox CA, Wahle E, Wickens M (1994) Nuclear polyadenylation factors recognize cytoplasmic polyadenylation elements. Genes Dev 8:1106–1116

Bond GL, Prives C, Manley JL (2000) Poly(A) polymerase phosphorylation is dependent on novel interactions with cyclins. Mol Cell Biol 20:5310–5320

Brown JW, Simpson CG (1998) Splice site selection in plant pre-mRNA splicing. Annu Rev Plant Physiol Plant Mol Biol 49:77–95

Brown PH, Tiley LS, Cullen BR (1991) Effect of RNA secondary structure on polyadenylation site selection. Genes Dev 5:1277–1284

Buratowski S (2005) Connections between mRNA 3' end processing and transcription termination. Curr Opin Cell Biol 17:257–261

Caballero JJ, Giron MD, Vargas AM, Sevillano N, Suarez MD, Salto R (2004) AU-rich elements in the mRNA 3'-untranslated region of the rat receptor for advanced glycation end products and their relevance to mRNA stability. Biochem Biophys Res Commun 319:247–255

Calvo O, Manley JL (2003) Strange bedfellows: polyadenylation factors at the promoter. Genes Dev 17:1321–1327

Calvo O, Manley JL (2005) The transcriptional coactivator PC4/Sub1 has multiple functions in RNA polymerase II transcription. EMBO J24:1009–1020

Chen S, Hyman LE (1998) A specific RNA-protein interaction at yeast polyadenylation efficiency elements. Nucleic Acids Res 26:4965–4974

Colgan DF, Murthy KG, Prives C, Manley JL (1996) Cell-cycle related regulation of poly(A) polymerase by phosphorylation. Nature 384:282–285

Dass B, Attaya EN, Michelle Wallace A, MacDonald CC (2001a) Overexpression of the CstF-64 and CPSF-160 polyadenylation protein messenger RNAs in mouse male germ cells. Biol Reprod 64:1722–1729

Dass B, McMahon KW, Jenkins NA, Gilbert DJ, Copeland NG, MacDonald CC (2001b) The gene for a variant form of the polyadenylation protein CstF-64 is on chromosome 19 and is expressed in pachytene spermatocytes in mice. J Biol Chem 276:8044–8050

Delaney KJ, Xu R, Zhang J, Li QQ, Yun KY, Falcone DL, Hunt AG (2006) Calmodulin interacts with and regulates the RNA-binding activity of an Arabidopsis polyadenylation factor subunit. Plant Physiol 140:1507–1521

Dettwiler S, Aringhieri C, Cardinale S, Keller W, Barabino SM (2004) Distinct sequence motifs within the 68–kDa subunit of cleavage factor Im mediate RNA binding, protein-protein interactions, and subcellular localization. J Biol Chem 279:35788–35797

Dichtl B, Aasland R, Keller W (2004) Functions for *S. cerevisiae* Swd2p in 3' end formation of specific mRNAs and snoRNAs and global histone 3 lysine 4 methylation. RNA 10:965–977

Dichtl B, Blank D, Sadowski M, Hubner W, Weiser S, Keller W (2002) Yhh1p/Cft1p directly links poly(A) site recognition and RNA polymerase II transcription termination. EMBO J 21:4125–4135

Dichtl B, Keller W (2001) Recognition of polyadenylation sites in yeast pre-mRNAs by cleavage and polyadenylation factor. EMBO J 20:3197–3209

Dickson KS, Bilger A, Ballantyne S, Wickens MP (1999) The cleavage and polyadenylation specificity factor in *Xenopus laevis* oocytes is a cytoplasmic factor involved in regulated polyadenylation. Mol Cell Biol 19:5707–5717

Dickson KS, Thompson SR, Gray NK, Wickens M (2001) Poly(A) polymerase and the regulation of cytoplasmic polyadenylation. J Biol Chem 276:41810–41816

Dominski Z, Yang XC, Marzluff WF (2005a) The polyadenylation factor CPSF-73 is involved in histone-pre-mRNA processing. Cell 123:37–48

Dominski Z, Yang XC, Purdy M, Wagner EJ, Marzluff WF (2005b) A CPSF-73 homologue is required for cell cycle progression but not cell growth and interacts with a protein having features of CPSF-100. Mol Cell Biol 25:1489–1500

Dong H, Deng Y, Chen J, Wang S, Peng S, Dai C, Fang Y, Shao J, Lou Y, Li D (2007) An exploration of 3'-end processing signals and their tissue distribution in *Oryza sativa*. Gene 389:107–113

Elliott BJ, Dattaroy T, Meeks-Midkiff LR, Forbes KP, Hunt AG (2003) An interaction between an Arabidopsis poly(A) polymerase and a homologue of the 100 kDa subunit of CPSF. Plant Mol Biol 51:373–384

Evans D, Perez I, MacMorris M, Leake D, Wilusz CJ, Blumenthal T (2001) A complex containing CstF-64 and the SL2 snRNP connects mRNA 3' end formation and trans-splicing in *C. elegans* operons. Genes Dev 15:2562–2571

Forbes KP. 2005. Characterization of plant polyadenylation trans-acting factors that modify poly(A) polymerase activity. Plant Physiology. Lexington, KY: University of Kentucky.

Forbes KP, Addepalli B, Hunt AG (2006) An Arabidopsis Fip1 homolog interacts with RNA and provides conceptual links with a number of other polyadenylation factor subunits. J Biol Chem 281:176–186

Giranton JL, Ariza MJ, Dumas C, Cock JM, Gaude T (1995) The S locus receptor kinase gene encodes a soluble glycoprotein corresponding to the SKR extracellular domain in *Brassica oleracea*. Plant J 8:827–834

Graber JH, Cantor CR, Mohr SC, Smith TF (1999) In silico detection of control signals: mRNA 3'-end-processing sequences in diverse species. Proc Natl Acad Sci USA 96:14055–14060

Graveley BR, Fleming ES, Gilmartin GM (1996) RNA structure is a critical determinant of poly(A) site recognition by cleavage and polyadenylation specificity factor. Mol Cell Biol 16:4942–4951

Gross S, Moore CL (2001) Rna15 interaction with the A-rich yeast polyadenylation signal is an essential step in mRNA 3'-end formation. Mol Cell Biol 21:8045–8055

Gunderson SI, Beyer K, Martin G, Keller W, Boelens WC, Mattaj LW (1994) The human U1A snRNP protein regulates polyadenylation via a direct interaction with poly(A) polymerase. Cell 76:531–541

Gunderson SI, Vagner S, Polycarpou-Schwarz M, Mattaj IW (1997) Involvement of the carboxyl terminus of vertebrate poly(A) polymerase in U1A autoregulation and in the coupling of splicing and polyadenylation. Genes Dev 11:761–773

Haas BJ, Delcher AL, Mount SM, Wortman JR, Smith RK, Jr., Hannick LI, Maiti R, Ronning CM, Rusch DB, Town CD, Salzberg SL, White O (2003) Improving the Arabidopsis genome annotation using maximal transcript alignment assemblies. Nucleic Acids Res 31:5654–5666

Hammell CM, Gross S, Zenklusen D, Heath CV, Stutz F, Moore C, Cole CN (2002) Coupling of termination, 3' processing, and mRNA export. Mol Cell Biol 22:6441–6457

Hansen WR, Barsic-Tress N, Taylor L, Curthoys NP (1996) The 3'-nontranslated region of rat renal glutaminase mRNA contains a pH-responsive stability element. Am J Physiol Renal Physiol 271:F126–F131

Henderson IR, Liu F, Drea S, Simpson GG, Dean C (2005) An allelic series reveals essential roles for FY in plant development in addition to flowering-time control. Development 132:3597–3607

Herr AJ, Molnar A, Jones A, Baulcombe DC (2006) Defective RNA processing enhances RNA silencing and influences flowering of Arabidopsis. Proc Natl Acad Sci USA 103:14994–15001

Hofmann I, Schnolzer M, Kaufmann I, Franke WW (2002) Symplekin, a constitutive protein of karyo- and cytoplasmic particles involved in mRNA biogenesis in *Xenopus laevis* oocytes. Mol Biol Cell 13:1665–1676

Hu J, Lutz CS, Wilusz J, Tian B (2005) Bioinformatic identification of candidate *cis*-regulatory elements involved in human mRNA polyadenylation. RNA 11:1485–1493

Hunt A (1994) Messenger RNA 3' end formation in plants. Annu Rev Plant Physiol Plant Mol Biol 45:47–60

Hunt AG, MacDonald MH (1989) Deletion analysis of the polyadenylation signal of a pea ribulose-1,5-bisphosphate carboxylase small-subunit gene. Plant Mol Biol 13:125–138

Iida K, Seki M, Sakurai T, Satou M, Akiyama K, Toyoda T, Konagaya A, Shinozaki K (2004) Genome-wide analysis of alternative pre-mRNA splicing in *Arabidopsis thaliana* based on full-length cDNA sequences. Nucleic Acids Res 32:5096–5103

Imai Y, Matsuo N, Ogawa S, Tohyama M, Takagi T (1998) Cloning of a gene, YT521, for a novel RNA splicing-related protein induced by hypoxia/reoxygenation. Brain Res Mol Brain Res 53:33–40

Ishikawa T, Yoshimura K, Tamoi M, Takeda T, Shigeoka S (1997) Alternative mRNA splicing of 3'-terminal exons generates ascorbate peroxidase isoenzymes in spinach (*Spinacia oleracea*) chloroplasts. Biochem J 328:795–800

Ji G, Zheng J, Shen Y, Wu X, Jiang R, Lin Y, Loke JC, Davis KM, Reese GJ, Li QQ (2007) Predictive modeling of plant messenger RNA polyadenylation sites. BMC Bioinformatics 8:43

Jin Y, Bian T (2004) Nontemplated nucleotide addition prior to polyadenylation: a comparison of Arabidopsis cDNA and genomic sequences. RNA10:1695–1697

Kashiwabara S, Noguchi J, Zhuang T, Ohmura K, Honda A, Sugiura S, Miyamoto K, Takahashi S, Inoue K, Ogura A, Baba T (2002) Regulation of spermatogenesis by testis-specific, cytoplasmic poly(A) polymerase TPAP. Science 298:1999–2002

Kashiwabara S, Zhuang T, Yamagata K, Noguchi J, Fukamizu A, Baba T (2000) Identification of a novel isoform of poly(A) polymerase, TPAP, specifically present in the cytoplasm of spermatogenic cells. Dev Biol 228:106–115

Kaufmann I, Martin G, Friedlein A, Langen H, Keller W (2004) Human Fip1 is a subunit of CPSF that binds to U-rich RNA elements and stimulates poly(A) polymerase. EMBO J 23:616–626

Kavanagh E, Buchert M, Tsapara A, Choquet A, Balda MS, Hollande F, Matter K (2006) Functional interaction between the ZO-1-interacting transcription factor ZONAB/DbpA and the RNA processing factor symplekin. J Cell Sci 119:5098–5105

Kessler MM, Henry MF, Shen E, Zhao J, Gross S, Silver PA, Moore CL (1997) Hrp1, a sequence-specific RNA-binding protein that shuttles between the nucleus and the cytoplasm, is required for mRNA 3'-end formation in yeast. Genes Dev 11:2545–2556

Kikuchi S, Satoh K, Nagata T, Kawagashira N, Doi K, Kishimoto N, Yazaki J, Ishikawa M, Yamada H, Ooka H, Hotta I, Kojima K, Namiki T, Ohneda E, Yahagi W, Suzuki K, Li CJ, Ohtsuki K, Shishiki T, Otomo Y, Murakami K, Iida Y, Sugano S, Fujimura T, Suzuki Y, Tsunoda Y, Kurosaki T, Kodama T, Masuda H, Kobayashi M, Xie Q, Lu M, Narikawa R, Sugiyama A, Mizuno K, Yokomizo S, Niikura J, Ikeda R, Ishibiki J, Kawamata M, Yoshimura A, Miura J, Kusumegi T, Oka M, Ryu R, Ueda M, Matsubara K, Kawai J, Carninci P, Adachi J, Aizawa K, Arakawa T, Fukuda S, Hara A, Hashizume W, Hayatsu N, Imotani K, Ishii Y, Itoh M, Kagawa I, Kondo S, Konno H, Miyazaki A, Osato N, Ota Y, Saito R, Sasaki D, Sato K, Shibata K, Shinagawa A, Shiraki T, Yoshino M, Hayashizaki Y, Yasunishi A (2003) Collection, mapping, and annotation of over 28,000 cDNA clones from japonica rice. Science 301:376–379

Kim M, Ahn SH, Krogan NJ, Greenblatt JF, Buratowski S (2004) Transitions in RNA polymerase II elongation complexes at the 3' ends of genes. EMBO J 23:354–364

Ko CH, Brendel V, Taylor RD, Walbot V (1998) U-richness is a defining feature of plant introns and may function as an intron recognition signal in maize. Plant Mol Biol 36:573–583

Kolev NG, Steitz JA (2005) Symplekin and multiple other polyadenylation factors participate in 3'-end maturation of histone mRNAs. Genes Dev 19:2583–2592

Kuhn U, Wahle E (2004) Structure and function of poly(A) binding proteins. Biochim Biophys Acta 1678:67–84
Le YJ, Kim H, Chung JH, Lee Y. (2001) Testis-specific expression of an intronless gene encoding a human poly(A) polymerase. Mol Cells 11:379–385
Lee YJ, Lee Y, Chung JH (2000) An intronless gene encoding a poly(A) polymerase is specifically expressed in testis. FEBS Lett 487:287–292
Legendre M, Gautheret D (2003) Sequence determinants in human polyadenylation site selection. BMC Genomics 4:7
Li Q, Hunt AG (1995) A near-upstream element in a plant polyadenylation signal consists of more than six nucleotides. Plant Mol Biol 28:927–934
Li Q, Hunt AG (1997) The polyadenylation of RNA in Plants. Plant Physiol 115:321–325
Licatalosi DD, Geiger G, Minet M, Schroeder S, Cilli K, McNeil JB, Bentley DL (2002) Functional interaction of yeast pre-mRNA 3′ end processing factors with RNA polymerase II. Mol Cell 9:1101–1111
Loke JC, Stahlberg EA, Strenski DG, Haas BJ, Wood PC, Li QQ. (2005) Compilation of mRNA polyadenylation signals in Arabidopsis revealed a new signal element and potential secondary structures. Plant Physiol 138:1457–1468
Lorković ZJ, Barta A (2002) Genome analysis: RNA recognition motif (RRM) and K homology (KH) domain RNA-binding proteins from the flowering plant *Arabidopsis thaliana*. Nucleic Acids Res 30:623–635
Lorković ZJ, Wieczorek Kirk DA, Lambermon MH, Filipowicz W (2000) Pre-mRNA splicing in higher plants. Trends Plant Sci 5:160–167
MacDonald CC, Wilusz J, Shenk T (1994) The 64-kilodalton subunit of the CstF polyadenylation factor binds to pre-mRNAs downstream of the cleavage site and influences cleavage site location. Mol Cell Biol 14:6647–6654
MacDonald MH, Mogen BD, Hunt AG (1991) Characterization of the polyadenylation signal from the T-DNA-encoded octopine synthase gene. Nucleic Acids Res 19:5575–5581
Macknight R, Bancroft I, Page T, Lister C, Schmidt R, Love K, Westphal L, Murphy G, Sherson S, Cobbett C, Dean C (1997) FCA, a gene controlling flowering time in Arabidopsis, encodes a protein containing RNA-binding domains. Cell 89:737–745
Mandel CR, Kaneko S, Zhang H, Gebauer D, Vethantham V, Manley JL, Tong L (2006) Polyadenylation factor CPSF-73 is the pre-mRNA 3′-end-processing endonuclease. Nature 444:953–956
Meeks LR (2005) Isolation and characterization of the four *Arabidopsis thaliana* poly(A) polymerase genes. Plant Physiology. Lexington, KY: University of Kentucky
Mendez R, Murthy KG, Ryan K, Manley JL, Richter JD (2000) Phosphorylation of CPEB by Eg2 mediates the recruitment of CPSF into an active cytoplasmic polyadenylation complex. Mol Cell 6:1253–1259
Meyers BC, Morgante M, Michelmore RW (2002) TIR-X and TIR-NBS proteins: two new families related to disease resistance TIR-NBS-LRR proteins encoded in Arabidopsis and other plant genomes. Plant J 32:77–92
Meyers BC, Vu TH, Tej SS, Ghazal H, Matvienko M, Agrawal V, Ning J, Haudenschild CD (2004) Analysis of the transcriptional complexity of *Arabidopsis thaliana* by massively parallel signature sequencing. Nat Biotechnol 22:1006–1011
Millevoi S, Geraghty F, Idowu B, Tam JL, Antoniou M, Vagner S (2002) A novel function for the U2AF 65 splicing factor in promoting pre-mRNA 3′-end processing. EMBO Rep 3:869–874
Millevoi S, Loulergue C, Dettwiler S, Karaa SZ, Keller W, Antoniou M, Vagner S (2006) An interaction between U2AF 65 and CF I(μ) links the splicing and 3′ end processing machineries. EMBO J 25:4854–4864
Miyamoto S, Chiorini JA, Urcelay E, Safer B (1996) Regulation of gene expression for translation initiation factor eIF-2 alpha: importance of the 3′ untranslated region. Biochem J 315:791–798

Mogen BD, MacDonald MH, Leggewie G, Hunt AG (1992) Several distinct types of sequence elements are required for efficient mRNA 3' end formation in a pea *rbcS* gene. Mol Cell Biol 12:5406–5414

Morlando M, Ballarino M, Greco P, Caffarelli E, Dichtl B, Bozzoni I (2004) Coupling between snoRNP assembly and 3' processing controls box C/D snoRNA biosynthesis in yeast. EMBO J 23:2392–2401

Morlando M, Greco P, Dichtl B, Fatica A, Keller W, Bozzoni I (2002) Functional analysis of yeast snoRNA and snRNA 3'-end formation mediated by uncoupling of cleavage and polyadenylation. Mol Cell Biol 22:1379–1389

Murthy KG, Manley JL (1995) The 160-kD subunit of human cleavage-polyadenylation specificity factor coordinates pre-mRNA 3'-end formation. Genes Dev 9:2672–2683

Nagasaki H, Arita M, Nishizawa T, Suwa M, Gotoh O (2005) Species-specific variation of alternative splicing and transcriptional initiation in six eukaryotes. Gene 364:53–62

Nedea E, He X, Kim M, Pootoolal J, Zhong G, Canadien V, Hughes T, Buratowski S, Moore CL, Greenblatt J (2003) Organization and function of APT, a subcomplex of the yeast cleavage and polyadenylation factor involved in the formation of mRNA and small nucleolar RNA 3'-ends. J Biol Chem 278:33000–33010

Niwa M, Rose SD, Berget SM (1990) In vitro polyadenylation is stimulated by the presence of an upstream intron. Genes Dev 4:1552–1559

Ohtsubo N, Iwabuchi M (1994) The conserved 3'-flanking sequence, AATGGAAATG, of the wheat histone H3 gene is necessary for the accurate 3'-end formation of mRNA. Nucleic Acids Res 22:1052–1058

Ok SH, Jeong HJ, Bae JM, Shin JS, Luan S, Kim KN (2005) Novel CIPK1-associated proteins in Arabidopsis contain an evolutionarily conserved C-terminal region that mediates nuclear localization. Plant Physiol 139:138–150

Proudfoot N (2004) New perspectives on connecting messenger RNA 3' end formation to transcription. Curr Opin Cell Biol 16:272–278

Qu X, Perez-Canadillas JM, Agrawal S, De Baecke J, Cheng H, Varani G, Moore C (2007) The C-terminal domains of vertebrate CstF-64 and its yeast orthologue Rna15 form a new structure critical for mRNA 3'-end processing. J Biol Chem 282:2101–2115

Quesada V, Macknight R, Dean C, Simpson GG (2003) Autoregulation of FCA pre-mRNA processing controls Arabidopsis flowering time. EMBO J 22:3142–3152

Razem FA, El-Kereamy A, Abrams SR, Hill RD (2006) The RNA-binding protein FCA is an abscisic acid receptor. Nature 439:290–294

Rothnie HM (1996) Plant mRNA 3'-end formation. Plant Mol Biol 32:43–61

Rothnie HM, Reid J, Hohn T (1994) The contribution of AAUAAA and the upstream element UUUGUA to the efficiency of mRNA 3'-end formation in plants. EMBO J 13:2200–2210

Rouget C, Papin C, Mandart E (2006) Cytoplasmic CstF-77 protein belongs to a masking complex with cytoplasmic polyadenylation element-binding protein in *Xenopus* oocytes. J Biol Chem 281:28687–28698

Ruegsegger U, Beyer K, Keller W (1996) Purification and characterization of human cleavage factor Im involved in the 3' end processing of messenger RNA precursors. J Biol Chem 271:6107–6113

Ruegsegger U, Blank D, Keller W (1998) Human pre-mRNA cleavage factor Im is related to spliceosomal SR proteins and can be reconstituted in vitro from recombinant subunits. Mol Cell 1:243–253

Ryan K, Calvo O, Manley JL (2004) Evidence that polyadenylation factor CPSF-73 is the mRNA 3' processing endonuclease. RNA 10:565–573

Sanfacon H, Brodmann P, Hohn T (1991) A dissection of the cauliflower mosaic virus polyadenylation signal. Genes Dev 5:141–149

Simpson CG, Jennings SN, Clark GP, Thow G, Brown JW (2004) Dual functionality of a plant U-rich intronic sequence element. Plant J 37:82–91

Simpson GG, Dijkwel PP, Quesada V, Henderson I, Dean C (2003) FY is an RNA 3' end-processing factor that interacts with FCA to control the Arabidopsis floral transition. Cell 113:777–787

Simpson GG, Filipowicz W (1996) Splicing of precursors to mRNA in higher plants: mechanism, regulation and sub-nuclear organisation of the spliceosomal machinery. Plant Mol Biol 32:1–41

Skadsen RW, Knauer NS (1995) Alternative polyadenylation generates three low-pI alpha-amylase mRNAs with differential expression in barley. FEBS Lett 361:220–224

Sorek R, Shamir R, Ast G (2004) How prevalent is functional alternative splicing in the human genome? Trends Genet 20:68–71

Stoilov P, Rafalska I, Stamm S (2002) YTH: a new domain in nuclear proteins. Trends Biochem Sci 27:495–497

Tacahashi Y, Helmling S, Moore CL (2003) Functional dissection of the zinc finger and flanking domains of the Yth1 cleavage/polyadenylation factor. Nucleic Acids Res 31:1744–1752

Tang G, Zhu X, Gakiere B, Levanony H, Kahana A, Galili G (2002) The bifunctional LKR/SDH locus of plants also encodes a highly active monofunctional lysine-ketoglutarate reductase using a polyadenylation signal located within an intron. Plant Physiol 130:147–154

Tantikanjana T, Nasrallah ME, Stein JC, Chen CH, Nasrallah JB (1993) An alternative transcript of the S locus glycoprotein gene in a class II pollen-recessive self-incompatibility haplotype of *Brassica oleracea* encodes a membrane-anchored protein. Plant Cell 5:657–666

Tian B, Hu J, Zhang H, Lutz CS (2005) A large-scale analysis of mRNA polyadenylation of human and mouse genes. Nucleic Acids Res 33:201–212

Tian B, Pan Z, Lee JY (2007) Widespread mRNA polyadenylation events in introns indicate dynamic interplay between polyadenylation and splicing. Genome Res 17:156–165

Touriol C, Morillon A, Gensac MC, Prats H, Prats AC (1999) Expression of human fibroblast growth factor 2 mRNA is post-transcriptionally controlled by a unique destabilizing element present in the 3'-untranslated region between alternative polyadenylation sites. J Biol Chem 274:21402–21408

Tzafrir I, Pena-Muralla R, Dickerman A, Berg M, Rogers R, Hutchens S, Sweeney TC, McElver J, Aux G, Patton D, Meinke D (2004) Identification of genes required for embryo development in Arabidopsis. Plant Physiol 135:1206–1220

Valentini SR, Weiss VH, Silver PA (1999) Arginine methylation and binding of Hrp1p to the efficiency element for mRNA 3'-end formation. RNA 5:272–280

Venkataraman K, Brown KM, Gilmartin GM (2005) Analysis of a noncanonical poly(A) site reveals a tripartite mechanism for vertebrate poly(A) site recognition. Genes Dev 19:1315–1327

Wahle E, Ruegsegger U (1999) 3'-End processing of pre-mRNA in eukaryotes. FEMS Microbiol Rev 23:277–295

Wallace AM, Dass B, Ravnik SE, Tonk V, Jenkins NA, Gilbert DJ, Copeland NG, MacDonald CC (1999) Two distinct forms of the 64,000 Mr protein of the cleavage stimulation factor are expressed in mouse male germ cells. Proc Natl Acad Sci USA 96:6763–6768

Wallace AM, Denison TL, Attaya EN, MacDonald CC (2004) Developmental distribution of the polyadenylation protein CstF-64 and the variant tauCstF-64 in mouse and rat testis. Biol Reprod 70:1080–1087

Wu C, Alwine JC (2004) Secondary structure as a functional feature in the downstream region of mammalian polyadenylation signals. Mol Cell Biol 24:2789–2796

Wu L, Ueda T, Messing J. (1993) 3'-end processing of the maize 27 kDa zein mRNA. Plant J 4:535–544

Wu L, Ueda T, Messing J (1994) Sequence and spatial requirements for the tissue- and species-independent 3'-end processing mechanism of plant mRNA. Mol Cell Biol 14:6829–6838

Xiao YL, Smith SR, Ishmael N, Redman JC, Kumar N, Monaghan EL, Ayele M, Haas BJ, Wu HC, Town CD (2005) Analysis of the cDNAs of hypothetical genes on Arabidopsis chromosome 2 reveals numerous transcript variants. Plant Physiol 139:1323–1337

Xing H, Mayhew CN, Cullen KE, Park-Sarge OK, Sarge KD (2004) HSF1 modulation of Hsp70 mRNA polyadenylation via interaction with symplekin. J Biol Chem 279:10551–10555

Xu R, Ye X, Quinn Li Q (2004) AtCPSF73-II gene encoding an Arabidopsis homolog of CPSF 73 kDa subunit is critical for early embryo development. Gene 324:35–45

Xu R, Zhao H, Dinkins RD, Cheng X, Carberry G, Li QQ. (2006) The 73 kD subunit of the cleavage and polyadenylation specificity factor (CPSF) complex affects reproductive development in Arabidopsis. Plant Mol Biol 61:799–815

Yan J, Marr TG. 2005. Computational analysis of 3'-ends of ESTs shows four classes of alternative polyadenylation in human, mouse, and rat. Genome Res 15:369–375

Yao Y, Song L, Katz Y, Galili G. (2002) Cloning and characterization of Arabidopsis homologues of the animal CstF complex that regulates 3' mRNA cleavage and polyadenylation. J Exp Bot 53:2277–2278

Zarudnaya MI, Kolomiets IM, Potyahaylo AL, Hovorun DM (2003) Downstream elements of mammalian pre-mRNA polyadenylation signals: primary, secondary and higher-order structures. Nucleic Acids Res 31:1375–1386

Zhao J, Hyman L, Moore C (1999) Formation of mRNA 3' ends in eukaryotes: mechanism, regulation, and interrelationships with other steps in mRNA synthesis. Microbiol Mol Biol Rev 63:405–445

Zhuang T, Kashiwabara S, Noguchi J and Baba T (2004) Transgenic expression of testis-specific poly(A) polymerase TPAP in wild-type and TPAP-deficient mice. J Reprod Dev 50:207–213

State of Decay: An Update on Plant mRNA Turnover

D.A. Belostotsky

Contents

Introduction 180
mRNA Degradation Pathways: An Overview 181
Plant mRNA Decay Pathways 183
   mRNA Decay-Initiating Events and Directionality of Decay 184
   Deadenylation 184
   Decapping 185
   Plant NMD 186
   Degradation of mRNA Fragments Resulting from RISC-Mediated Cleavage 187
   Degradation of mRNA via Depurination: A Unique Case of Pokeweed
      Antiviral Protein 187
Insights into Plant mRNA Decay from Transcriptome Profiling Studies 188
   Global View of Unstable mRNAs in WT Plants 188
   Transcriptome Profiling of *xrn4* Mutants 189
   Transcriptome Profiling of *parn* Mutant 190
   Arabidopsis Exosome 190
   Regulated mRNA Decay in Plants 192
   Outlook 194
References 195

Abstract Proper degradation of plant messenger RNA is crucial for the maintenance of cellular and organismal homeostasis, and it must be properly regulated to enable rapid adjustments in response to endogenous and external cues. Only a few dedicated studies have been done so far to address the fundamental mechanisms of mRNA decay in plants, especially as compared with fungal and mammalian model systems. Consequently, our systems-level understanding of plant mRNA decay remains fairly rudimentary. Nevertheless, a number of serendipitous findings in recent years have reasserted the central position of the regulated mRNA decay

---

D.A. Belostotsky
School of Biological Sciences, University of Missouri-Kansas City, Kansas City,
MO 64110, USA
e-mail: belostotskyd@umkc.edu

in plant physiology. In addition, the meteoric rise to prominence of the plant small RNA field has spawned a renewed interest in the general plant mRNA turnover pathways. Combined with the advent of widely accessible microarray platforms, these advances allow for a renewed hope of rapid progress in our understanding of the fundamental rules governing regulated mRNA degradation in plants. This chapter summarizes recent findings in this field.

Introduction

Messenger RNA stability is one of the key parameters that determine the level of expression of all genes at steady-state conditions, and it plays a particularly important role in rapid adjustments of the cellular status in response to external and endogenous signals. A number of excellent reviews are available that discuss the fundamental aspects of mRNA stability (e.g., Coller and Parker 2004; Meyer et al. 2004; Ross 1995; Wilusz et al. 2001), and several articles are specifically devoted to mRNA stability in plants (Cheng and Chen 2004; Fedoroff 2002; Gutierrez et al. 1999). Therefore, this chapter focuses mostly on the recent findings in this area not covered by earlier reviews. Organellar and most of the RNAi-related topics have been reviewed recently (Bollenbach et al. 2004; Vaucheret 2006; Vazquez 2006) and are beyond the scope of this chapter.

Despite the rapid progress of plant and particularly Arabidopsis functional genomics and systems biology, studies of plant mRNA stability are still lagging behind other systems, such as mammalian and fungal model organisms. However, this state of affairs is likely to change in the next few years, because of several factors. First, there has been a great deal of renewed interest in RNA metabolism in plants in general, mainly propelled by the small RNA field. Second, new and powerful tools have become available (and moreover widely accessible) to many researchers, such as whole genome oligonucleotide microarrays, powerful bioinformatics software packages, as well as the versatile inducible and repressible expression systems. Third, a number of the mutant screen-based studies in diverse areas of plant biology during the past few years fortuitously identified several key mRNA turnover factors. Thus there are good reasons to anticipate a renaissance in the field of plant mRNA stability.

One technical note is also warranted. Most discussions of the quantitative studies of mRNA decay make certain implicit assumptions and/or have limitations that should be kept in mind. Typically, the mRNA decay measurements that are reported in the literature assume a single-hit model, so that the rate of change in mRNA concentration at any time point (dC/dt) is assumed to follow first-order kinetics. While more rigorous (and complicated) treatments are available that account for the multistep nature of mRNA degradation (Cao and Parker 2001, 2003), they are rarely used. Another caveat concerns the widespread practice of measuring the mRNA decay rates by following the disappearance of mRNA after inhibiting transcription by pharmacological means. However, one should keep in mind that prolonged treatment with global transcriptional inhibitors (such as actinomycin D, cordycepin, or DRB) can lead to secondary effects, for example, due to the depletion of the mRNAs that encode regulatory factors and effectors of mRNA stability. Therefore, such measurements are most valid at the early time points of treatment, that is, for the most unstable messages. On the other hand, gene-specific transcriptional inhibition bypasses these sorts of artifacts, but requires engineering of specialized reporter constructs. Moreover, the gene-specific transcriptional inhibition achievable with even the most sophisticated regulated promoter systems is rarely complete.

mRNA Degradation Pathways: An Overview

Eukaryotic mRNAs are generally believed to undergo degradation through a defined sequence of steps, that is, along specific *pathways*. Two major pathways have been described for degradation of normal mRNAs, that is, those without any structural defects (Fig. 1). The first pathway, termed deadenylation-dependent

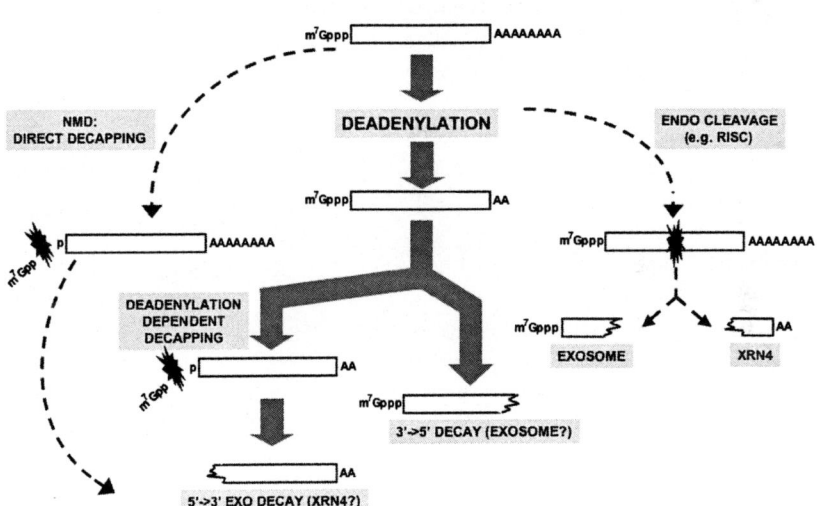

Fig. 1 Summary of eukaryotic mRNA decay pathways, with emphasis on known mRNA decay events and factors from plants. Main turnover pathways for degradation of mRNAs without any structural defects are deadenylation-dependent decapping and deadenylation-dependent exosomal decay (*thick gray arrows*). The role of XRN4 in the deadenylation-dependent decapping pathway in plants may be minor, and the role of the plant exosome in deadenylation-dependent 3'-5' decay has not been defined (*question marks*). The mechanism of NMD in plants is not known. NMD proceeds via direct decapping in yeast (*dashed arrow* on the left) but is initiated by endonucleolytic cleavage in flies. Endonucleolytic mRNA cleavage by RISC in plants is followed by exosomal and XRN4-mediated decay of upstream and downstream fragments, respectively; products of other endonucleolytic mRNA cleavage events may also be degraded similarly.

decapping, is initiated by the shortening of the 3'-poly(A) tail by one of several deadenylating enzymes. Following this step, decapping enzyme removes the 5'-m^{7}G cap and thereby clears the access for the processive 5'-3' exoribonuclease of the Xrn1 family. Alternatively, mRNA can be degraded after deadenylation in the 3'-5' direction by the exosome complex (discussed in detail below). The relative contribution of these two mRNA termini-initiated pathways to the overall turnover of bulk mRNA is not completely clear even in yeast (see, e.g., He et al. 2003) and, moreover, can be species specific. Less frequently, mRNA decay can be initiated by internal endonucleolytic cleavage(s), for example by miRNA- or siRNA-programmed RISC. Experimental evidence suggests that after the internal cleavage, the 5' fragment can be degraded in the 3'-5' direction by exosome and the 3' fragment degraded in the 5'-3' direction by Xrn1. In addition, several variant degradation pathways have evolved to detect and destroy various types of aberrant mRNAs. The nonsense-mediated decay (NMD) eliminates transcripts containing premature stop codons, while the nonstop decay degrades the mRNAs that lack a termination codon altogether. Finally, a so-called no-go decay pathway acts to eliminate the translationally stalled mRNAs.

This brief overview circumscribes the three access points for the exonucleases that processively degrade the body of mRNA (3' end, 5' end or internal cleavage site). It is notable that for normal mRNAs, not only the 3' end but also the 5' end-initiated route as well requires the shortening and/or complete removal of the poly(A) tail (hereafter collectively termed "deadenylation"). This implies communication between the 5' and the 3' ends of mRNA, which is also reflected in the switch of the site of action during deadenylation-dependent decapping pathway from the 3' end to the 5' end. The mechanistic basis of this communication largely resides in the physical interaction of the poly(A) binding protein (PABP) with the translation initiation factor eIF4G (Jacobson 1996; Kahvejian et al. 2001; Wells et al. 1998). Deadenylation eliminates the binding site for the PABP on the mRNA, leading to destabilization of the interactions of eIF4G and its partner, cap binding protein eIF4E, with the 5'-cap, and hence facilitating the mRNA decapping.

Deadenylation often serves as a rate-determining step of the mRNA decay process as a whole, and computational modeling of mRNA decay in yeast suggests that altering the rate of deadenylation is a highly powerful way to manipulate the overall steady-state level of a given transcript (Cao and Parker 2001). Deadenylation is also subject to regulation. Rapid mRNA deadenylation is often dependent on specific, portable sequence determinants found in the mRNA 3'-UTR, such as AU-rich elements in mammals (Chen et al. 1995) or DST sequences in plants (Gil and Green 1996; Sullivan and Green 1996). Such elements most likely act as binding sites for specific *trans*-acting factors (sequence-specific RNA binding proteins). Binding of such factors can modulate deadenylation in several ways: via direct stimulation and/or recruitment of deadenylases, via altering mRNA secondary structure, via displacing or remodeling of protecting RNA binding protein(s), for example, PABP, or by even targeting PABP for proteolytic degradation. Moreover, deadenylation can be also regulated by extracellular stimuli acting through signal transduction pathways (Vasudevan and Peltz 2001; Winzen et al. 1999). Requirement for deadenylation can be bypassed in certain situations, such as during NMD in yeast (Muhlrad and Parker 1994). In this case, decapping occurs without prior deadenylation, triggered by an abnormal RNP structure resulting from improper termination at a premature stop codon.

One recent and exciting development in the field of mRNA turnover has been the realization that its reactions are spatially compartmentalized, rather than randomly distributed. First, many (but not all) of the enzymes of mRNA turnover in yeast and mammalian cells colocalize in a small number of discrete cytoplasmic processing bodies (P-bodies, reviewed by Eulalio et al. 2007). Second, mRNA decay intermediates that are artificially stabilized by introduction of a *cis*-acting structural block that inhibits exonucleases likewise accumulate in the P-bodies. Third, mutating the key 5'-3' exonucleolytic enzyme in yeast (Xrn1) leads to both accumulation of mRNA in P-bodies and a dramatic change in the P-body morphology (Sheth and Parker 2003). P-bodies also represent a physical embodiment of an important and widespread cross talk between the processes of mRNA decay and translational control, for example, translationally inhibited transcripts that dissociate from the polysomes likewise move to the P-bodies (Coller and Parker 2005; Teixeira et al. 2005). In addition, P-bodies are strongly implicated in miRNA-mediated gene silencing in mammalian cells (see, e.g. Pillai et al. 2005). Among the unresolved issues in P-body biology are the relative contributions of localized and free cytoplasmic mRNA degradation to the overall flux of turnover, the exact relationship between the P-bodies and related stress granules, as well as the dynamics of the factors and mRNA substrates exchange in and out of P-bodies (Kedersha et al. 2005). It is difficult to tell, for example, whether the P-bodies represent reasonably permanent, pre-existing structures that become populated with the mRNA substrates and degradative enzymes during decay, or result predominantly from de novo formation.

Finally, a cautionary note is warranted concerning the seductively elegant view of defined mRNA decay pathways occurring as linear sequences of discrete steps. Complex macromolecular processes usually possess a certain degree of inbuilt stochasticity, for example, as has been recently shown for transcription (Chubb et al. 2006), and it is likely that mRNA turnover is no exception. However, because very little is known about plant mRNA decay pathways, for the purpose of this discussion we will also follow this simplified deterministic approach, and only mention the few instances where randomized order of events is supported by experimental evidence.

Plant mRNA Decay Pathways

Information on the pathways of plant mRNA decay remains fragmentary, and in no case has the complete sequence of mRNA degradative events been worked out. Hence, the ensuing overview focuses mostly on the key steps of plant mRNA degradation, considered separately.

mRNA Decay-Initiating Events and Directionality of Decay

It is commonly assumed that both 5'-3' and 3'-5' mRNA decay pathways are operating in plant cells, but in fact there are very few examples in which the directionality and the sequence of steps of decay have been at least partially delineated. This is largely due to the absence of reliable methods of capturing mRNA decay intermediates. In yeast, this is commonly accomplished by inserting into the reporter mRNA a polyG tract that forms G-quartet structure and hence blocks both 5'-3' and 3'-5' exonucleolytic decay, thus allowing visualization of the degradation intermediates. For poorly understood reasons, this approach has not been fruitful in plant systems (Gutierrez et al. 1999). However, in a few exceptional cases the degradation intermediates in plants are detectable. One example is the *PHYA* mRNA in oat, where naturally occurring incompletely degraded intermediates can be visualized by Northern blotting (Higgs and Colbert 1994). Their structure suggests that the *PHYA* mRNA decay most likely proceeds in both 5'-3' and 3'-5' directions (although alternative interpretations cannot be ruled out). On the other hand, the degradation of the soybean Rubisco *SRS4* mRNA is initiated by endonucleolytic cleavages, when assayed in vitro in soybean as well as in *Arabidopsis* extracts (Tanzer and Meagher 1995). Importantly, the pattern of cleavages appears identical in vivo, indicating that the in vitro conditions faithfully recapitulate the physiological degradation pathway. SRS4 mRNA cleavage events occur stochastically and are contingent on neither prior mRNA deadenylation nor decapping. The cut sites likely act as entry points for the subsequent exonucleolytic 3'-5' and 5'-3' removal of the upstream and downstream cleavage products, respectively. Two other well-understood examples of mRNA decay initiated by endonucleolytic cleavages involve regulated cotranslational degradation of Arabidopsis CGS mRNA as well as the endonucleolytic mRNA cleavage by miRNA-programmed RISC, as discussed below.

Deadenylation

Removal of the poly(A) tail is often the initiating and rate-limiting step of mRNA decay in mammals and yeast, but this has not been proven in plants. In fact, for the most part the role of deadenylation during mRNA decay in higher plants has been inferred, rather than directly demonstrated. In pollinated tobacco pistils, some stylar mRNAs undergo poly(A) shortening, which in some but not all cases leads to the degradation of the body of the mRNA (Wang et al., 1996). The cis-acting signal for the rice αAmy3 α-amylase mRNA destabilization has been mapped to its 3'-UTR, and likely acts through accelerated deadenylation (Chan and Yu 1998); the poly(A) tail length of the αAmy3 is also modulated during development (Lue and Lee 1994).

One of the factors responsible for plant mRNA deadenylation, a broadly conserved poly(A) ribonuclease PARN (AtPARN), has been characterized (Reverdatto et al. 2004). The null allele of AtPARN is embryo lethal. Interestingly, this arrest of embryogenesis is associated with elongation of poly(A) tails on only some, but not all, mRNAs in embryos. This finding, as well as the results of in vitro assays with recombinant enzyme and mutants thereof, support the notion that it is a bona fide deadenylase, but they also indicate a complex division of labor and/or redundancy relationships among the multiple plant deadenylation enzymes. In fact, the essentiality of the AtPARN is remarkable given that deadenylases in Arabidopsis are potentially represented by as many as 19 distinct proteins: 2 PARN-like, 6 CCR4, and 11 POP2 homologs. The CCR4 and POP2 proteins act as a heterodimer and play the most dominant role in mRNA deadenylation in most species (Tucker et al. 2001; Yamashita et al. 2005). On the other hand, PARN has been implicated in modulating the mRNA poly(A) tail length during *Xenopus* oocyte maturation (Kim and Richter 2006) and in deadenylation of select mammalian mRNAs (Lai et al. 2003; Tran et al. 2004), but it has not been shown to make a significant contribution to mRNA turnover in somatic cells. Moreover, PARN is totally absent from *Drosophila* and *S. cerevisiae* genomes. Overall, the identity of the enzyme(s) responsible for the bulk mRNA deadenylation in plants remains unresolved.

Decapping

Decapping of an mRNA molecule irreversibly commits it to fast decay, and therefore it plays a central role in the 5'-3' directed mRNA degradation pathway. Decapping of mRNA has additional important consequences, including inhibition of translation (Sonenberg 1988), and also may possibly facilitate the recruitment of RNA-dependent RNA polymerases (Axtell et al. 2006; Belostotsky 2004; Gazzani et al. 2004). Decapping of mRNA is catalyzed by the conserved enzyme Dcp2 in association with Dcp1 and, in mammals but not yeast, another protein, HEDLS (Fenger-Gron et al. 2005; Lykke-Andersen and Wagner 2005; van Dijk et al. 2002; Wang et al. 2002). Mammalian Dcp2 acts not only as the catalytic subunit of the complex but is also responsible for the interactions with the RNA-binding recruiting factors, such as TTP (Fenger-Gron et al. 2005). On the other hand, decapping by a distinct scavenger decapping enzyme (DcpS) takes place in the context of the 3'-5' mRNA decay pathway (Liu et al. 2002). Accordingly, DcpS has a pronounced preference for the short capped oligonucleotides (as opposed to full-length capped mRNA, which is preferred by Dcp2) (Gu et al. 2004). Plants apparently lack DcpS, indicating that the terminal removal of the 5'-cap in the course of 3'-5' decay is either unnecessary or carried out by a different activity.

An unexpected entry point into the dissection of plant mRNA decapping apparatus arose from a screen for vein patterning defects that culminated in the identification of *VCS (VARICOSE)* gene, encoding a homolog of HEDLS (Deyholos et al. 2003). Subsequent studies led to isolation of plant homologs of Dcp2 and Dcp1 (Xu et al. 2006; Goeres et al. 2007). Decapping assays with recombinant proteins as well as protein-protein interaction studies provided unambiguous evidence that DCP2, DCP1, and VCS constitute the essential components of plant decapping complex. Null alleles of these factors severely inhibit plant development and are seedling lethal. Curiously, vascular patterning and differentiation is abnormal in all three mutants, although the mechanistic basis of this common phenotype remains unclear. A survey of changes in mRNA levels due to the inhibition of decapping in *dcp2*, *dcp1*, and *vcs* mutants showed a surprisingly high degree of specificity (Goeres et al. 2007). Only a subset of the sampled mRNAs overaccumulated (in the 5′-capped form, as expected), and to a different degree (Xu et al. 2006). This finding parallels the limited and highly specific effect of mutations in XRN4 (encoding the plant equivalent of the yeast and mammalian Xrn1, discussed below) on only a small number of mRNAs, and stands in contrast with much broader effects of the respective mutations in yeast (He et al. 2003). This suggests that the cast of players responsible for the bulk mRNA decay in plants is still very incomplete.

Plant NMD

For historical and idiosyncratic reasons, NMD in plants has been investigated in perhaps more detail than the degradation pathways of normal mRNA. NMD has evolved to detect and destroy the transcripts bearing premature termination codons (PTCs). NMD involves three interacting proteins, UPF1, UPF2, and UPF3, that are highly conserved, yet the rules of recognition of the NMD substrates are different in mammals vs. flies and yeast. In mammalian cells the stop codon is defined as PTC if it occurs >50 nt upstream from the exon-exon junctions that are marked by the deposition of a multiprotein exon junction complex (EJC; reviewed in Rehwinkel et al. 2006). The core components of EJC directly interact with the NMD effector proteins UPF1, -2, and -3 (Gehring et al. 2003). On the other hand, in yeast (Amrani et al. 2004) and flies (Behm-Ansmant et al. 2007) the PTC recognition occurs when the translational termination event takes place far upstream of the natural 3′-UTR and is recognized as abnormal because it precludes interactions between the terminating ribosome and PABP bound to the poly(A) tail. Hence, PTC recognition in yeast and *Drosophila* is independent of splicing and EJC.

Plant genomes encode homologs of the UPF proteins, and plant UPF1 and UPF3 have been directly implicated in NMD (Arciga-Reyes et al. 2006; Hori and Watanabe 2005; Yoine et al. 2006a, 2006b). Plant genomes encode many of the EJC components, as well (Belostotsky and Rose 2005); nevertheless, many observations of NMD in plants were made with intronless transcripts (Dickey et al. 1994; Petracek et al. 2000; van Hoof and Green 1996; Voelker et al. 1990). This is inconsistent with a mammalian-like mode of PTC recognition. Indeed, direct experimental probing showed that plant UPF1 is required for NMD of both spliced and unspliced mRNAs (Arciga-Reyes et al. 2006), and insertion of an intron 80 nt downstream of the stop codon does not destabilize the reporter transcript in transgenic plants (Rose 2004). On the other hand, agroinfiltration-based transient assays show that introns inserted sufficiently far downstream of the stop codon can trigger NMD, as do long 3′-UTRs (Kertesz et al. 2006). Studies of the waxy gene in rice also show evidence for a mechanistic link between splicing and NMD, although of a distinct nature from mammals, because the intron located upstream, rather than downstream, of PTC modulates NMD efficiency in this system (Isshiki 2001). It should be noted that only a limited number of NMD reporter constructs have been tested so far, either in transient or transgenic assays, while the long 3′-UTRs, as well as of introns in the 3′-UTRs of plant genes are quite common. Therefore, broader dedicated studies are needed to resolve this issue. One attractive strategy would be to use microarrays to investigate the potential endogenous targets of NMD in the Arabidopsis *upf* mutants. Such line of inquiry is even more exciting in light of the fact that Arabidopsis *upf1* and *upf3* mutants exhibit a range of unexpected vegetative and floral abnormalities, including jagged leaves, late flowering, fused flowers, and seedling lethality. This not only indicates that NMD pathway in plants has endogenous substrates, but also shows that their downregulation is essential for normal development. Moreover, the *upf1* mutants are also impaired in gene silencing by an inverted repeat transgene, suggesting an interesting intersection of NMD and silencing pathways (Arciga-Reyes et al. 2006). A parallel link has been reported in *C. elegans* (Domeier et al. 2000).

Degradation of mRNA Fragments Resulting from RISC-Mediated Cleavage

Downstream products resulting from the cleavage of mRNA by miRNA-programmed RISC bear an exposed uncapped 5′-PO_4 end and are degraded by Xrn4, the functional equivalent of the Xrn1 in yeast (Souret et al. 2004). However, only in about half of the examined miRNA targets the downstream mRNA cleavage products appeared to be sensitive to the Xrn4 activity (Souret et al. 2004). The basis of this selectivity in the Xrn4 substrate choice remains unknown. The matching upstream RISC cleavage products possess the 5′-cap but lack the poly(A) tail, and are likely to be decayed by the exosome complex. This has not yet been directly demonstrated for the miRNA-programmed RISC, but is clearly the case for the siRNA-programmed RISC in *Drosophila* (Orban and Izaurralde 2005) and *Chlamydomonas* (Ibrahim et al. 2006). Moreover, the exosome-mediated removal of the upstream cleavage product in *Chlamydomonas* is apparently stimulated by untemplated oligoadenylation by an unconventional poly(A) polymerase that is similar to the yeast enzymes Trf4/Trf5 (Ibrahim et al. 2006).

Degradation of mRNA via Depurination: A Unique Case of Pokeweed Antiviral Protein

The single-chain ribosome inactivating protein from pokeweed *Phytolacca americana* is a N-glycosidase that specifically depurinates the conserved sarcin/ricin loop in the 25S rRNA. PAP [for pokeweed antiviral protein, i.e. not to be confused with poly(A) polymerase] also possesses an antiviral activity that can be genetically uncoupled from rRNA depurination (Tumer et al. 1997). In addition, PAP destabilizes its own mRNA in vivo (Parikh et al. 2002). PAP is able to specifically and tightly bind to the mRNA 5'-cap and depurinate mRNA in vitro at multiple but specific positions downstream from the cap (Hudak et al. 2002; the cap itself is neither cleaved nor depurinated). While all of the above activities require the N-glycosidase active site, they can be mutationally separated; for example, the RNA binding domain mutant of PAP can depurinate rRNA but is unable to down-regulate its own mRNA (Parikh et al. 2002). A truncated variant of the ribosomal protein L3 delta suppresses the autoregulation of PAP mRNA in vivo: In its presence, the PAP mRNA is stabilized but the L3 delta mRNA is destabilized (Di and Tumer 2005). The mechanistic basis of this phenomenon remains to be elucidated, as are the reasons for preferential destabilization of the PAP mRNA over other cellular mRNA in vivo and the downstream events following the mRNA depurination. The depurination-initiated mRNA decay mechanism so far appears to be unique to this agriculturally important protein.

Insights into Plant mRNA Decay from Transcriptome Profiling Studies

Experience has shown that identification of novel plant mRNA decay factors by forward genetic screens is extremely difficult, likely because of their redundancy and/or lethality (Johnson et al. 2000). On the other hand, microarray-based profiling of mutations in the known mRNA decay genes not only promises rapid advances but also offers a systems-level view of the landscape of plant mRNA degradation as a whole. Moreover, microarrays can also be used for global monitoring of mRNA decay rates in WT plants in combination with treatments with global transcriptional inhibitors. Several such studies have been conducted to date.

Global View of Unstable mRNAs in WT Plants

Gutierrez et al. used a partial cDNA array (covering ~7,800 annotated genes) and an oligo(dT)-primed target from Arabidopsis seedlings treated with cordycepin, a global transcriptional inhibitor that acts as an elongating chain terminator (Gutierrez et al. 2002). Notwithstanding the aforementioned caveats that are applicable to all global inhibitor studies, this analysis provides a valuable glimpse of Arabidopsis "mRNA degradome." About 1% of all messages appear unstable, decaying with half-lives of <60 min. A comparison of their structural features, and specifically of their 3' UTRs, does not reveal any obvious common clues to the defining features of instability, suggesting that many diverse determinants are at play here, rather than just a few common ones. mRNAs encoding transcription factors were overrepresented among the unstable messages. This was not unexpected, because instability of transcription factor mRNAs allows for a rapid reprogramming of cellular gene expression profiles in response to changing conditions. In addition, a disproportionally large fraction of unstable transcripts appear to be regulated by touch and by circadian rhythms. Touch responses must be rapid by their very nature, and cycling of mRNA levels implicit in circadian rhythms likewise demands fast and efficient clearance of the preexisting transcripts; thus both of these findings make immediate biological sense. This study remains virtually the only attempt to date to define the plant mRNA turnover rates globally, and follow-ups are sure to be informative (Narsai et al. 2007).

Transcriptome Profiling of xrn4 Mutants

The Arabidopsis homolog of Xrn1, the major exonuclease of the 5'-3' mRNA decay pathway in yeast, has been isolated and characterized in four independent studies. First, Xrn4 was the focus of a dedicated study by the Green laboratory (Kastenmayer and Green 2000; Souret et al. 2004). They have shown that Xrn4 functions in the cytoplasm and complements at least some of the 5'-3' mRNA decay defects of the yeast *xrn1* strain, as well as characterized its substrates with partial cDNA microarray, including some miRNA-targeted transcripts as discussed above. Independently, the Sablowski group (Gazzani et al. 2004) zeroed in on XRN4 in the course of a suppressor screen to find mutations that reverse the inhibition of cotyledon and leaf development caused by ectopic overexpression of glucocorticoid-regulated transcription factor STM-GR, which activates meristematic developmental program in response to dexamethasone. The reversion of the STM-GR dependent phenotype in the *xrn4* background was due to an increased degradation of the *STM-GR* mRNA via RNA silencing pathway that required RNA-dependent RNA polymerase RDR6. Hence, XRN4 likely degrades the RNA species that, if allowed to accumulate, becomes an efficient template for RdRP. This interpretation is also consistent with the two-hit model of the RdRP-dependent siRNA biogenesis that was recently proposed (Axtell et al. 2006). Third, the Ecker laboratory has independently isolated XRN4 as a result of positional cloning of ethylene-insensitive gene *ein5*. Consistent with this finding, *ein5* plants show reduced induction of ethylene-responsive genes (Olmedo et al. 2006). Similar results were obtained independently by Genschik and colleagues (Potuschak et al. 2006).

Yet despite its involvement in diverse aspects of plant biology, knocking out Arabidopsis XRN4 causes no apparent phenotype under normal growth conditions and does not affect the decay rate of the known unstable transcripts (Souret et al. 2004). Although subsequent experiments identified several internally cleaved mRNAs as Xrn4 substrates as discussed above, the total number of mRNAs upregulated in the mutant plants out of 15,000 ESTs represented on the array was shockingly small (14 genes). The Ecker lab used whole-genome tiling microarrays to undertake the comprehensive molecular characterization of *xrn4* deficiency (Olmedo et al. 2006). Consistent with the identification of XRN4 via an ethylene-insensitive mutation, mutant plants show reduced induction of ethylene-responsive genes. Two of the upregulated transcripts, *EBF1* and *EBF2*, encode the F-box proteins that target for degradation the main transcriptional activator of ethylene responses (EIN3). This finding provides an elegant explanation for the role of Xrn4 in ethylene responses. More significant for the purposes of this review is the fact that only ~1.5% of all mRNAs were upregulated in *xrn4* mutants. This number is comparable to the number of mRNAs upregulated in the *ein2* plants that are impaired in the function of a membrane protein of ethylene signal transduction pathway that has no general effect on gene expression. The results of the two independent studies showing that only a small fraction of the transcriptome is affected by XRN4 function are in stark contrast with the situation in yeast, where the 5'-3' decay by Xrn1 is the major pathway of mRNA turnover and its inactivation severely impairs growth. Hence, the questions of the significance of 5'-3' decay for the bulk plant mRNA turnover, as well as the roles of the plant XRN homologs in it may be well worth revisiting.

Transcriptome Profiling of parn Mutant

Hirayama and colleagues (Nishimura et al. 2005) independently isolated the gene encoding mRNA deadenylase AtPARN in the course of the study of the mutation in Arabidopsis that causes hypersensitivity of seed germination to ABA. Interestingly, they found that the mRNA encoding AtPARN itself is upregulated by ABA, salt, and osmotic stress, suggesting that its deadenylase function becomes particularly important under such conditions. They also conducted a transcriptome study of the AtPARN partial loss of function mutant, using a cDNA array covering some 7,000 genes. It showed that AtPARN deficiency results in upregulated expression of a suite of genes that is similar, though not identical, to the gene set that is induced by ABA. Together, these results suggest that one of the functions of AtPARN may be to fine-tune the strength of the transcriptome response to the stress that is induced by ABA and related conditions. It is also notable that only some ~150 genes were overexpressed (using a rather generous 1.5-fold cutoff). In light of the potentially extensive redundancy of plant deadenylases such a relatively minor effect is not as surprising as in the case of XRN4. On the other hand, the fact that a comparable number of genes were also underexpressed in the mutant plants raises a concern that secondary effects may have obscured the true significance of the AtPARN activity.

Arabidopsis Exosome

The problem of secondary effects can be at least partially bypassed by employing conditional mutants. As opposed to the constitutive loss of function alleles, conditional alleles can be examined during early time points after the phenotype is triggered, that is, before the contribution of secondary effects to the overall changes in the transcriptome becomes significant. This strategy was adopted by the author's laboratory during the study of Arabidopsis exosome (Chekanova et al. 2007).

Exosome is a nuclear and cytoplasmic macromolecular complex that mediates essential reactions of RNA maturation and degradation with 3'-5' directionality. This remarkable macromolecular machine is universally essential for viability and carries out three very different kinds of reactions: (a) degradation of some RNA species, including messenger RNAs in the course of their normal turnover, (b) surgically-precise processing of various stable structural RNAs in the course of their biogenesis, and (c) proofreading of several types of RNA, such that the molecules that are not properly processed, folded, and/or assembled into the appropriate RNP particles are identified and destroyed. During such proofreading reactions, exosome interfaces with the auxiliary factor called TRAMP (for Trf4/5-Air1/2-Mtr4 polyadenylation complex) whose Trf4 (or Trf5) subunit adds an oligo(A) tail to the selected RNA substrates, which marks them for degradation (reviewed in Houseley et al. 2006). The versatility of exosome function necessitates sophisticated substrate discrimination mechanisms that are understood in only a few select cases. A comprehensive definition of the substrates of exosome has not been described in multicellular eukaryotes, and is incomplete even in yeast.

Nuclear and cytoplasmic forms of yeast exosome share 10 common components. The six RNase PH-domain proteins of the exosome core, Rrp41, Rrp42, Rrp43, Rrp45, Rrp46, and Mtr3, are organized into a hexameric ring that is capped on one side by the three subunits containing S1 and KH domains (Rrp40, Rrp4, and Csl4) that contribute to the structural integrity of the complex and to its RNA binding activity. In yeast, every subunit of the exosome core is essential for viability, and genetic depletion of any subunit triggers highly similar spectra of defects in RNA processing (Allmang et al. 1999a, 1999b). The recent X-ray crystallographic study of human exosome indicates that its structural integrity obligatorily depends on the presence of all nine core subunits of the complex (Liu et al. 2006). However, despite the presence of the six RNase PH domains in the yeast exosome core, the role of these proteins appears to be structural rather than catalytic (Dziembowski et al. 2007). On the other hand, plant exosome is special in possessing at least one catalytically active RNase PH domain subunit, RRP41 (Chekanova et al. 2000).

To assess the contribution of the individual exosome subunits to its multiple functions, including cytoplasmic mRNA turnover (probably the most prominent degradative activity of exosome), we engineered an inducible RNAi of *RRP4* or *RRP41*, by expressing the segments of respective cDNAs as a pair of inverted repeats separated under the control of estradiol-regulated chimeric transactivator XVE (Zuo et al. 2000). Addition of estradiol rapidly induces expression of the hairpin RNAs and elicits posttranscriptional silencing of respective targets in vivo. Such genetic depletion of RRP4 or RRP41 by RNAi was deployed in conjunction with Affymetrix whole-genome tiling microarrays to globally identify the exosome-regulated RNA species. In the first series of experiments, oligo(dT) primed target was used to interrogate the array probes, which limited the focus of the study to the polyadenylated

RNA species. Nonetheless, it revealed numerous novel exosome substrates, including multiple stable structural RNA species, pri-miRNAs, tandem repeat-associated transcripts serving as siRNA precursors, as well as numerous noncoding RNAs including a novel class of noncoding transcripts colinear with the 5′ ends of known mRNAs. The abundance of the polyadenylated stable RNAs that become upregulated in the exosome-depleted seedlings indicates that plants widely employ polyadenylation- and exosome-mediated RNA quality control (Chekanova et al. 2007).

Three aspects of this tiling microarray study are most interesting in the context of this chapter. First, as in the case of XRN4 and PARN, only a small number of mRNAs were upregulated. However, this is probably because priming of the target with oligo(dT) in these experiments precluded the detection of deadenylated mRNA decay intermediates, and hence should not be interpreted as evidence against any major role of exosome in mRNA turnover (this is currently being addressed with random-primed microarray targets). Notably, intronless genes were overrepresented among the upregulated mRNAs, suggesting that many members of this group may correspond to processed pseudogenes, whose transcripts are degraded posttranscriptionally by a distinct mechanism that involves exosome-mediated deadenylation as well as decay. The precedent for exosome-mediated deadenylation is provided by the nonstop decay pathway in yeast, in which exosome first removes the poly(A) tail and then proceeds to degrade the rest of the message in the 3′-5′ direction, that is, without involvement of any specialized deadenylase (Frischmeyer et al. 2002; van Hoof et al. 2002). Second, some mRNAs showed subunit-specific responses. For example, glycosyltransferase At5g54060 mRNA is upregulated upon the depletion of RRP4 but not RRP41; conversely, β-galactosidase At5g56870 mRNA is upregulated upon the depletion of RRP41 but not of RRP4. Although the mechanistic basis of this specificity remains unclear, these examples are instructive in demonstrating functional nonequivalence of exosome subunits and are consistent with the distinct phenotypes of respective knockouts (*rrp41* null is female gametophyte lethal, while *rrp4* null is embryo lethal). Third, several mRNAs exhibited extensions beyond their annotated 3′ ends in the exosome-depleted seedlings, without any effect on their overall levels. Yeast exosome is known to degrade aberrant readthrough mRNAs that arise as a result of mutations in the 3′ end processing apparatus (Torchet et al. 2002) or natural readthrough events (Vasiljeva and Buratowski 2006). These observations indicate that such a mechanism is also employed in plant cells, although only for a limited number of transcripts. An interesting alternative possibility is that exosome participates in the 3′ end formation, rather than degradation, of some RNAP II transcription units.

Regulated mRNA Decay in Plants

Many plant genes have been shown, and many more are suspected, to be regulated at the level of mRNA of decay in response to various stimuli. Several such cases have been reviewed previously (Gutierrez et al. 1999), including pea ferredoxin

Fed-1 mRNA that is subject to a translation-coupled stabilization by light (Petracek et al. 1997) and also possesses a constitutive darkness instability element (Bhat et al. 2004), the rice αAmy3 destabilized by carbon source [high sucrose (Chan and Yu 1998; Sheu et al. 1996)], and soybean β-expansin transcript that is strongly stabilized in response to cytokinin (Downes and Crowell 1998). The stability of mRNA encoding a cell wall protein in common bean is sharply reduced in response to exposure to the fungal elicitor (Zhang and Mehdy 1994; Zhang et al. 1993) and correlates with binding of a 50-kDa protein to a U-rich site in the mRNA 3' UTR. Stability of the transcripts of light-harvesting, chlorophyll-binding (Lhcb) gene family in Arabidopsis and pea is regulated in response to a single blue light pulse and mediated by a compact (64 nt), portable *cis*-acting element in the mRNA 5'-UTR (Anderson et al. 1999). The destabilization-inducing pathway requires the function of phototropin PHOT1 and its interacting partner NPH3 (Folta and Kaufman 2003), but the mechanistic basis of accelerated decay remains unknown.

Degradation of cystathionine γ-synthase (CGS) in Arabidopsis represents an exceptionally beautiful mechanism of regulated mRNA decay. CGS1 mRNA, encoding an enzyme that catalyzes the first committed step of methionine biosynthesis, is subject to a negative feedback control by methionine and its derivatives. CGS1 downregulation in high methionine is coincident with the appearance of the 5'-truncated mRNA degradation product. Interestingly, the *cis*-acting determinant triggering the CGS1 mRNA decay resides in a short amino acid sequence segment of CGS1 protein called MTO. That the peptide, rather than the mRNA segment encoding it, acts as a decay determinant suggested that CGS1 mRNA degradation occurs cotranslationally (Chiba et al. 1999). This effect has been recapitulated in a wheat germ in vitro translation system responsive to *S*-adenosyl-l-methionine (SAM), a direct metabolite of methionine (Chiba et al. 2003). Subsequent toeprint analysis (a.k.a. primer extension inhibition) showed that SAM triggers translational arrest on the CGS1 mRNA template, with the MTO peptide stuck in the ribosome exit channel (Onouchi et al. 2005). This translational arrest then results in the degradation of the mRNA molecule on which the ribosome has stalled, by as yet unknown means and by an unknown nuclease. This example, in addition to its elegance as a biological regulatory mechanism, is also instructive in demonstrating the power of in vitro systems (generally almost unexploited) in the analyses of mRNA decay pathways.

Microarray analysis is another tool that undoubtedly should be applied more broadly to elucidate the global scope of the regulation of mRNA stability in plants. One instructive example stems from the study by Green and colleagues cited above that revealed the enrichment of circadian clock-controlled mRNAs among the unstable Arabidopsis transcripts. Subsequent follow-up studies revealed an even more complex picture (Lidder et al. 2005). Not only the CCL and SEN1 transcripts are regulated by circadian clock, but their decay rates vary depending on the time of day, so that mRNAs are more stable in the morning than in the afternoon. These differences in stability persist upon the transfer of plants into the continuous light, and hence must be truly circadian regulated and not simply be under the diurnal control by changing light conditions.

Furthermore, these mRNAs appear to be targeted by the mRNA degradation pathway that is mediated by a *cis*-acting instability determinant called DST. This conserved bipartite sequence, initially identified in the 3′-UTRs of small auxin up RNAs (SAURs) (McClure et al. 1989), remains the best characterized instability determinant in plants (Gil and Green 1996; Sullivan and Green 1996). Although the *trans*-acting factor(s) mediating DST action remain to be identified, the mutant defective in DST-mediated decay pathway is available (Johnson et al. 2000), and the half-life of CCL mRNA is altered in *dst1* mutant plants. Furthermore, the *dst1* mutant exhibits an altered circadian regulation at the whole plant level, as evidenced by changes in the oscillatory cotyledon movement, a classical circadian phenotype (Lidder et al. 2005). It will be interesting to identify those circadian regulated transcript(s) whose misregulated stability is responsible for this phenotype.

Outlook

Despite the continuing progress in studies of plant mRNA turnover, one major predicament of this field remains to be a lack of understanding of just how the "average" plant mRNA is turned over. A few exceptional cases that are better understood in mechanistic detail all fall under the rubric of "unusual." While this undoubtedly makes such systems biologically interesting, this obvious gap in the basic understanding of mRNA turnover cannot be ignored. Mutational inactivation of several of the "usual suspects" such as PARN and XRN4 resulted in surprisingly specific molecular signatures, as evidenced by microarray analyses. Likewise, inactivation of the decapping enzyme resulted in upregulation of only a subset of tested mRNAs. These results suggest that the often-cited yeast paradigm, according to which mRNA degradation occurs predominantly by deadenylation-dependent decapping, may not hold in plants. We may be surprised to encounter novel pathways and even plant-unique factors that contribute in significant ways to the regulation of plant mRNA stability. Among the important tasks for the future will be a more comprehensive global definition of the plant mRNA half-lives with whole-genome microarrays, in WT plants as well as in the mutants affected in the various factors of mRNA decay, along with the rigorous dissection of the sequence of events in the degradation of specific mRNA. Finally, we need deeper investigation of the compartments responsible for mRNA degradation reactions (such as P-bodies) by proteomic and cell biological approaches. The combination of rekindled interest in the mechanisms of mRNA turnover and the development of powerful technologies leads to the eager anticipation of rapid progress in this important field.

Acknowledgements Research on plant RNA metabolism in the author's laboratory is supported by USDA (Grant 2003-35304-13210), NSF (Grant MCB 0424651), and BARD (Grant US375605).

References

Allmang C, Kufel J, Chanfreau G, Mitchell P, Petfalski E, Tollervey D (1999a) Functions of the exosome in rRNA, snoRNA and snRNA synthesis. EMBO J 18:5399–5410

Allmang C, Petfalski E, Podtelejnikov A, Mann M, Tollervey D, Mitchell P (1999b) The yeast exosome and human PM-Scl are related complexes of $3'\rightarrow 5'$ exonucleases. Genes Dev 13:2148–2158

Amrani N, Ganesan R, Kervestin S, Mangus DA, Ghosh S, Jacobson A (2004) A faux 3'-UTR promotes aberrant termination and triggers nonsense-mediated mRNA decay. Nature 432:112–118

Anderson MB, Folta K, Warpeha K M, Gibbons J, Gao J, Kaufman LS (1999) Blue light-directed destabilization of the pea Lhcb1*4 transcript depends on sequences within the 5' untranslated region. Plant Cell 11:1579–1590

Arciga-Reyes L, Wootton L, Kieffer M, Davies B (2006) UPF1 is required for nonsense-mediated mRNA decay (NMD) and RNAi in Arabidopsis. Plant J 47:480–489

Axtell MJ, Jan C, Rajagopalan R, Bartel DP (2006) A two-hit trigger for siRNA biogenesis in plants. Cell 127:565–577

Behm-Ansmant I, Gatfield D, Rehwinkel J, Hilgers V, Izaurralde E (2007) A conserved role for cytoplasmic poly(A)-binding protein 1 (PABPC1) in nonsense-mediated mRNA decay. EMBO J 26:1591–1601

Belostotsky DA (2004) mRNA turnover meets RNA interference. Mol Cell 16:498–500

Belostotsky DA, Rose AB (2005) Plant gene expression in the age of systems biology: integrating transcriptional and post-transcriptional events. Trends Plant Sci 10:347–353

Bhat S, TangL, Krueger AD, Smith CL, Ford SR, Dickey LF, Petracek ME (2004) The Fed-1 (CAUU)4 element is a 5' UTR dark-responsive mRNA instability element that functions independently of dark-induced polyribosome dissociation. Plant Mol Biol 56:761–773

Bollenbach TJ, Schuster G, Stern DB (2004) Cooperation of endo- and exoribonucleases in chloroplast mRNA turnover. Prog Nucleic Acid Res Mol Biol 78:305–337

Cao D, Parker R (2003) Computational modeling and experimental analysis of nonsense-mediated decay in yeast. Cell 113:533–545

Cao D, Parker R (2001) Computational modeling of eukaryotic mRNA turnover. RNA 7:1192–1212

Chan MT, Yu SM (1998) The 3' untranslated region of a rice alpha-amylase gene functions as a sugar-dependent mRNA stability determinant. Proc Natl Acad Sci USA 95:6543–6547

Chen C-Y A, Xu N, Shyu A-B (1995) mRNA decay mediated by two distinct AU-rich elements from c-fos and granulocyte-macrophage colony-stimulating factor transcripts: different deadenylation kinetics and uncoupling from translation. Mol Cell Biol 15:5777–5788

Cheng Y, Chen X (2004) Posttranscriptional control of plant development. Curr Opin Plant Biol 7:20–25

Chekanova JA, Shaw RJ, Wills MA, Belostotsky DA (2000) Poly(A) tail-dependent exonuclease AtRrp41p from Arabidopsis thaliana rescues 5.8 S rRNA processing and mRNA decay defects of the yeast ski6 mutant and is found in an exosome-sized complex in plant and yeast cells. J Biol Chem 275:33158–33166

Chekanova JA, Gregory BD, Reverdatto SV, Chen H, Kumar R, Hooker T, Yazaki J, LI P, Skiba NP, Peng Q, Alonso JM, Brukhin V, Grossniklaus U, Ecker JR, Belostotsky DA (2007) Genome-wide high-resolution mapping of exosome substrates reveals hidden features in the *Arabidopsis* transcriptome. Cell 131: 1340–1353.

Chiba Y, Ishikawa M, Kijima F, Tyson RH, Kim J, Yamamoto A, Nambara E, Leustek T, Wallsgrove RM, Naito S (1999) Evidence for autoregulation of cystathionine gamma-synthase mRNA stability in arabidopsis. Science:1371–1374

Chiba Y, Sakurai R, Yoshino M, Ominato K, Ishikawa M, Onouchi H, Naito S (2003) *S*-adenosyl-l-methionine is an effector in the posttranscriptional autoregulation of the cystathionine gamma-synthase gene in Arabidopsis. Proc Natl Acad Sci USA 100:10225–10230

Chubb JR, Trcek T, Shenoy SM, Singer RH (2006) Transcriptional pulsing of a developmental gene. Curr Biol 16:1018–1025

Coller J, Parker R (2004) Eukaryotic mRNA decapping. Annu Rev Biochem 7:861–890

Coller J, Parker R (2005) General translational repression by activators of mRNA decapping. Cell 122:875–886

Deyholos MK, Cavaness GF, Hall B, King E, Punwani J, Van Norman J, Sieburth LE (2003) VARICOSE, a WD-domain protein, is required for leaf blade development. Development 130:6577–6588

Di R, Tumer NE (2005) Expression of a truncated form of ribosomal protein L3 confers resistance to pokeweed antiviral protein and the *Fusarium* mycotoxin deoxynivalenol. Mol Plant Microbe Interact 18:762–770

Dickey LF, Nguyen T, Allen GC, Thompson WF (1994) Light modulation of ferredoxin mRNA abundance requires an open reading frame. Plant Cell 6:1171–1176

Domeier ME, Morse DP, Knight SW, Portereiko M, Bass BL, Mango SE (2000) A link between RNA interference and nonsense-mediated decay in *Caenorhabditis elegans*. Science 289:1928–1931

Downes BP, Crowell DN (1998) Cytokinin regulates the expression of a soybean beta-expansin gene by a post-transcriptional mechanism. Plant Mol Biol 37:437–444

Dziembowski A, Lorentzen E, Conti E, Seraphin B (2007) A single subunit, Dis3, is essentially responsible for yeast exosome core activity. Nat Struct Mol Biol 14:15–22

Eulalio A, Behm-Ansmant I, Izaurralde E (2007) P bodies: at the crossroads of post-transcriptional pathways. Nat Rev Mol Cell Biol. 8:9–22

Fedoroff NV (2002) RNA-binding proteins in plants: the tip of an iceberg? Curr Opin Plant Biol 5:452–459

Fenger-Gron M, Fillman C, Norrild B, Lykke-Andersen J (2005) Multiple processing body factors and the ARE binding protein TTP activate mRNA decapping. Mol Cell 20:905–915

Folta KM, Kaufman LS (2003) Phototropin 1 is required for high-fluence blue-light-mediated mRNA destabilization. Plant Mol Biol 51:609–618

Frischmeyer PA, van Hoof A, O'Donnell K, Guerrerio AL, Parker R, Dietz HC (2002) An mRNA surveillance mechanism that eliminates transcripts lacking termination codons. Science 295, 2258–2261

Gazzani S, Lawrenson T, Woodward C, Headon D, Sablowski R (2004) A link between mRNA turnover and RNA interference in Arabidopsis. Science 306:1046–1048

Gehring NH, Neu-Yilik G, Schell T, Hentze MW, Kulozik AE (2003) Y14 and hUpf3b form an NMD-activating complex. Mol Cell 11:939–949

Goeres DC, Van Norman JMV, Zhang W, Fauver NA, Spencer ML, Sieburth LE (2007) Components of the *Arabidopsis* mRNA Decapping Complex Are Required for Early Seedling Development. The Plant Cell 19:1549–1564

Gil P, Green PJ (1996) Multiple regions of the Arabidopsis SAUR-AC1 gene control transcript abundance: the 3′ untranslated region functions as an mRNA instability determinant. EMBO J 15:1678–1686

Gu M, Fabrega C, Liu SW, Liu H, Kiledjian M, Lima CD (2004) Insights into the structure, mechanism, and regulation of scavenger mRNA decapping activity. Mol Cell 14:67–80

Gutierrez RA, Ewing RM, Cherry JM, Green PJ (2002) Identification of unstable transcripts in Arabidopsis by cDNA microarray analysis: rapid decay is associated with a group of touch- and specific clock-controlled genes. Proc Natl Acad Sci USA 99:11513–11518

Gutierrez R A, MacIntosh GC, Green PJ (1999) Current perspectives on mRNA stability in plants: multiple levels and mechanisms of control. Trends Plant Sci 4:429–438

He F, Li X, Spatrick P, Casillo R, Dong S, Jacobson A (2003) Genome-wide analysis of mRNAs regulated by the nonsense-mediated and 5′ to 3′ mRNA decay pathways in yeast. Mol Cell 12:1439–1552

Higgs DC, Colbert JT (1994) Oat phytochromeA mRNA degradation appears to occur via two distinct pathways. Plant Cell 6:1007–1019

Hori K, Watanabe Y (2005) UPF3 suppresses aberrant spliced mRNA in Arabidopsis. Plant J 43:530–540

Houseley J, Lacava J, Tollervey D (2006) RNA-quality control by the exosome. Nat Rev Mol Cell Biol 7:529–539

Hudak KA, Bauman JD, Tumer NE (2002) Pokeweed antiviral protein binds to the cap structure of eukaryotic mRNA and depurinates the mRNA downstream of the cap. RNA 8:1148–1159

Ibrahim F, Rohr J, Jeong W J, Hesson J, Cerutti H (2006) Untemplated oligoadenylation promotes degradation of RISC-cleaved transcripts. Science 314:1893

Isshiki M, Yamamoto Y, Satoh H, Shimamoto K (2001) Nonsense-mediated decay of mutant waxy mRNA in rice. Plant Physiol 125:1388–1398

Jacobson A (1996) Poly(A) metabolism and translation: the closed loop model. In Translational control, J. W. B. Hershey, Mathews, M.B., Sonenberg, N., ed. (CSHL Press), pp. 451–480

Johnson MA, Perez-Amador MA, Lidder P, Green PJ (2000) Mutants of Arabidopsis defective in a sequence-specific mRNA degradation pathway. Proc Natl Acad Sci USA 97:13991–13996

Kahvejian A, Roy G, Sonenberg N (2001) The mRNA closed-loop model: the function of PABP and PABP-interacting proteins in mRNA translation. Cold Spring Harb Symp Quant Biol 66:293–300

Kastenmayer JP, Green PJ (2000) Novel features of the XRN-family in Arabidopsis: evidence that AtXRN4, one of several orthologs of nuclear Xrn2p/Rat1p, functions in the cytoplasm. Proc Natl Acad Sci USA 97:13985–13990

Kedersha N., Stoecklin G, Ayodele M, Yacono P, Lykke-Andersen J, Fritzler MJ, Scheuner D, Kaufman RJ, Golan DE, Anderson P (2005) Stress granules and processing bodies are dynamically linked sites of mRNP remodeling. J Cell Biol 169:871–884

Kertesz S, Kerenyi Z, Merai Z, Bartos I, Palfy T, Barta E, Silhavy D (2006) Both introns and long 3'-UTRs operate as *cis*-acting elements to trigger nonsense-mediated decay in plants. Nucleic Acids Res 34:6147–6157

Kim JH, Richter JD (2006) Opposing polymerase-deadenylase activities regulate cytoplasmic polyadenylation. Mol Cell 24:173–183

Lai WS, Kennington EA, Blackshear PJ (2003) Tristetraprolin and its family members can promote the cell-free deadenylation of AU-rich element-containing mRNAs by poly(A) ribonuclease. Mol Cell Biol 23:3798–3812

Lidder P, Gutierrez RA, Salome PA, McClung CR, Green PJ (2005) Circadian control of messenger RNA stability. Association with a sequence-specific messenger RNA decay pathway. Plant Physiol 138:2374–2385

Liu, H, Rodgers ND, Jiao X, Kiledjian M (2002) The scavenger mRNA decapping enzyme DcpS is a member of the HIT family of pyrophosphatases. EMBO J 21:4699–4708

Liu Q, Greimann JC, Lima CD (2006) Reconstitution, activities, and structure of the eukaryotic RNA exosome. Cell 127:1223–1237

Lue M-Y, Lee H-t (1994) Poly(A) tail shortening of alpha-amylase mRNAs in vegetative tissues of *Oryza sativa*. Biochem Biophys Res Commun 202:1031–1037

Lykke-Andersen J, Wagner E (2005) Recruitment and activation of mRNA decay enzymes by two ARE-mediated decay activation domains in the proteins TTP and BRF-1. Genes Dev 19:351–361

McClure BA, Hagen G, Brown CS, Gee MA, Guilfoyle TJ (1989) Transcription, organization, and sequence of an auxin-regulated gene cluster in soybean. Plant Cell 1:229–239

Meyer S, Temme C, Wahle E (2004) Messenger RNA turnover in eukaryotes: pathways and enzymes. Crit Rev Biochem Mol Biol 39:197–216

Muhlrad D, Parker R (1994) Premature translational termination triggers mRNA decapping. Nature 370:578–581

Narsai R, Howell KA, Millar AH, Nicholas O'Toole N, Ian Small I, Whelan J (2007) Genome-Wide Analysis of mRNA Decay Rates and Their Determinants in Arabidopsis thaliana. Plant Cell 19:3418–3436

Nishimura N, Kitahata N, Seki M, Narusaka Y, Narusaka M, Kuromori T, Asami T, Shinozaki K, Hirayama T (2005) Analysis of ABA hypersensitive germination2 revealed the pivotal functions of PARN in stress response in Arabidopsis. Plant J 44:972–984

Olmedo G, Guo H, Gregory BD, Nourizadeh SD, Aguilar-Henonin L, Li H, An F, Guzman P, Ecker JR (2006) ETHYLENE-INSENSITIVE5 encodes a 5′→3′ exoribonuclease required for regulation of the EIN3-targeting F-box proteins EBF1/2. Proc Natl Acad Sci USA 103:13286–13293

Onouchi H, Nagami Y, Haraguchi Y, Nakamoto M, Nishimura Y, Sakurai R, Nagao N, Kawasaki D, Kadokura Y, Naito S (2005) Nascent peptide-mediated translation elongation arrest coupled with mRNA degradation in the CGS1 gene of Arabidopsis. Genes Dev 19:1799–1810

Orban TI, Izaurralde E (2005) Decay of mRNAs targeted by RISC requires XRN1, the Ski complex, and the exosome. RNA 11:459–469

Parikh BA, Coetzer C, Tumer NE (2002) Pokeweed antiviral protein regulates the stability of its own mRNA by a mechanism that requires depurination but can be separated from depurination of the alpha-sarcin/ricin loop of rRNA. J Biol Chem 277:41428–41437

Petracek ME, Dickey LF, Huber SC, Thompson WF (1997) Light-regulated changes in abundance and polyribosome association of ferredoxin mRNA are dependent on photosynthesis. Plant Cell 9:2291–2300

Petracek ME, Nuygen T, Thompson WF, Dickey LF (2000) Premature termination codons destabilize ferredoxin-1 mRNA when ferredoxin-1 is translated. Plant J 21:563–569

Pillai RS, Bhattacharyya SN, Artus CG, Zoller T, Cougot N, Basyuk E, Bertrand E, Filipowicz W (2005) Inhibition of translational initiation by Let-7 MicroRNA in human cells. Science 309:1573–1576

Potuschak T, Vansiri A, Binder BM, Lechner E, Vierstra RD, Genschik P (2006) The exoribonuclease XRN4 is a component of the ethylene response pathway in Arabidopsis. Plant Cell 18:3047–3057

Rehwinkel J, Raes J, Izaurralde E (2006) Nonsense-mediated mRNA decay: Target genes and functional diversification of effectors. Trends Biochem Sci 31, 639–646.

Reverdatto SV, Dutko JA, Chekanova JA, Hamilton DA, Belostotsky DA (2004) mRNA deadenylation by PARN is essential for embryogenesis in higher plants. RNA 10:1200–1214

Rose AB (2004) The effect of intron location on intron-mediated enhancement of gene expression in Arabidopsis. Plant J 40:744–751

Ross J. (1995) mRNA stability in mammalian cells. Microbiol Rev 59, 423–450

Sheth U, Parker R (2003) Decapping and decay of messenger RNA occur in cytoplasmic processing bodies. Science 300:805–808

Sheu JJ, Yu TS, Tong WF, Yu SM (1996) Carbohydrate starvation stimulates differential expression of rice alpha-amylase genes that is modulated through complicated transcriptional and posttranscriptional processes. J Biol Chem 271:26998–27004

Sonenberg N (1988) Cap-binding proteins of eukaryotic messenger RNA: functions in initiation and control of translation. Prog Nucl Acid Res Mol Biol 35:173–207

Souret FF, Kastenmayer JP, Green PJ (2004) AtXRN4 degrades mRNA in Arabidopsis and its substrates include selected miRNA targets. Mol Cell 15:173–183

Sullivan ML, Green P J (1996) Mutational analysis of the DST element in tobacco cells and transgenic plants: identification of residues critical for mRNA instability. RNA 2:308–315

Tanzer MM, Meagher RB (1995) Degradation of the soybean ribulose-1,5-bisphosphate carboxylase small-subunit mRNA, SRS4, initiates with endonucleolytic cleavage. Mol Cell Biol 15:6641–6652

Teixeira D, Sheth U, Valencia-Sanchez MA, Brengues M, Parker R (2005) Processing bodies require RNA for assembly and contain nontranslating mRNAs. RNA 11:371–382

Torchet C, Bousquet-Antonelli C, Milligan L, Thompson E, Kufel J, Tollervey D (2002) Processing of 3′-extended read-through transcripts by the exosome can generate functional mRNAs. Mol Cell 9:1285–1296

Tran H, Schilling M, Wirbelauer C, Hess D, Nagamine Y (2004) Facilitation of mRNA deadenylation and decay by the exosome-bound, DExH protein RHAU. Mol Cell 13:101–111

Tucker M, Valencia-Sanchez MA, Staples RR, Chen J, Denis CL, Parker R (2001) The transcription factor associated Ccr4 and Caf1 proteins are components of the major cytoplasmic mRNA deadenylase in *Saccharomyces cerevisiae*. Cell 104:377–386

Tumer NE, Hwang DJ, Bonness M (1997) C-terminal deletion mutant of pokeweed antiviral protein inhibits viral infection but does not depurinate host ribosomes. Proc Natl Acad Sci USA 94:3866–3871

van Dijk E, Cougot N, Meyer S, Babajko S, Wahle E, Seraphin B (2002) Human Dcp2: a catalytically active mRNA decapping enzyme located in specific cytoplasmic structures. EMBO J 21:6915–6924

van Hoof A, Frischmeyer PA, Dietz HC, Parker R (2002). Exosome-mediated recognition and degradation of mRNAs lacking a termination codon. Science 295:2262–2264

van Hoof A, Green PJ (1996) Premature nonsense codons decrease the stability of phytohemagglutinin mRNA in a position-dependent manner. Plant J 10:415–424

Vasiljeva L, Buratowski S (2006) Nrd1 interacts with the nuclear exosome for 3' processing of RNA polymerase II transcripts. Mol Cell 21:239–248

Vasudevan S, Peltz SW (2001) Regulated ARE-mediated mRNA decay in *Saccharomyces cerevisiae*. Mol Cell 7:1191–1200

Vaucheret H (2006) Post-transcriptional small RNA pathways in plants: mechanisms and regulations. Genes Dev 20:759–771

Vazquez F (2006) Arabidopsis endogenous small RNAs: highways and byways. Trends Plant Sci 11:460–468

Voelker TA, Moreno J, Chrispeels MJ (1990) Expression analysis of a pseudogene in transgenic tobacco: A frameshift mutation prevents mRNA accumulation. Plant Cell 2:255–261

Wang H, Wu H-μ, Cheung AY (1996) Pollination induces mRNA poly(A) tail shortening and cell deterioration in flower transmitting tissue. Plant J 9:715–727.

Wang, Z, Jiao X, Carr-Schmid A, Kiledjian M (2002) The hDcp2 protein is a mammalian mRNA decapping enzyme. Proc Natl Acad Sci USA 99:12663–12668

Wells SE, Hillner PE, Vale RD, Sachs AB (1998) Circularization of mRNA by eukaryotic translation initiation factors. Mol Cell 2:135–140

Wilusz CJ, Wormington M, Peltz SW (2001) The cap-to-tail guide to mRNA turnover. Nat Rev Mol Cell Biol 2:237–246

Winzen R, Kracht M, Ritter B, Wilhelm A, Chen CY, Shyu AB, Muller M, Gaestel M, Resch K, Holtmann H (1999) The p38 MAP kinase pathway signals for cytokine-induced mRNA stabilization via MAP kinase-activated protein kinase 2 and an AU-rich region-targeted mechanism. EMBO J 18:4969–4980

Xu J, Yang JY, Niu QW, Chua NH (2006) Arabidopsis DCP2, DCP1, and VARICOSE form a decapping complex required for postembryonic development. Plant Cell 18:3386–3398

Yamashita A, Chang TC, Yamashita Y, Zhu W, Zhong Z, Chen CY, Shyu AB (2005) Concerted action of poly(A) nucleases and decapping enzyme in mammalian mRNA turnover. Nat Struct Mol Biol 12:1054–1063

Yoine M, Nishii T, Nakamura K (2006a) Arabidopsis UPF1 RNA helicase for nonsense-mediated mRNA decay is involved in seed size control and is essential for growth. Plant Cell Physiol 47:572–580

Yoine M., Ohto MA, Onai K, Mita S, Nakamura K (2006b) The lba1 mutation of UPF1 RNA helicase involved in nonsense-mediated mRNA decay causes pleiotropic phenotypic changes and altered sugar signalling in Arabidopsis. Plant J 47:49–62

Zhang S, Mehdy MC (1994) Binding of a 50-kD protein to a U-rich sequence in an mRNA encoding a proline-rich protein that is destabilized by fungal elicitor. Plant Cell 6:135–145

Zhang S, Sheng J, Liu Y, Mehdy MC (1993) Fungal elicitor-induced bean proline-rich protein mRNA down-regulation is due to destabilization that is transcription and translation dependent. Plant Cell 5:1089–1099

Zuo J, Niu QW, Chua NH (2000) An estrogen receptor-based transactivator XVE mediates highly inducible gene expression in transgenic plants. Plant J 24:265–273

Regulation of Flowering Time by RNA Processing

L.C. Terzi, G.G. Simpson(✉)

Contents

Introduction 202
Regulated Gene Expression Controls Arabidopsis Flowering Time 202
Alternative Polyadenylation in the Control of Flowering Time 204
FY and FCA 204
    Alternative Processing of *FCA* Pre-mRNA Has a Functional Consequence
    for Flowering Time Control 206
    The Alternative Processing of *FCA* Pre-mRNA Is Temporally and Spatially
        Controlled 207
    How do FCA and FY Control *FLC* Expression? 207
Alternative Splicing in the Control of Flowering Time 208
A Connection Between Alternative Splicing and Flowering Time Regulation
by Ambient Temperature 209
Blocking mRNA Export Affects Flowering Time 210
microRNAs in the Control of Flowering Time 211
    microRNAs Regulating Flowering Time 212
Factors Connecting Other RNA Processing Events to Flowering Time Control 213
Perspective 214
References 215

Abstract Plants control the time at which they flower by integrating environmental cues such as day length and temperature with an endogenous program of development. Flowering time is a quantitative trait and a model for how precision in gene regulation is delivered. In this review, we reveal that flowering time control is particularly rich in RNA processing-based gene regulatory phenomena. We review those factors which function in conserved RNA processing events like alternative 3′ end formation, splicing, RNA export and miRNA biogenesis and how they affect flowering time. Likewise, we review the novel plant-specific RNA-binding proteins identified as regulators of flowering time control. In addition, we add to the network of flowering time control pathways, information on alternative processing of flowering time gene pre-mRNAs. Finally, we describe new approaches to dissect the mechanisms which underpin this control.

G.G. Simpson
Division of Plant Sciences, College of Life Sciences, Dundee University at SCRI, Invergowrie, Scotland, UK and Genetics Programme, SCRI, Invergowrie, Scotland, UK
e-mail: g.g.simpson@dundee.ac.uk

Introduction

The output of Arabidopsis genetic screens has given us unbiased access to the genes required to control complex plant-specific processes. It is striking that, to date, factors associated with RNA processing have emerged from screens for regulators of flowering time more often than from screens for any other aspect of plant biology. Here, we review the RNA-binding proteins and miRNAs required to regulate flowering time and give a perspective on how to understand the mechanism by which they act.

Regulated Gene Expression Controls Arabidopsis Flowering Time

Plants control the time at which they flower in order to ensure that they flower in conditions favourable for reproductive success. This is an inherently complex process which involves the integration of responses to environmental cues with an endogenous program of development. Arabidopsis flowering is promoted in response to environmental cues such as day length, light quality and quantity and temperature (Ausin et al. 2005; Samach and Wigge 2005; Thomas 2006). In addition, flowering is promoted in response to endogenous signals, such as the phytohormone gibberellic acid. The promotion of flowering in response to these environmental and endogenous cues is mediated by genetically separable pathways. The output of these diverse signalling pathways is integrated through their convergence at a common set of genes, collectively termed floral pathway integrators (FPI). In turn, FPI factors promote the activation of the floral meristem identity genes and, hence, the onset of flower formation.

A key feature of the network of pathways which control flowering is floral repression. A potent repressor of flowering in Arabidopsis is the MADS box transcription factor FLC, which functions to inhibit the upregulation of FPI genes. Arabidopsis flowering is exquisitely sensitive to the level of FLC expression. When FLC is expressed at high levels, the repression it mediates effectively prevents Arabidopsis from responding to conditions otherwise favourable for flowering. As a result, precision (and robustness) in the control of *FLC* expression is delivered by a large number of factors which either promote or prevent the accumulation of *FLC* mRNA. For example, components of the genetically defined autonomous pathway of flowering time control prevent *FLC* mRNA accumulation and in doing so enable the FPI genes to respond to the pathways which promote flowering.

Flowering time is a quantitative and adaptive trait and clearly precision and modification of gene regulation underpins this. While flowering in response to day length, mediated by the photoperiod pathway, is governed by a transcription factor cascade, the regulation of *FLC* depends on multiple factors which appear not to regulate each other but are predicted to control *FLC* through different mechanisms dependent on chromatin modification or RNA processing. Most flowering time genes exhibit alternative splicing or alternative polyadenylation of their pre-mRNAs (see Fig. 1).

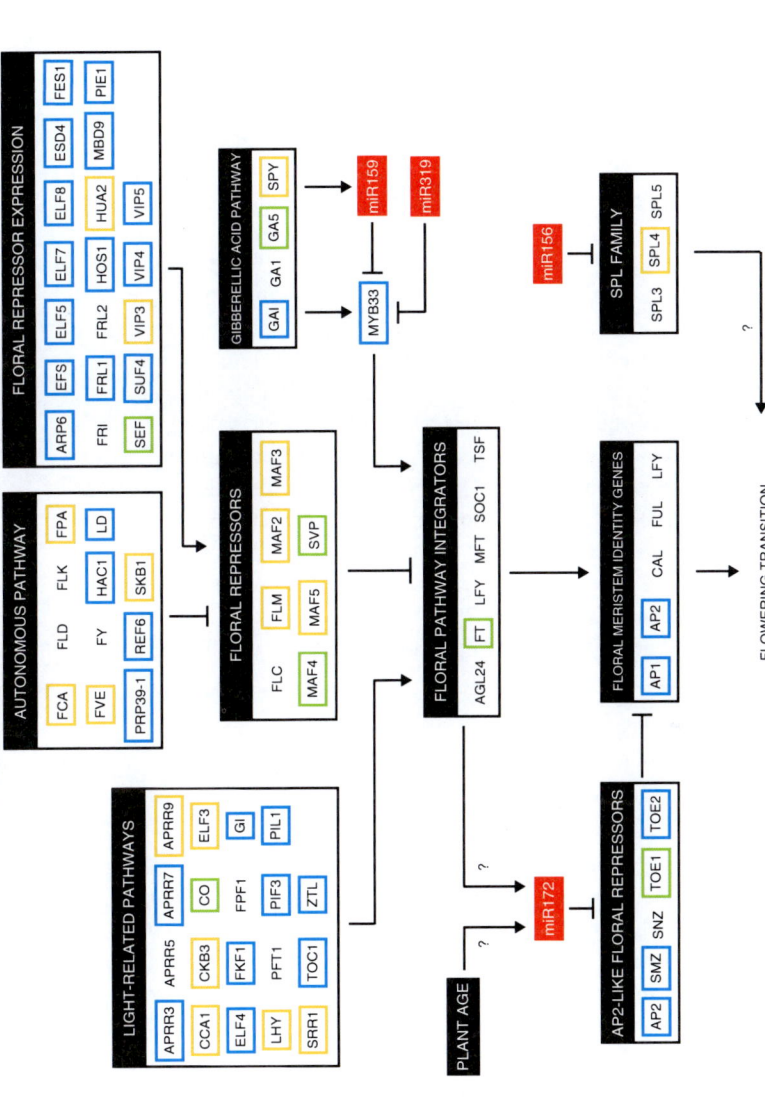

Fig. 1 Alternative processing of Arabidopsis flowering time genes. Genes controlling flowering time are grouped together in the pathways in which they function (the precise inter-relationship of genes within each pathway is not shown). *Arrowheads* identify activities promoting the expression of other genes controlling flowering, while *T-bars* illustrate repressive activities. Genes for which evidence has been found for alternative pre-mRNA splicing are boxed in *green*, alternative polyadenylation are boxed in *blue* and those genes for which evidence exists for both types of alternative pre-mRNA processing are boxed in *orange*. The targets of miRNAs (*red boxes*) are indicated. Evidence of alternative splicing was derived from AtGDB (http://www.plantgdb.org/AtGDB/index.php) and alternative polyadenylation from MPSS data (http://mpss.udel.edu/at/?). Alternative polyadenylation evidence is derived from MPSS signatures widely spaced in the genes and does not include the heterogeneity found in the conventional 3′ UTRs. *Question marks* refer to disagreement in the literature about control of miR172 expression or unknown targets of SPL genes.

Alternative Polyadenylation in the Control of Flowering Time

Mutations in genes that encode factors required for cleavage and polyadenylation of pre-mRNA exhibit flowering time defects (Table 1, Fig. 1) (see the chapter by A. G. Hunt, this volume). David Baulcombe and co-workers identified several mutations in factors required for 3' end formation in a screen for factors affecting RNA silencing. The *ENHANCED SILENCING PHENOTYPE* (*ESP*) mutants *ESP1*, *ESP4* and *ESP5* are likely components of the core cleavage and polyadenylation machinery (Herr et al. 2006). *ESP5* encodes the orthologue of mammalian cleavage and polyadenylation specificity factor (CPSF) 100, a key component of the CPSF complex participating in cleavage and polyadenylation of pre-mRNA (reviewed in Proudfoot 2004). ESP5 protein interacts with ESP4 (Herr et al. 2006), a homologue of another mammalian CPSF component, Symplekin/Pta1 (Proudfoot 2004). ESP5 also interacts with CPSF160, poly(A) polymerase (Elliott et al. 2003) and CPSF73 (Herr et al. 2006), the pre-mRNA 3' end endonuclease (Mandel et al. 2006). *ESP1* encodes a protein resembling mammalian cleavage stimulatory factor (CstF) 64 but may not be the Arabidopsis orthologue. This is because another Arabidopsis gene encodes a protein more closely related to CstF64 with the equivalent protein domains. In contrast, ESP1 lacks the N-terminal RNA recognition motif (RRM), an RNA-binding domain, of CstF64. Consistent with a role for *ESP1*, *ESP4* and *ESP5* in pre-mRNA 3' end formation, loss-of-function mutations in these genes result in increased transcriptional read-through at the end of RNA transcripts (Herr et al. 2006). Loss-of-function *esp1, 4* and *5* mutants all flower early, but the molecular basis governing this phenotype has not yet been established (see below).

FY and FCA

Loss-of-function mutations in FY, another factor implicated in Arabidopsis pre-mRNA 3' end processing, also affects flowering time. However, in contrast to mutations in *esp1, 4* and *5,* which flower early, loss-of-function *fy* mutations flower late. The reason for this difference is not clear, but it may relate to the different Arabidopsis backgrounds in which these mutants were isolated. Late-flowering *fy* mutants have been identified in Columbia and Landsberg *erecta* backgrounds whereas the *esp* mutants were isolated in a C24 background. *FY* functions in the autonomous pathway to prevent the accumulation of *FLC* mRNA. *FY* encodes a protein which is highly conserved in all eukaryotes for which sequence is available (Ohnacker et al. 2000; Simpson et al. 2003). FY comprises seven repeats of the WD protein interaction domain and it is this part of the protein which is highly conserved in other eukaryotes. Of the FY-related proteins, only *Saccharomyces cerevisiae*

Table 1 Flowering time phenotype caused by altered expression of RNA processing factors

Gene product	Locus	Flowering time pathway	Mutant phenotype	Over-expressor phenotype	Molecular function	FLC regulation
ABH1	At2g13540	n.a.	Early	n.d.	Cap binding	Yes
DCL1	At1g01040	n.a.	Late	n.d.	miRNA biogenesis	n.d.
ELF5	At5g62640	Floral repressor expression	Early	Wild type	Splicing	Yes
ESP1	At1g73840	n.a.	Early	n.d.	Polyadenylation	n.d.
ESP3	At1g32490	n.a.	Early	n.d.	Splicing	n.d.
ESP4	At5g01400	n.a.	Early	n.d.	Polyadenylation	n.d.
ESP5	At5g23880	n.a.	Early	n.d.	Polyadenylation	n.d.
FCA	At4g16280	Autonomous	Late	Early	Polyadenylation	Yes
FES1	At2g33835	FRIGIDA	Early	n.d.	n.d.	Yes
FLK	At3g04610	Autonomous	Late	Early	n.d.	Yes
FPA	At2g43410	Autonomous	Late	Early	n.d.	Yes
FY	At5g13480	Autonomous	Late	n.d.	Polyadenylation	Yes
HEN1	At4g20410	n.a.	Late	n.d.	miRNA biogenesis	n.d.
HUA2	At5g23150	Floral repressor expression	Early	n.d.	n.d.	Yes
HYL1	At1g09700	n.a.	Late	n.d.	miRNA biogenesis	n.d.
PRP39-1	At1g04080	Autonomous	Late	n.d.	Splicing	Yes
SE (SERRATE)	At2g27100	n.a.	Early	n.d.	miRNA biogenesis	n.d.
SR30	At1g09140	n.a.	n.d.	Late	Splicing	n.d.
U2AF35a	At1g27650	n.a.	Late	n.d.	Splicing	n.d.
U2AF35b	At5g42820	n.a.	Late	n.d.	Splicing	Yes
UPF1	At5g47010	n.a.	Late	n.d.	NMD	n.d.
SAR1	At1g33410	n.a.	Early	n.d.	mRNA nuclear export	n.d.
SAR3/PRE	At1g80680	n.a.	Early	n.d.	mRNA nuclear export	n.d.
LOS4	At3g53110	n.a.	Early	n.d.	mRNA nuclear export	No

Refer to Fig. 1 for the classification of flowering time pathways. See text for details
n.d., not determined. n.a., not assigned

Pfs2p has an established function as an essential component of the pre-mRNA 3' end processing machinery (Ohnacker et al. 2000). Pfs2p is stably associated with a complex of cleavage factor II-polyadenylation factor I (CFII-PFI) and interacts with a component of cleavage factor Ia (CFIA). Pfs2p therefore bridges CFII-PFI and CFIA and, consistent with this, is required for both the cleavage and polyadenylation reactions of 3' end formation (Ohnacker et al. 2000). Does FY function in 3' end formation in Arabidopsis? Two pieces of evidence indicate that it might. First, FY is required for alternative 3' end formation of *FCA* pre-mRNA (see below): In the absence of FY, promoter proximal 3' end formation within FCA intron 3 occurs less often (Simpson et al. 2003). Second, FY was found to be stably associated with a complex purified from Arabidopsis comprised of homologues of the cleavage and polyadenylation machinery: CPSF100/ESP5, CPSF160, CPSF73 and symplekin/ESP4 (Herr et al. 2006).

Like *FY*, *FCA* functions in the autonomous pathway to prevent the accumulation of *FLC* mRNA. In contrast to the evolutionarily conserved sequence of *FY*, *FCA* encodes a novel, plant-specific RNA-binding protein (Macknight et al. 1997; Quesada et al. 2003). FCA physically interacts with FY (Simpson et al. 2003). While *S. cerevisiae* Pfs2p is comprised of only repeats of the WD protein interaction domain, related sequences in higher eukaryotes have evolved extended and divergent carboxy termini (Henderson et al. 2005). Higher plant FY homologues possess PPLP motifs in this region and FY binds the FCA WW domain through these PPLP sequences (Henderson et al. 2005; Simpson et al. 2003).

Alternative Processing of FCA Pre-mRNA Has a Functional Consequence for Flowering Time Control

FCA expression is post-transcriptionally regulated through alternative polyadenylation and this has a functional consequence for flowering time control. Three major transcripts are produced as a result of alternative processing of the relatively large (2.2 kb) intron 3: (a) *FCAγ*, fully spliced *FCA* mRNA, (b) *FCAα*, a transcript in which all introns are excised except intron 3, (3) *FCAβ*, a transcript formed from premature cleavage and polyadenylation within intron 3. Therefore, *FCA* pre-mRNA is alternatively spliced and alternatively polyadenylated. Only the fully spliced γ isoform is active in flowering time control. This pattern of alternative processing is conserved in FCA homologues in other plant species (Lee et al. 2005).

The functional significance of this alternative processing can at least partly be explained by the autoregulatory activity of FCA. Quesada et al. (2003) revealed that elevated levels of *FCAγ* expressed from a transgene promote cleavage and polyadenylation within intron 3 of endogenous *FCA* pre-mRNA. This serves to autoregulate active FCA expression, because as a result, less fully spliced endogenous *FCAγ* accumulates. Consistent with this, loss-of-function *fca* mutants exhibit reduced accumulation of *FCAγ*. The presence of introns effectively limits the accumulation of active FCA expression through the autoregulatory function of FCA. This has a functional consequence for flowering time control because if FCA is expressed from a transgene with its native promoter, but lacking introns, the resulting plants flower early (Quesada et al. 2003).

FY is required for FCA autoregulation. Less *FCA* pre-mRNA is cleaved and polyadenylated within intron 3 in loss-of-function *fy* mutants and as a result more of the fully spliced *FCA*γ transcript accumulates (Quesada et al. 2003; Simpson et al. 2003). In addition, the enhanced promotion of cleavage and polyadenylation within intron 3 witnessed in wild-type or *fca-1* mutant plants over-expressing *FCA*γ from a transgene is not seen in an *fy* mutant background. How might FCA and FY function to autoregulate FCA expression? FCA could interfere with 3' end formation in the conventional FCA 3' UTR and as a result cause cleavage and polyadenylation to occur at the site within intron 3 by default. Alternatively, FCA could more actively promote cleavage and polyadenylation within intron 3. As *FCA* autoregulation does not depend upon its native 3' UTR (Quesada et al. 2003), the former possibility, that FCA more actively promotes 3' end processing within intron 3, appears likely. A straightforward model would suggest that FCA binds to its own RNA close to this site and through its interaction with FY brings in the 3' end formation processing machinery. This model remains untested.

The Alternative Processing of FCA Pre-mRNA Is Temporally and Spatially Controlled

The autoregulation of FCA does not appear to yield uniform homeostasis of active FCA expression as the processing of *FCA* intron 3 is under developmental control. Macknight et al (1997) fused the sequences of FCA's promoter through exon 5 to the β-glucuronidase (*GUS*) reporter gene. If *FCA* pre-mRNA was cleaved and polyadenylated within intron 3 or if intron 3 was retained, no GUS activity would be detected. Instead, GUS activity would only be detected if intron 3 was correctly spliced. These plants showed elevated GUS expression at days 4–6 after germination, with a simultaneous upregulation of GUS activity in the shoot and root apex. The other major location of elevated GUS activity was at the sites of lateral root emergence. This indicates that at a certain time and sites of development, the FCA autoregulation functions can be altered. This appears to be most closely associated with regions of active cellular proliferation.

How do FCA and FY Control FLC Expression?

The late flowering phenotype of *fy* mutants is not due to a genome-wide misregulation of 3' end formation leading to compromised gene expression and hence generally slowed development. Instead, the late flowering phenotype of loss-of-function *fy* mutants can be explained exclusively by the mis-regulation of *FLC*, as a null allele of *flc* suppresses the late flowering phenotype of *fy* mutants (Henderson et al. 2005). The directness by which FCA and FY control *FLC* expression is unknown. The simplest interpretation of the autoregulation function of FCA is that FCA binds its own pre-mRNA and recruits the 3′ end processing machinery to an otherwise weak poly A site. Since FCA requires FY for *FLC* repression, the same model could apply for FCA's role in flowering time control. However, no alternatively polyadenylated transcripts of *FLC* have been reported to accumulate, let alone in a manner dependent on FCA or FY. A second possibility is that FCA and FY regulate the expression of an intermediate—a factor that also controls *FLC* expression. This might even include loci adjacent to *FLC* which appear to be co-regulated with *FLC* since loss-of-function *fca-1* mutants show elevated levels of *FLC* and the adjacent gene *UPSTREAM OF FLC* (*UFC*: At5g10150), raising the possibility that the effects of FCA on *FLC* are an indirect consequence of regulating other genes. However, based on reporter gene analysis, Sheldon et al. (2002) concluded that the minimal *cis* element of *FLC* required for *FCA*-mediated repression which they tested comprised the region that encompassed *FLC* exon 1, intron 1 and exon 2.

The functional equivalent of Pfs2p identified in extracts which carry out cleavage and polyadenylation within mammalian extracts was proposed to be a protein known as CstF50 (Ohnacker et al. 2000). Like Pfs2p, CstF50 comprises seven WD repeats (although the sequence identity to Pfs2p is very low). In addition, mammalian CstF50 and *S. cerevisiae* Pfs2p physically interact with related proteins which function in 3′ end formation: mammalian CstF50 with CPSF73 and CstF77, and *S. cerevisiae* Pfs2p with the yeast homologues of these proteins YSH1 and RNA14, respectively. Interestingly, however, higher eukaryotes encode genes related to both Pfs2p/FY and CstF50, while *S. cerevisiae* only encodes Pfs2p. This raises the possibility that divergent complexes affecting constitutive and regulated 3′ end formation may have evolved in higher eukaryotes. Null alleles of FY are lethal, but it is not clear whether this is as a result of defective constitutive 3′ end formation, the regulation of 3′ end processing of an essential gene or another function of FY. This indicates that the study of RNA processing in flowering time control also has implications for understanding regulated RNA processing in other eukaryotes.

Alternative Splicing in the Control of Flowering Time

Viable mutations in multiple splicing factors disrupt flowering time control (Table 1). These factors are predicted to affect different steps of the splicing reaction. For example, PRP39-1 and SR30 affect 5′ splice site selection.

Loss of function *prp39-1* mutants flower late. *PRP39-1* prevents the accumulation of *FLC* mRNA, indicating that, like FCA and FY, it acts within the autonomous pathway of flowering time control (Wang et al. 2007). *PRP39-1* encodes a protein which is associated with the U1 snRNP and is required for U1 snRNP assembly on the 5′ splice site (Chen et al. 2006; Lockhart and Rymond 1994).

SR30 encodes an SR protein, a family of proteins involved in constitutive and alternative splicing of pre-mRNA, which is related to the mammalian splicing factor SF2/ASF (Reddy 2004) (see the chapters by A. Barta et al. and G. S. Ali and A. S. N. Reddy, this volume). SR30 affects the alternative splicing of endogenous plant genes in a concentration-dependent manner (Lopato et al. 1999) although targets connected to flowering time control have not yet been identified. Over-expression of SR30 results in late flowering (Table 1). At the protein level, SR30 participates in a network of splicing factor interactions through its interaction with SRZ33 (At2g37340) (Lopato et al. 2002), a plant-specific SR protein affecting alternative splicing *in vivo* (Kalyna et al. 2003). SRZ21 (At1g23860), SRZ22 (At4g31580) and SR33 (At1g55310) interact with SRZ33 (Lopato et al. 2002) and with one of the three U1 snRNP-specific proteins, U1-70K, (Golovkin and Reddy 1998, 1999).

Flowering time may be regulated by factors affecting other steps of the splicing reaction. The assembly of the U2 snRNP at the branch point is facilitated by U2AF, a heterodimeric splicing factor composed of a large (U2AF65) and a small (U2AF35) subunit (Lorković et al. 2000). U2AF35 is encoded by two genes in Arabidopsis: *U2AF35a* and *U2AF35b* (Wang and Brendel 2006). RNAi and antisense knockdown of the expression of both genes cause late flowering associated with elevated levels of mRNA encoding the floral repressor, *FLC* (Wang and Brendel 2006). This might partly be explained by a compromise of FCA function in these U2AF35 knockdown lines as cryptic and aberrant splicing of *FCA* pre-mRNA was found in these plants (Wang and Brendel 2006).

Lesions in *ELF5* causes early flowering partially due to decreased expression of *FLC* (Noh et al. 2004). ELF5 is predicted to interact with the U5 snRNP since it is similar to metazoan NpwBP. NpwBP interacts with the WW domain of Npw38 in mouse (Komuro et al. 1999), the specific partner of the U5 snRNP component U5-15kDa/dim1p (Waragai et al. 2000).

ESP3 encodes a homologue of yeast PRP2, an essential DEAH RNA helicase involved in pre-mRNA splicing (Teigelkamp et al. 1994; Vijayraghavan et al. 1989). PRP2 is required for the progression through the first transesterification step of splicing (Kim and Lin, 1996). *esp3* mutants are early flowering, but the alterations in RNA processing conferring this phenotype are not yet known (Herr et al. 2006). Consistent with a role for ESP3 in pre-mRNA splicing, *FCA* pre-mRNA processing was affected in *esp3* mutants (Herr et al. 2006): The $FCA\alpha$ isoform, retaining intron 3 of *FCA* (Quesada et al. 2003), was more abundant in the mutant.

A Connection Between Alternative Splicing and Flowering Time Regulation by Ambient Temperature

Ambient temperature is an important environmental regulator of flowering time (Blazquez et al. 2003; Lee et al. 2007). Elevated temperatures promote flowering in Arabidopsis and even a modest increase in growth temperature (from 23°C to

27°C), is a potent inducer of flowering (Balasubramanian et al. 2006). The molecular basis underpinning this response has received relatively little attention until now. However, the recent characterization of the flowering response to ambient temperature of different Arabidopsis mutants and ecotypes has led to the idea that there is a genetic basis to this aspect of flowering time control. As with other aspects of floral promotion, elevated levels of *FLC* inhibit thermal induction of flowering. However, thermosensitivity does not require *FLC* (Balasubramanian et al. 2006). Instead, another repressor of flowering, *FLM* (At1g77080), is involved (Balasubramanian et al. 2006; Werner et al. 2005), as a genetic analysis of ecotypes defective in thermal response identified deletions in *FLM*.

The thermal induction of flowering is as effective in accelerating flowering time as day length (a shift from non-inductive short-day growth conditions to long days). However, thermal induction is genetically distinct from the photoperiod pathway. Consistent with this, a genomic analysis of gene expression revealed distinct patterns in response to thermal and photoperiod induction (Balasubramanian et al., 2006). Interestingly, significant changes in gene expression identified upon thermal induction of flowering were in RNA processing factors. The mRNA expression of several SR proteins, *SR30, SRZ33, SRZ21* and *SR33*, was increased upon thermal induction, indicating that alternative splicing may be an important feature of this response (Balasubramanian et al. 2006). Significantly, alternatively spliced isoforms of *FLM* mRNA can be detected *in vivo* (Scortecci et al. 2001) and thermal induction triggers a shift in isoform ratios, with the major isoform decreasing upon temperature increase (Balasubramanian et al. 2006). Thermal induction also affects the splicing pattern of another floral repressor, *MAF2* (At5g65050) (Balasubramanian et al. 2006). The functional significance of this alternative splicing to flowering time control in response to ambient temperature has yet to be assessed. Genetic analysis suggests that FLM functions closely with SHORT VEGETATIVE PHASE (SVP) (Lee et al. 2007; Scortecci et al. 2003), another repressor of flowering. *SVP* pre-mRNA is also alternatively spliced, although it is not yet known whether this is affected by ambient temperature.

Blocking mRNA Export Affects Flowering Time

Once pre-mRNA molecules are spliced and polyadenylated, they must be exported to the cytoplasm through the nuclear pore complex (NPC), a process tightly linked with gene expression (Cole and Scarcelli 2006; Sommer and Nehrbass 2005). Failure to ensure efficient nuclear export of mRNA affects flowering time as revealed by the phenotype of mutants disrupted in genes encoding proteins associated with the nuclear pore complex: SAR1/NUP160, SAR3 and LOS4. *SAR1* and *SAR3* encode nucleoporins (Dong et al. 2006; Parry et al. 2006) (see the chapter by V. Chinnusamy et al., this volume), the building blocks of the NPC (Lim and Fahrenkrog 2006). Nucleoporins are organized in sub-complexes within the NPC: The vertebrate homologues of SAR1 and SAR3 belong to the NUP107–160 complex (Loiodice et al. 2004), the core element of the central framework essential for NPC assembly (Lim and Fahrenkrog 2006). In agreement with a central role of SAR1 and SAR3 in NPC assembly, mRNA export was greatly affected in the corresponding mutants (Dong et al. 2006; Parry et al. 2006). Mutations in these NPC components flower early (Table 1). The flowering time gene(s) affected by mRNA export is currently unknown.

LOS4 encodes a putative DEAD box RNA helicase enriched at the nuclear rim, suggesting it is involved in mRNA export (Gong et al. 2002, 2005). Consistent with this, mRNA accumulates in the nuclei of *los4* mutants. *FLC* expression is not greatly affected in *los4* mutants (Gong et al. 2005) and thus the genes which affect flowering time and which depend on *LOS4* for their expression are presently unknown.

microRNAs in the Control of Flowering Time

RNA processing factors involved in microRNA (miRNA) biogenesis affect flowering time. miRNAs are riboregulators of gene expression involved in developmental processes in various multi-cellular organisms. They are 20- to 24-nt, small noncoding RNAs serving as guides to the RNA-induced silencing complex (RISC) in the identification and silencing of target mRNAs. Specificity is provided by base pairing of the miRNA to miRNA target sequence, leading to translational repression or degradation of mRNA (Zhang et al. 2007). miRNAs are usually encoded by independent transcription units. However, miRNA genes produce long, pri-miRNA molecules which must be processed to yield the mature miRNA (Kim 2005). In agreement with miRNAs regulating flowering time (see below), mutations in genes participating in the biogenesis of miRNA exhibit flowering time defects (Golden et al. 2002; Vazquez et al. 2004).

The first step of miRNA biogenesis, the synthesis of the pri-miRNA transcript from its miRNA gene, shares high similarity with genes encoding mRNA (Jones-Rhoades et al. 2006): miRNA genes are most likely transcribed by RNA polymerase II (Megraw et al. 2006; Xie et al. 2005) and may be capped (Xie et al. 2005), polyadenylated (Aukerman and Sakai 2003; Kurihara and Watanabe 2004) and spliced (Sunkar et al. 2005). This initial mRNA-like processing of pri-miRNA highlights the similarity between pri-miRNA and pre-mRNA. Whether it is required for downstream miRNA biogenesis is currently unclear.

Further miRNA processing requires the cropping of a pri-miRNA stem-loop structure into a pre-miRNA and the dicing of the pre-miRNA to a miRNA/miRNA* duplex (Chen 2005; Kim 2005). Both processes occur in Arabidopsis nuclei and require the activity of a Dicer-like protein (DCL1), a dsRNA-binding protein (HYL1) and a methyltransferase (HEN1) (Chen 2005). Mutants in the corresponding genes affect flowering time (Golden et al. 2002; Vazquez et al. 2004). *SERRATE (SE)* is required for plant microRNA processing (Grigg et al. 2005; Lobbes et al. 2006; Yang et al. 2006) and, therefore, *se* mutants accumulate less miRNAs (Lobbes et al. 2006; Yang et al. 2006). *se* mutants compromise the *FRIGIDA (FRI)*-mediated enhancement of *FLC* expression (Bezerra et al. 2004), thereby accelerating the transition to flowering. This suggests that the expression of negative regulators of *FLC* is controlled by miRNA-based pathway(s). A reduction of *FRI*-mediated expression of *FLC* is also observed in *abh1* mutants (Bezerra et al. 2004). *ABH1* encodes the large subunit of the eukaryotic nuclear mRNA cap-binding complex (Hugouvieux et al. 2001). It has, therefore, been proposed that the mRNAs of key regulators of flowering time may be highly sensitive to a loss of *ABH1*. Alternatively, since pri-miRNAs are capped, it is feasible that a loss of *ABH1* function prevents their accumulation and/or processing. It will therefore be interesting to determine the levels of miRNAs in *abh1* mutants.

microRNAs Regulating Flowering Time

Arabidopsis miRNAs (Rajagopalan et al. 2006) regulate multiple developmental events (Chen 2005). The involvement of three miRNA families (miR172, miR159/miR319 and miR156) in flowering time regulation has recently been demonstrated (Jones-Rhoades et al. 2006).

The miR172 family contains four members in Arabidopsis: miR172a-1, miR172a-2, miR172b and miR172c (Park et al. 2002). Plants over-expressing miR172a-1 or miR172a-2 flower extremely early (Aukerman and Sakai, 2003) since these miRNAs prevent the expression of repressors of flowering belonging to a subfamily of *APETALA2* (*AP2*)-like transcription factor genes: *AP2, TARGET OF EAT 1* (*TOE1*), *TARGET OF EAT 2* (*TOE2*), *SCHLAFMÜTZE* (*SMZ*) and *SCHNARCHZAPFEN* (*SNZ*) (Aukerman and Sakai 2003; Chen 2004; Schmid et al. 2003). There is no consensus in the literature regarding the control of miR172 expression. Whereas Schmid et al. (2003) reported that miR172a-2 expression depends on day length, *CO* and *FT*, Aukerman and Sakai found that miR172a-2 accumulates with increasing plant age and is not affected by a mutation in *CO* (Aukerman and Sakai 2003).

The overexpression of *miR319* (also called miR-JAW) results in several developmental defects, including a delay in flowering (Palatnik et al. 2003). miR319 shares at least one mRNA target (*MYB33*) with miR159a, as predicted from their sequence homology. *MYB33* is required for the expression of the floral pathway integrator *LFY*. It was previously shown that *MYB33* mRNA was directly cleaved by miR159a (Achard et al. 2004), resulting in a delay in flowering time specifically in short days. In contrast, over-expression of miR319 delays flowering in long days, suggesting different regulation of these two miRNA family members. miR159 expression is promoted by gibberellin (Achard et al. 2004). It will therefore be interesting to analyse the effect of different growth conditions and/or treatments on the expression of miR319. Other genes of the TCP transcription factor family are down-regulated after over-expression of *miR319* and it remains possible that some of these might affect flowering time (Palatnik et al. 2003).

miR156b over-expression results in late flowering in long days and down-regulates the expression of 10 members of the *SQUAMOSA PROMOTER BINDING PROTEIN LIKE* (SPL) family containing target sites for miR156 (Schwab et al. 2005). Consistent with this, over-expression of *SPL3* promotes early flowering (Cardon et al. 1997) and *SPL3* mRNA levels increase during the transition to flowering (Cardon et al. 1997) and in plants exposed to a photo-inductive stimulus (Schmid et al. 2003). The regulation of *SPL3* (as well as *SPL4* and *SPL5*) by *miR156* has recently been described *in vivo*. Using miR156a-insensitive variants of *SPL3* mRNA, Wu and Poethig (2006) showed that, in short days, *miR156* regulated the temporal expression of *SPL3* early in development. *miR156* expression is consistent through the juvenile vegetative phase and then decreases during the transition to the adult vegetative phase. Concomitantly, *SPL3* expression is not detected in juvenile plants and increases during the transition to adult. A *miR156*-insensitive variant of *SPL3* escapes *miR156* regulation and can be detected in the juvenile phase of vegetative development. This study demonstrated that *miR156* regulation of *SPL3* expression is first involved in vegetative phase transition during *Arabidopsis* development. Whether and how it extends to the regulation of flowering time is unclear at the moment.

Factors Connecting Other RNA Processing Events to Flowering Time Control

Other RNA processing factors are known to control *FLC* expression. However, a molecular function has not yet been assigned to them. *FPA* and *FLK* belong to the autonomous pathway and encode proteins with RRM and KH domains, respectively (Mockler et al. 2004; Schomburg et al. 2001). *FES1* encodes a CCCH-type zinc finger protein and is required for *FRI*-mediated enhancement of *FLC* expression (Schmitz et al. 2005). CCCH-type zinc fingers have been associated with RNA binding (Carballo et al. 1998; Cheng et al. 2003), although this has not yet been shown for FES1. *HUA2* is required for the expression of the floral repressors, *FLC* and the *MAFs* (Doyle et al. 2005). HUA2 contains an RPR domain, a motif found in proteins involved in RNA metabolism (Doerks et al. 2002), and has been shown to promote splicing of *AGAMOUS* pre-mRNA intron 2 (Cheng et al. 2003). A reduction of the *FRI*-mediated enhancement of *FLC* expression is also observed in *abh1* mutants (Bezerra et al. 2004). *ABH1* encodes the large subunit of the eukaryotic nuclear mRNA cap-binding complex (Hugouvieux et al. 2001), suggesting that CBC80 is required for *FLC* expression. The cap can stimulate splicing, polyadenylation and nuclear export. It is not known at what level CBC80 affects *FLC* expression or even if this is a direct effect on *FLC* pre-mRNA.

Transcription and pre-mRNA processing can incorporate errors or generate aberrant RNAs. RNA surveillance mechanisms scan pre-mRNAs and target such RNAs for decay. For example, mRNAs with premature termination codons are identified and degraded by a process known as nonsense-mediated decay. This process not only functions in RNA quality control but can be dynamically incorporated into regulated gene expression. Viable Arabidopsis mutations in the nonsense-mediated decay factor UPF1 flower late (Arciga-Reyes et al. 2006) although the affected RNA targets which influence flowering time have not yet been identified. UPF1 also appears to be required for Arabidopsis RNAi (Arciga-Reyes et al. 2006) and so at this stage it is not clear through which function UPF1 affects flowering.

The expression of *FLC* and other *MAF* floral repressors is dependent on Arabidopsis homologues of the RNA polymerase II-associated factor complex (PAF1c) (Quesada et al. 2005). This complex functions as a passive elongation factor facilitating the recruitment of chromatin modifiers and RNA processing factors to transcribed genes. Loss of PAF1c function in yeast affects the expression of only a subset of genes, apparently through aberrant 3' end formation leading to RNA degradation (Penheiter et al. 2005). This function may be significant in Arabidopsis flowering time control in light of the fact that many of the factors in the autonomous pathway which antagonize the function of PAF1c in promoting *FLC* expression are RNA processing factors.

Perspective

The inherent complexity and importance of regulating flowering time make it no surprise that diverse mechanisms of gene regulation are involved in this process. While we do not yet know all of the details in the mechanism of flowering time control, the established and accepted concept of genetically separable pathways controlling flowering provides a valuable framework for understanding this. With respect to the role of RNA processing in this process, we generally lack two key pieces of information for the mutants and factors affecting flowering: (a) What are the changes in gene expression resulting from mutations in genes that affect flowering time? (b) What are the immediate targets of the RNA processing factors involved?

These are tractable problems. We have compiled here the flowering time genes which exhibit alternative processing of pre-mRNA and which are targets of miRNAs (Fig. 1). Whole genome TILING arrays, direct sequencing, or flowering time gene-specific analysis of alternative RNA processing in mutants affecting flowering time should reveal where in the network of flowering time control perturbations are found. In addition, the ability to use whole genome TILING arrays and direct sequencing in combination with the development of *in vivo* cross-linking and immunoprecipitation technologies in Arabidopsis will allow us to address the directness by which identified RNA-binding proteins regulate flowering.

Such an analysis also requires the characterisation of the functional consequence of particular RNA processing events for flowering time control. So far, this has only been clearly established for *FCA* pre-mRNA and certain miRNA targets. However, once these are established they will generate reporters for new Arabidopsis genetic screens with the potential to specifically dissect signal transduction pathways leading to regulated RNA processing affecting this developmental switch.

Flowering time control presents an especially attractive model system to study the mechanisms and relevance of RNA processing to regulated gene expression. The genetic tractability and quantitative nature of flowering time control can let us reveal how precision in gene regulation is delivered. It can let us determine how conserved factors thought to function in constitutive gene expression, like CBC80, PRP-39 or CPSF100 selectively affect gene expression. In addition, this is a field where we can understand how plant-specific RNA processing factors like FCA and FPA function, potentially revealing new ways to regulate gene expression at the RNA level not yet identified or present in other organisms.

Acknowledgements Our work is supported by the Scottish Executive Environment and Rural Affairs Department and by BBSRC Grant BB/D000653/1.

References

Achard P, Herr, A Baulcombe DC, Harberd NP (2004) Modulation of floral development by a gibberellin-regulated microRNA. Development 131:3357–3365
Arciga-Reyes L, Wootton L, Kieffer M, Davies B (2006) UPF1 is required for nonsense-mediated mRNA decay (NMD) and RNAi in Arabidopsis. Plant J 47:480–489
Aukerman MJ, Sakai H (2003) Regulation of flowering time and floral organ identity by a MicroRNA and its APETALA2-like target genes. Plant Cell 15:2730–2741
Ausin I, Alonso-Blanco C, Martinez-Zapater JM (2005) Environmental regulation of flowering. Int J Dev Biol 49:689–705
Balasubramanian S, Sureshkumar S, Lempe J, Weigel D (2006) Potent induction of *Arabidopsis thaliana* flowering by elevated growth temperature. PLoS Genetics 2:e106
Bezerra I C, Michaels SD, Schomburg FM, Amasino RM (2004) Lesions in the mRNA cap-binding gene ABA HYPERSENSITIVE 1 suppress FRIGIDA-mediated delayed flowering in Arabidopsis. Plant J 40:112–119
Blazquez MA, Ahn JH, Weigel D (2003) A thermosensory pathway controlling flowering time in *Arabidopsis thaliana*. Nat Genet 33:168–171
Carballo E, Lai WS, Blackshear PJ (1998) Feedback inhibition of macrophage tumor necrosis factor-alpha production by tristetraprolin. Science 281:1001–1005
Cardon GH, Hohmann S, Nettesheim K, Saedler H, Huijser P. (1997) Functional analysis of the *Arabidopsis thaliana* SBP-box gene SPL3: a novel gene involved in the floral transition. Plant J 12:367–377
Chen X (2004) A MicroRNA as a translational repressor of APETALA2 in arabidopsis flower development. Science 303:2022–2025
Chen X (2005) microRNA biogenesis and function in plants. FEBS Lett 579:5923–5931
Chen YI, Maika SD, Stevens SW (2006) Epitope tagging of proteins at the native chromosomal loci of genes in mice and in cultured vertebrate cells. J Mol Biol 361:412–419
Cheng Y, Kato N, Wang W, Li J, Chen X (2003) Two RNA binding proteins, HEN4 and HUA1, act in the processing of AGAMOUS pre-mRNA in *Arabidopsis thaliana*. Dev Cell 4:53–66
Cole CN, Scarcelli JJ (2006) Transport of messenger RNA from the nucleus to the cytoplasm. Curr Opin Cell Biol 18:299–306
Doerks T, Copley RR, Schultz J, Ponting CP, Bork P (2002) Systematic identification of novel protein domain families associated with nuclear functions. Genome Res 12:47–56
Dong C H, Hu X, Tang W, Zheng X, Kim YS, Lee BH, Zhu JK (2006) A putative Arabidopsis nucleoporin, AtNUP160, is critical for RNA export and required for plant tolerance to cold stress. Mol Cell Biol 26:9533–9543

Doyle MR, Bizzell CM, Keller MR, Michaels SD, Song J, Noh YS, Amasino RM (2005). HUA2 is required for the expression of floral repressors in *Arabidopsis thaliana*. Plant J 41:376–385

Elliott BJ, Dattaroy T, Meeks-Midkiff LR, Forbes KP, Hunt AG (2003). An interaction between an Arabidopsis poly(A) polymerase and a homologue of the 100 kDa subunit of CPSF. Plant Mol Biol 51:373–384

Golden TA, Schauer SE, Lang JD, Pien S, Mushegian AR, Grossniklaus U, Meinke DW, Ray A (2002) SHORT INTEGUMENTS1/SUSPENSOR1/CARPEL FACTORY, a Dicer homolog, is a maternal effect gene required for embryo development in Arabidopsis. Plant Physiol 130:808–822

Golovkin M, Reddy AS (1999) An SC35-like protein and a novel serine/arginine-rich protein interact with Arabidopsis U1-70K protein. J Biol Chem 274:36428–36438

Golovkin M Reddy ASN (1998) The plant U1 small nuclear ribonucleoprotein particle 70 K protein interacts with two novel serine/arginine-rich proteins. Plant Cell 10:1637–1648

Gong Z, Dong CH, Lee H, Zhu J, Xiong L, Gong, D, Stevenson, B, Zhu JK (2005) A DEAD box RNA helicase is essential for mRNA export and important for development and stress responses in Arabidopsis. Plant Cell 17:256–267

Gong Z Lee H, Xiong L, Jagendorf A, Stevenson B, Zhu J-K (2002) RNA helicase-like protein as an early regulator of transcription factors for plant chilling and freezing tolerance. Proc Natl Acad Sci USA 99:11507–11512

Grigg SP, Canales C, Hay A, Tsiantis M (2005) SERRATE coordinates shoot meristem function and leaf axial patterning in Arabidopsis. Nature 437:1022–1026

Henderson IR, Liu F, Drea S, Simpson GG, Dean C (2005) An allelic series reveals essential roles for FY in plant development in addition to flowering-time control. Development 132:3597–3607

Herr AJ, Molnar A, Jones A, Baulcombe DC (2006) Defective RNA processing enhances RNA silencing and influences flowering of Arabidopsis. Proc Natl Acad Sci USA 103:14994–15001

Hugouvieux V, Kwak JM, Schroeder JI (2001) An mRNA cap binding protein, ABH1, modulates early abscisic acid signal transduction in Arabidopsis. Cell 106:477–487

Jones-Rhoades MW, Bartel DP, Bartel, B. (2006). MicroRNAS and their regulatory roles in plants. Annu Rev Plant Biol 57:19–53

Kalyna M, Lopato S, Barta A (2003). Ectopic expression of atRSZ33 reveals its function in splicing and causes pleiotropic changes in development. Mol Biol Cell 14:3565–3577

Kim SH, Lin RJ (1996). Spliceosome activation by PRP2 ATPase prior to the first transesterification reaction of pre-mRNA splicing. Mol Cell Biol 16:6810–6819

Kim VN (2005) MicroRNA biogenesis: coordinated cropping and dicing. Nat Rev Mol Cell Biol 6:376–385

Komuro A, Saeki M, Kato S (1999) Association of two nuclear proteins, Npw38 and NpwBP, via the interaction between the WW domain and a novel proline-rich motif containing glycine and arginine. J Biol Chem 274:36513–36519

Kurihara Y, Watanabe Y (2004) Arabidopsis micro-RNA biogenesis through Dicer-like 1 protein functions. Proc Natl Acad Sci USA 101:12753–12758

Lee JH, Cho YS, Yoon HS, Suh MC, Moon J, Lee I, Weigel D, Yun CH, Kim JK (2005) Conservation and divergence of FCA function between Arabidopsis and rice. Plant Mol Biol 58:823–838

Lee JH, Yoo SJ, Park SH, Hwang, I, Lee JS, Ahn JH (2007) Role of SVP in the control of flowering time by ambient temperature in Arabidopsis. Genes Dev 21:397–402

Lim RY, Fahrenkrog B (2006) The nuclear pore complex up close. Curr Opin Cell Biol 18:342–347

Lobbes D, Rallapalli G, Schmidt DD, Martin C, Clarke J (2006) SERRATE: a new player on the plant microRNA scene. EMBO Rep 7:1052–1058

Lockhart SR, Rymond BC (1994) Commitment of yeast pre-mRNA to the splicing pathway requires a novel U1 small nuclear ribonucleoprotein polypeptide, Prp39p. Mol Cell Biol 14:3623–3633

Loiodice I, Alves A, Rabut G, Van Overbeek M, Ellenberg J, Sibarita JB, Doye V (2004) The entire Nup107–160 complex, including three new members, is targeted as one entity to kinetochores in mitosis. Mol Biol Cell 15:3333–3344

Lopato S, Forstner C, Kalyna M, Hilscher J, Langhammer U, Indrapichate K, Lorković Z J, Barta A (2002) Network of interactions of a novel plant-specific Arg/Ser-rich protein, atRSZ33, with atSC35-like splicing factors. J Biol Chem 277:39989–39998

Lopato S, Kalyna M, Dorner S, Kobayashi R, Krainer AR, Barta A (1999) atSRp30, one of two SF2/ASF-like proteins from *Arabidopsis thaliana*, regulates splicing of specific plant genes. Genes Dev 13:987–1001

Lorković ZJ, Wieczorek Kirk DA, Lambermon MH, Filipowicz W (2000) Pre-mRNA splicing in higher plants. Trends Plant Sci 5:160–167

Macknight R, Bancroft I, Page T, Lister C, Schmidt R, Love K, Westphal L, Murphy G, Sherson S, Cobbett C, Dean C (1997) FCA, a gene controlling flowering time in Arabidopsis, encodes a protein containing RNA-binding domains. Cell 89:737–745

Mandel CR, Kaneko S, Zhang H, Gebauer D, Vethantham V, Manley JL, Tong L (2006) Polyadenylation factor CPSF-73 is the pre-mRNA 3′-end-processing endonuclease. Nature 444:953–956

Megraw M, Baev V, Rusinov V, Jensen ST, Kalantidis K, Hatzigeorgiou AG (2006) MicroRNA promoter element discovery in Arabidopsis. RNA 12:1612–1619

Mockler TC, Yu X, Shalitin D, Parikh D, Michael TP, Liou J, Huang J, Smith Z, Alonso JM, Ecker JR et al. (2004). Regulation of flowering time in Arabidopsis by K homology domain proteins. Proc Natl Acad Sci USA 101:12759–12764

Noh Y.-S, Bizzell CM, Noh B, Schomburg FM, Amasino RM (2004) EARLY FLOWERING 5 acts as a floral repressor in Arabidopsis. Plant J 38:664–672

Ohnacker M, Barabino SM, Preker PJ, Keller W (2000) The WD-repeat protein pfs2p bridges two essential factors within the yeast pre-mRNA 3′-end-processing complex. EMBO J 19:37–47

Palatnik JF, Allen E, Wu X, Schommer C, Schwab R, Carrington JC, Weigel D (2003) Control of leaf morphogenesis by microRNAs. Nature 425:257–263

Park W, Li J, Song R, Messing J, Chen X (2002) CARPEL FACTORY, a Dicer homolog, and HEN1, a novel protein, act in microRNA metabolism in *Arabidopsis thaliana*. Curr Biol 12:1484–1495

Parry G, Ward S, Cernac A, Dharmasiri S, Estelle M (2006) The Arabidopsis SUPPRESSOR OF AUXIN RESISTANCE proteins are nucleoporins with an important role in hormone signaling and development. Plant Cell 18:1590–1603

Penheiter KL, Washburn TM, Porter SE, Hoffman MG, Jaehning JA (2005) A posttranscriptional role for the yeast Paf1-RNA polymerase II complex is revealed by identification of primary targets. Mol Cell 20:213–223

Proudfoot N (2004) New perspectives on connecting messenger RNA 3′ end formation to transcription. Curr Opin Cell Biol 16:272–278

Quesada V, Dean C, Simpson GG (2005) Regulated RNA processing in the control of Arabidopsis flowering. Int J Dev Biol 49:773–780

Quesada V, Macknight R, Dean C, Simpson GG (2003) Autoregulation of FCA pre-mRNA processing controls Arabidopsis flowering time. EMBO J 22:3142–3152

Rajagopalan R, Vaucheret H, Trejo J, Bartel DP (2006) A diverse and evolutionarily fluid set of microRNAs in *Arabidopsis thaliana*. Genes Dev 20:3407–3425

Reddy ASN (2004) Plant serine/arginine-rich proteins and their role in pre-mRNA splicing. Trends Plant Sci 9:541–547

Samach A, Wigge PA (2005) Ambient temperature perception in plants. Curr Opin Plant Biol 8:483–486

Schmid M, Uhlenhaut NH, Godard F, Demar M, Bressan R, Weigel D, Lohmann JU (2003) Dissection of floral induction pathways using global expression analysis. Development 130:6001–6012

Schmitz RJ, Hong L, Michaels S, Amasino RM (2005) FRIGIDA-ESSENTIAL 1 interacts genetically with FRIGIDA and FRIGIDA-LIKE 1 to promote the winter-annual habit of Arabidopsis thaliana. Development 132:5471–5478

Schomburg FM, Patton DA, Meinke DW, Amasino RM (2001) FPA, a gene involved in floral induction in Arabidopsis, encodes a protein containing RNA-recognition motifs. Plant Cell 13:1427–1436

Schwab R, Palatnik JF, Riester M, Schommer C, Schmid M, Weigel D (2005). Specific effects of microRNAs on the plant transcriptome. Dev Cell 8:517–527

Scortecci K, Michaels SD, Amasino RM (2003) Genetic interactions between FLM and other flowering-time genes in *Arabidopsis thaliana*. Plant Mol Biol 52:915–922

Scortecci KC, Michaels SD, Amasino RM (2001) Identification of a MADS-box gene, FLOWERING LOCUS M that represses flowering. Plant J 26:229–236

Sheldon CC, Conn AB, Dennis ES, Peacock WJ (2002) Different regulatory regions are required for the vernalization-induced repression of FLOWERING LOCUS C and for the epigenetic maintenance of repression. Plant Cell 14:2527–2537

Simpson GG, Dijkwel PP, Quesada V, Henderson I, Dean C (2003) FY Is an RNA 3′ end-processing factor that interacts with FCA to control the Arabidopsis floral transition. Cell 113:777–787

Sommer P, Nehrbass U (2005) Quality control of messenger ribonucleoprotein particles in the nucleus and at the pore. Curr Opin Cell Biol 17:294–301

Sunkar R, Girke T, Jain PK, Zhu J-K (2005) Cloning and characterization of MICRORNAS from rice. Plant Cell 17:1397–1411

Teigelkamp S, McGarvey M, Plumpton M, Beggs JD (1994) The splicing factor PRP2, a putative RNA helicase, interacts directly with pre-mRNA. EMBO J 13:888–897

Thomas B (2006) Light signals and flowering. J Exp Bot 57:3387–3393

Vazquez F, Gasciolli V, Crete P, Vaucheret H. (2004) The nuclear dsRNA binding protein HYL1 is required for microRNA accumulation and plant development, but not posttranscriptional transgene silencing. Curr Biol 14:346–351

Vijayraghavan U, Company M, Abelson J (1989) Isolation and characterization of pre-mRNA splicing mutants of *Saccharomyces cerevisiae*. Genes Dev 3:1206–1216

Wang BB, and Brendel V (2006) Molecular characterization and phylogeny of U2AF35 homologs in plants. Plant Physiol 140:624–636

Wang C, Tian Q, Hou Z, Mucha M, Aukerman M, Olsen OA (2007) The *Arabidopsis thaliana* AT PRP39-1 gene, encoding a tetratricopeptide repeat protein with similarity to the yeast pre-mRNA processing protein PRP39, affects flowering time. Plant Cell Rep 26:1357–1366

Waragai M, Junn E, Kajikawa M, Takeuchi S, Kanazawa I, Shibata M, Mouradian MM, Okazawa H (2000) PQBP-1/Npw38, a nuclear protein binding to the polyglutamine tract, interacts with U5-15kD/dim1p via the carboxyl-terminal domain. Biochem Biophys Res Commun 273:592–595

Werner JD, Borevitz JO, Uhlenhaut NH, Ecker JR, Chory J, Weigel D (2005) FRIGIDA-independent variation in flowering time of natural *Arabidopsis thaliana* accessions. Genetics 170:1197–1207

Wu G, Poethig RS (2006) Temporal regulation of shoot development in *Arabidopsis thaliana* by miR156 and its target SPL3. Development 133:3539–3547

Xie Z, Allen E, Fahlgren N, Calamar A, Givan SA, Carrington JC (2005) Expression of Arabidopsis MIRNA genes. Plant Physiol 138:2145–2154

Yang L, Liu Z, Lu F, Dong A, Huang H (2006) SERRATE is a novel nuclear regulator in primary microRNA processing in Arabidopsis. Plant J 47:841–850

Zhang B, Wang Q, Pan X (2007) MicroRNAs and their regulatory roles in animals and plants. J Cell Physiol 210:279–289

Alternative Splicing in Plant Defense

W. Gassmann

Contents

Introduction	220
Plant Resistance Proteins	220
Resistance Protein Classes	220
Distinct CNL and TNL *R* Gene Structures	221
R Gene-Dependent Signaling	221
Alternative Splicing of CNL Gene Transcripts	222
Alternative Splicing of TNL Gene Transcripts	223
Conservation of Alternative ORFs via Different Alternative Splicing Mechanisms	223
Requirement for Alternative TNL *R* Gene Transcripts	224
Possible Functions of Alternative TNL Transcript Products	226
Possible Mechanisms of Alternative Splicing Regulation	227
Coupling of Transcription and Alternative Splicing	227
Nonsense-Mediated Decay	228
Future Directions	228
References	229

Abstract Plant resistance proteins directly or indirectly perceive the presence of pathogen virulence factors and trigger an effective form of plant immunity that often includes programmed host cell death. Because the activation of resistance proteins has the potential to be detrimental to the plant, this process is tightly regulated on multiple levels. Several resistance genes have been shown to be alternatively spliced. Depending on the resistance gene, alternative transcripts are thought to limit the expression of R proteins or encode truncated R proteins with a positive role in defense activation. In addition, *R* gene alternative splicing is dynamic during the defense response. Possible mechanisms of *R* gene alternative splicing regulation and how alternative *R* gene transcripts fit into the current view of resistance protein-mediated defense responses are discussed.

W. Gassmann
Division of Plant Sciences and C.S. Bond Life Sciences Center, University of Missouri, Columbia, MO 65211-7310, USA
e-mail: gassmannw@missouri.edu

Introduction

Plant defense against microbial pathogens comes in many forms and at many levels. All plants are resistant to all pathogens at some basal level, but this is not always sufficient to prevent pathogen colonization and growth. A very effective and genetically well-defined form of plant defense is mediated by highly specific resistance (*R*) genes that elicit defense responses upon perception of so-called pathogen avirulence (*avr*) genes, originally described in the gene-for-gene hypothesis (Flor 1971). Over the last decades a large body of research has established that these *avr* genes encode effector proteins whose original function is to promote pathogen fitness and aggressiveness (Abramovitch et al. 2006; Grant et al. 2006). Plants in turn evolved R proteins that directly or indirectly detect the presence of these pathogen effector proteins. Therefore, this resistance mechanism has recently been named effector-triggered immunity (ETI) to more clearly capture the evolutionary and functional basis of gene-for-gene resistance (Chisholm et al. 2006; Jones and Dangl 2006).

Activation of R proteins by an avirulent pathogen usually results in localized plant programmed cell death (Goodman and Novacky 1994; Greenberg and Yao 2004). Constitutive activation of defense responses leads to pleiotropic and deleterious phenotypes. Consequently, ETI must be tightly regulated, both positively and negatively, to provide a fine balance between rapid response to pathogens and prevention of host cell damage (McDowell and Simon 2006). The data available to date suggest that alternative splicing of *R* gene transcripts is one facet of the fine-tuning of R protein activity, either by impacting the total amount of R protein that is being synthesized or by encoding R protein variants. Although alternative splicing of transcripts from defense genes other than *R* genes has been reported (Li and Howe 2001; Shigeoka et al. 2002), the functional significance of the defense response of these genes and their alternative transcripts is not easily determined. Therefore, this review focuses on the alternative splicing of *R* gene transcripts, where some progress has been made in measuring the relevance of alternative splicing in plant defense.

Plant Resistance Proteins

Resistance Protein Classes

Plant R proteins come in many forms. To date, five distinct classes of predicted R proteins have been described, with additional proteins classified as R proteins that do not fall into any of the five major classes (Martin et al. 2003; Nimchuk et al. 2003). Most characterized plant *R* genes encode predicted proteins with nucleotide binding site (NBS) and leucine-rich repeat (LRR) motifs thought to be important for regulation of activation and protein-protein interaction specificity, respectively (Tameling et al. 2002; Martin et al. 2003; Belkhadir et al. 2004; Dodds et al. 2006; McDowell and Simon 2006; Tameling et al. 2006). The NBS-LRR R proteins can be grouped into two main families based on their N-terminal domains (Meyers et al. 1999). One family possesses a coiled-coil (CC) domain (also called a leucine zipper domain). The other contains a Toll/interleukin-1 receptor (TIR) domain that shows sequence similarity to the cytosolic domain of Drosophila Toll and mammalian interleukin-1 transmembrane receptors, proteins known to be involved in the animal innate immune response (Martin et al. 2003; Jones and Dangl 2006). These two families of *R* genes are not equally distributed in the plant kingdom: Whereas the majority of predicted Arabidopsis *R* genes belong to the TIR-NBS-LRR (TNL) family, this family of genes appears to be absent in monocots (Meyers et al. 1999). In general, CC-NBS-LRR (CNL) and TNL proteins are predicted to be cytosolic, that is, the majority do not contain predicted transmembrane domains or organellar targeting signals. For a few R proteins, however, distinct and dynamic subcellular localizations have been established experimentally (Boyes et al. 1998; Deslandes et al. 2003; Burch-Smith et al. 2007; Shen et al. 2007).

Distinct CNL and TNL R Gene Structures

Based on a thorough analysis of the Arabidopsis *R* gene complement, Meyers et al. divided each of the CNL and TNL families of *R* genes into groups based on protein motifs and gene structure (Meyers et al. 2003). In the context of alternative splicing, it is noteworthy that the best-understood group of Arabidopsis *R* genes, the CNL genes *RPS2* (Bent et al. 1994), *RPM1* (Grant et al. 1995), and *RPS5* (Warren et al. 1998), are encoded by intronless genes. Although transcript variants based on alternative polyadenylation sites in the maize *Rp1D* gene, which has an intronless open reading frame (ORF), have been detected (Ayliffe et al. 2004), no alternative transcripts have ever been reported for the intronless Arabidopsis CNL genes. Arabidopsis CNL genes of known function with introns include the two genes *RPP8* and *RPP7*, which recognize specific strains of the oomycete pathogen *Hyaloperonospora parasitica* (formerly called *Peronospora parasitica*, hence *RPP* genes) (McDowell et al. 1998; Eulgem et al. 2007). Known CNL *R* genes from other plant species also contain introns and display diverse gene structures. In contrast, TNL genes from many plant species that have been functionally characterized to date show a highly conserved gene structure, with the first and second exons encoding the TIR and the NBS domain, respectively. This conservation of gene structure and its possible functional significance are discussed further below.

R Gene-Dependent Signaling

The genetics of Arabidopsis defense signaling in general distinguishes between the two major families of resistance genes (Aarts et al. 1998). Disease resistance conferred by members of the CNL family of bacterial resistance genes is strongly dependent on a functional *NDR1* allele (non-race-specific disease resistance) (Century et al. 1995, 1997) and unaffected by mutations in *EDS1* (enhanced disease susceptibility) (Parker et al. 1996; Falk et al. 1999), whereas TNL genes depend on a functional *EDS1* allele and are unaffected by mutations in *NDR1*. The structural basis for the dichotomy between *EDS1*- and *NDR1*-dependent resistance genes and the homology of R protein TIR domains to the cytosolic signaling domains of animal innate immunity transmembrane receptors suggests that the putative N-terminal domain of resistance proteins (TIR or CC) interacts with downstream signaling elements. However, not all R proteins fit into this scheme, suggesting that additional factors determine R protein signaling pathways (McDowell et al. 2000; Bittner-Eddy and Beynon 2001; Chandra-Shekara et al. 2004).

Additional genes that play a role in regulating disease resistance signaling by multiple *R* genes include *PBS2/RAR1* (Warren et al. 1999; Muskett et al. 2002; Tornero et al. 2002) and *SGT1* (Austin et al. 2002; Tör et al. 2002), which encode co-chaperones of R proteins that appear to mainly affect R protein stability (Shen et al. 2003; Bieri et al. 2004). Additional results with site-directed mutagenesis experiments between RAR1-dependent and -independent barley MLA R proteins suggest that, beyond the absolute amount of R protein, RAR1 is important in regulating the amount of active R protein (Halterman and Wise 2004, 2006). RAR1 and SGT1 physically interact with components of the ubiquitin ligase complex, suggesting that protein degradation may be part of R protein regulation (Austin et al. 2002; Azevedo et al. 2002; Liu et al. 2002).

The heat shock protein HSP90 has emerged as a major interactor of several R proteins (Hubert et al. 2003; Lu et al. 2003; Takahashi et al. 2003; Liu et al. 2004). HSP90 interacts with RAR1 and SGT1 and similarly regulates the stability of R proteins. It has therefore been suggested that HSP90 together with RAR1 and SGT1 hold R proteins in a signaling-competent but downregulated state before activation by the avirulence signal (Belkhadir et al. 2004). Importantly in the context of this review, the main function of these proteins is to regulate the amount of active R proteins in the cell. In other words, changes in the amount of active R protein either slow down the response or activate it inappropriately, highlighting the quantitative nature of the *R* gene-mediated defense response (Tao et al. 2003; Caldo et al. 2004; Caldo et al. 2006). The underlying hypothesis of alternative splicing studies of *R* gene studies fits with this conclusion. The available evidence so far suggests that with CNL genes alternative splicing regulates the expression and hence the amount of R proteins, whereas with TNL genes the expression of distinct R protein variants is dynamically controlled.

Alternative Splicing of CNL Gene Transcripts

Alternative splicing of CNL genes has been most carefully studied with the barley *Mla* genes that control ETI to the barley powdery mildew fungus *Blumeria graminis* f. sp. *hordei* (Bgh). Full-length and truncated *Mla* alleles are clustered in a complex locus (Wei et al. 2002). Across different barley varieties, approximately 30 *Mla* alleles with distinct recognition specificities have been mapped to this locus, and several *Mla* alleles have been cloned (Halterman et al. 2001, 2003; Zhou et al. 2001; Shen et al. 2003; Halterman and Wise 2004). A first indication of *Mla* transcript alternative splicing was discovered in the course of cloning *Mla6* (Halterman et al. 2001; Shen et al. 2003; Halterman and Wise 2004). An *Mla6*-like gene with an almost identical 5'-untranslated region (UTR) sequence and a severely truncated

ORF was revealed by cDNA sequencing. While *Mla6* itself appeared constitutively spliced, the truncated cDNAs provided evidence for two alternatively spliced introns within the conserved 5'-UTR, whereas *Mla6* cDNAs showed constitutive splicing of the second intron but not the first predicted intron. However, the full-length *Mla6* gene was sufficient to provide *Mla6* function in a transient assay system (Halterman et al. 2001).

The structure of the *Mla13* allele is very similar to that of *Mla6*, including the presence of two introns in the 5'-UTR. Here, it was shown that the single *Mla13* gene gives rise to at least five distinct transcripts that differ in the splicing of the two 5'-UTR introns and the presence or absence of a GTG trimer at the 3' end of the first exon, indicating alternative 5' splice sites in intron 1(Halterman et al. 2003). The different combinations of these features determine the presence or absence of small upstream ORFs (uORFs) between 10 and 70 amino acids in length that could determine efficiency of translation (Halterman et al. 2003). Interestingly, it was found that *Mla13* was induced by the avirulent Bgh isolate, and that the ratio of the different *Mla13* transcripts changed during the resistance response.

The hypothesis that the *Mla13* 5'-UTR influences protein expression was tested experimentally by fusing a firefly luciferase reporter cDNA to the genomic 5'-UTR or one of the five 5'-UTRs found in *Mla13* cDNAs (Halterman and Wise 2006). In addition, the start codons of the uORFs were mutated individually or in combination. In vitro transcribed cRNA was introduced into oat protoplasts to measure the efficiency of protein translation. Inactivation of uORF translation by mutation of the associated start codons indeed led to an increase in reporter protein synthesis by 50% with the 5'-UTR sequences that are present in the two most abundant transcripts during the resistance response, and by 150% if all uORF start codons were mutated in the genomic 5'-UTR. This provided evidence that the 5'-UTR and its associated uORFs downregulate protein expression (Halterman and Wise 2006). However, no difference in reporter protein activity was measured when comparing the five naturally occurring 5'-UTRs with intact uORF start codons. In other words, a switch from one transcript to another during the defense response would not change MLA13 protein expression levels if luciferase activity in oat protoplasts accurately measures MLA13 synthesis in barley cells. In conclusion, alternative splicing of the *Mla13* 5'-UTR likely establishes a constant ceiling for MLA13 synthesis, despite the induction of alternative transcripts that without uORFs would lead to a possibly deleterious overaccumulation of MLA13.

Alternative Splicing of TNL Gene Transcripts

Conservation of Alternative ORFs via Different Alternative Splicing Mechanisms

Members of the TNL family have been functionally characterized from a large number of diverse dicotyledonous plant species and include the tobacco *N* gene specifying resistance to tobacco mosaic virus (Whitham et al. 1994), the flax *L6* and *M* genes for resistance to the fungal pathogen *Melampsora lini* (Lawrence et al. 1995; Anderson et al. 1997), *RPP5* (Parker et al. 1997) and members of the *RPP1* gene cluster (Botella et al. 1998) from Arabidopsis conferring resistance to the oomycete pathogen *Hyaloperonospora parasitica*, Arabidopsis *RPS4* conferring resistance to *Pseudomonas syringae* pv. tomato expressing *avrRps4* (Gassmann et al. 1999), Arabidopsis *RAC1* against the fungal pathogen *Albugo candida* (Borhan et al. 2004), and the tomato *Bs4* gene against *Xanthomonas campestris* pv. vesicatoria expressing *avrBs4* (Schornack et al. 2004).

Despite their diverse origins and the pathogens that they recognize, these TNL genes share a conserved gene structure at their 5' ends, and in most cases alternative transcripts arising from these genes have been described. Remarkably, the mechanism of alternative transcript generation varies among TNL genes and includes alternative exons, alternative 5' or 3' splice site selection, and intron retention. Yet the encoded predicted proteins are very similar and consist of truncated TIR-NBS (TN) proteins. In the case of *N*, the putative truncated N protein arises through the alternative splicing of a 70-bp alternative exon within intron 3 (Whitham et al. 1994). Truncated L6 arises through inefficient splicing of intron 3, which contains an in-frame stop codon (Lawrence et al. 1995). Likewise, truncated RPS4 is encoded by alternative transcripts with retained introns 2 or 3 (Gassmann et al. 1999; Zhang and Gassmann 2003). A splicing-independent mechanism was observed with *RPP5*, where a predicted truncated RPP5 protein with identical amino acid sequence in the TIR and almost complete NBS domain is encoded by a genetically linked member of the *RPP5* gene family with an inserted retroelement (Parker et al. 1997; Noël et al. 1999). No alternative splicing was observed in *RPP5* itself. A second group of TNL genes represented by *RAC1* generate alternative transcripts encoding TIR-only proteins, via retention of intron 1 (Borhan et al. 2004). It should be noted that based on the scanning ribosome model of eukaryotic transcript translation, which entails that the ribosome binds the transcript cap at the 5'-end and slides along the mRNA until it encounters a start codon, a TNL transcript with retained intron 1 is more likely to encode a TIR protein rather than the suggested NBS-LRR protein.

Requirement for Alternative TNL R Gene Transcripts

The production of alternative transcripts that encode truncated TIR or TN proteins by different mechanisms is intriguing and suggests that these putative proteins fulfill an important function. The requirement for alternative transcripts in resistance function was experimentally tested with several TNL genes, with mixed results. First, it was found that a tobacco *N* transgene with the native *N* promoter lacking the alternative exon within the third intron did not provide full *N* function when stably transformed into susceptible *nn* tobacco plants, whereas removing any of the other three *N* gene introns did not compromise function. Susceptible plants transformed with a genomic *N* clone or genes lacking any of the other introns were fully resistant. In contrast, TMV-induced HR was delayed sufficiently on primary leaves of plants with the *N* gene lacking the alternative exon so that TMV spread systemically, causing a spreading systemic HR (Dinesh-Kumar and Baker 2000). Complete removal of intron 3 had similar effects, while a cDNA representing the alternative *N* transcript was fully nonfunctional (systemic mosaic symptoms and no spreading HR). Coexpression of regular and alternative cDNAs was no better than expression of the regular cDNA alone. A closer determination of the requirements for full function in a single *N* transgene clone identified intron 3, and genomic 3' sequences that are transcribed and regulate *N* alternative splicing, as essential for rapid HR to TMV that prevents systemic spread of the virus. Taken together, these results showed that a transgene encoding full-length N protein alone was not sufficient to achieve full resistance to TMV.

The fact that coexpression of cDNAs for the regular and alternative *N* transcripts also did not confer complete resistance suggested that the ratio of the two transcripts was an additional important feature of *N* gene regulation. Indeed, the ratio of *N* gene transcripts for full-length and truncated protein changed during the infection process: Whereas the regular *N* transcript dominated before TMV infection with a ratio of the regular to the alternative transcript of approximately 28:1, the alternative transcript dominated 1:23 6 h after TMV inoculation. By 9 h, the transcript ratio had returned to resting levels (Dinesh-Kumar and Baker 2000).

Similar results were found with the Arabidopsis *RPS4* gene, with some distinctions to *N*. The function of *RPS4* alternative transcripts, which are generated by intron 2 or 3 retention and splicing of a cryptic intron in exon 3, was tested by stably expressing intron-deprived transgenes in a susceptible Arabidopsis line. Surprisingly, removal of just one intron from the transgene was sufficient to abolish *RPS4* function completely (Zhang and Gassmann 2003). Removal of a single intron did not impact the splicing of the remaining introns, so either removal of a single intron sufficiently upsets a fine balance of transcript ratios or individual transcripts fulfill different functions. Interestingly, the function of intron-deprived transgenes was complemented by the presence of a second, truncated, *RPS4* transgene, directly showing a requirement for more than one *RPS4* transcript for function (Zhang and Gassmann 2003). Quantification of transcript abundances showed that only one of five experimentally verified alternative *RPS4* transcripts, with intron 3 retained, is transiently upregulated during the defense response, showing that this system is very specific and under dynamic regulation (Zhang and Gassmann 2007).

Conversely, transgenic flax plants stably expressing an intronless *L6* gene incapable of giving rise to alternative transcripts showed full resistance (Ayliffe et al. 1999). The authors speculated that the putative function of a truncated L6 protein could be substituted by truncated proteins of the highly related *M* locus, members of which are also alternatively spliced. However, more recent results with transiently coexpressed *L6* cDNA and the cognate avirulence gene *AvrL6* in tobacco, which gave an HR, make this unlikely (P. Dodds, personal communication). Likewise, a *Bs4* cDNA transiently expressed in tomato was also functional in recognizing *avrBs4* (Schornack et al. 2004). Recently, Jones and coworkers showed an *avrRps4*-independent hypersensitive response in tobacco when *RPS4* was overexpressed

(Zhang et al. 2004). Since this was also observed with a cDNA, they concluded that in this system no alternative transcripts are required. This is in contrast to the complete abolition of *RPS4* function of intron-deprived *RPS4* transgenes expressed from their native promoters in stable Arabidopsis transgenics (Zhang and Gassmann 2003). On balance, therefore, alternative transcripts in some cases serve a regulatory role in appropriate and accelerated activation of defense responses that might not be evident when transiently (over)expressing *R* gene cDNAs. The quantitative nature of the *R* gene-mediated defense response makes it difficult to interpret negative results regarding the function of alternative transcripts that were obtained with nonnative *R* gene expression levels.

Possible Functions of Alternative TNL Transcript Products

A truncated TN R protein could function as an adaptor protein analogous to the animal TIR domain-containing protein MyD88 that functions in innate immunity downstream of interleukin-1 and Toll-like receptor proteins (O'Neill 2000; Meyers et al. 2002; Tauszig-Delamasure et al. 2002). In an alternative model, the truncated TN protein interacts with the full-length TNL protein to disrupt intramolecular interactions that downregulate TNL protein activation. Disruption of self-inhibition in a subset of resistance complexes could prime the system for efficient activation upon pathogen recognition (Zhang and Gassmann 2003). Support for the model that R proteins are self-inhibited comes from many studies. In domain swap experiments between the tomato root-knot nematode and potato aphid resistance gene *Mi* and a closely related homolog, it was found that chimeric R proteins became constitutively active (Hwang et al. 2000). These authors proposed that resistance proteins are self-inhibited by interactions between their N- and C-terminal domains, and that this self-inhibition is relieved by interaction with a pathogen effector protein or host protein modified by pathogen invasion (Hwang et al. 2000).

Intriguingly, Moffett and coworkers showed that the Rx protein can be provided in two parts, a CC-NBS and an LRR part (Moffett et al. 2002). Interestingly, these two partial Rx proteins were found to interact in plant cells, and this interaction was disrupted by the cognate avirulence protein, the coat protein of potato virus X (Moffett et al. 2002). Additional support for self-inhibition is accumulating through the study of point mutations in R proteins that render them constitutively active (Bendahmane et al. 2002; Shirano et al. 2002; Zhang et al. 2003). These point mutations may disrupt the inter- or intramolecular interactions that keep resistance proteins in a downregulated state (Belkhadir et al. 2004; McDowell and Simon 2006).

In addition to the above models, and not mutually exclusive, the TIR or TN protein alone may trigger the resistance response more potently than the full-length protein. In this model, TIR and TN proteins are sufficient for defense activation and represent the actual signal moiety of the full-length protein. Unlike the full-length protein, TIR or TN proteins are difficult to control and are only produced once the pathway has been activated by the full-length protein. Several studies using transient expression of R protein subdomains have shown that the N-terminal domains of R proteins are sufficient to trigger cell death (Zhang et al. 2004; Ade et al. 2007). Rather than interacting with the full-length R protein, the TIR or TN proteins would directly engage with the downstream signaling machinery to activate resistance. It is also noteworthy that the Arabidopsis genome contains a large number of genes that encode predicted TIR and TN proteins, although the function of these genes is unknown (Meyers et al. 2002).

An additional role for truncated R proteins may arise from the documented dynamic subcellular localization of full-length R proteins (Boyes et al. 1998; Deslandes et al. 2003; Burch-Smith et al. 2007; Shen et al. 2007). Truncated R proteins could be localized to a different compartment from the full-length protein. Subcellular localization of R proteins and their potential corresponding truncated products arising from alternative splicing needs to be verified experimentally, because so far R protein localization is not easily predicted or defined by clear localization signals.

Possible Mechanisms of Alternative Splicing Regulation

The observation that *Mla13*, *N*, and *RPS4* transcript profiles are dynamic and induced by the cognate avirulent pathogen suggests that alternative splicing of *R* genes is temporally regulated and dynamic; however, very little is known about the underlying mechanisms. Some splice factors themselves are stress regulated, thus possibly providing a direct mechanism for dynamic *R* gene transcript alternative splicing during the defense response (Reddy 2004). These changes in splicing factors are described in other chapters (see the chapters by A. Barta et al. and G. S. Ali and A. S. N. Reddy, this volume). Here we discuss additional mechanisms that may have a function in regulating *R* gene expression.

Coupling of Transcription and Alternative Splicing

Several proteins associated with gene transcription regulate the incidence of alternative splicing of mammalian transcripts, suggesting that transcription and transcript processing are linked (Kornblihtt 2005). In humans, the splicing machinery recognizes exons (exon definition), and the predominant form of alternative splicing is exon skipping (Gupta et al. 2004). Interestingly, in humans and yeast it was found that the elongation rate of RNA polymerase II can be the most important factor in determining the ratio of transcripts with skipped exons (Kornblihtt 2005). The faster RNA polymerase II elongates a transcript, the higher the likelihood of exon skipping. If intron definition is the predominant form of intron recognition in Arabidopsis (Wang and Brendel 2006), then intron retention of weakly recognized introns would be expected to increase with higher transcription rates, analogous to increased exon skipping in humans.

Intriguingly, it was found that intron retention and overall mRNA levels of *RPS4* are coordinately regulated during the resistance response: As *RPS4* mRNA levels are induced, the amount of the alternative transcript with intron 3 retained increases (Zhang and Gassmann 2007). Intron retention was reported as the major mechanism of alternative transcript generation on a genomewide basis in Arabidopsis (Ner-Gaon et al. 2004; Wang and Brendel 2006). Interestingly, stress-response genes were overrepresented in the group of transcripts that showed intron retention (Ner-Gaon et al. 2004). Since stress-response genes by definition are upregulated during the stress response, this property may explain a number of instances of dynamically regulated alternative splicing. Gene induction is possibly a general feature of *R* gene regulation and was also found for *N* (Levy et al. 2004) and *Mla13* (Halterman et al. 2003).

Nonsense-Mediated Decay

An additional process that is likely to play a role in alternative *R* gene transcript accumulation is nonsense-mediated decay (NMD) (see the chapter by D. A. Belostotsky, this volume). This mRNA quality control mechanism is widespread among eukaryotes and degrades mRNAs with premature stop codons (Maquat 2004). Notably, alternative transcripts of TNL genes contain premature stop codons and are generally of very low abundance before infection (Dinesh-Kumar and Baker 2000) (Zhang and Gassmann 2007). Mutations in proteins in the NMD pathway such as UPF1 and UPF3 were shown to stabilize mRNAs containing premature stop codons (Hori and Watanabe 2005; Arciga-Reyes et al. 2006). Conceivably, downregulation of the NMD pathway during the defense response could lead to accumulation of alternative *R* gene transcripts. On the other hand, not all *RPS4* alternative transcripts were upregulated during the defense response, even though all alternative transcripts contain premature stop codons. Therefore, it is most likely a combination of increased alternative transcript generation and increased transcript stability that leads to the observed dynamic regulation of *R* gene transcripts.

Future Directions

To date, numerous studies have documented the generation of alternative *R* gene transcripts. In the case of TNL genes, these transcripts are predicted to encode truncated R proteins. A major challenge will be to detect these putative truncated proteins and to determine their function. Transient expression of truncated RPS4 has shown that the TIR and TIR-NBS domains alone are able to induce an *avr-Rps4*-independent HR when overexpressed. It is therefore not trivial to detect these proteins in vivo in the context of a natural infection process, which is necessary to determine the function of truncated R proteins in activating or modulating the resistance response. For CNL genes, it will be important to document additional examples where alternative transcripts with differing uORFs limit R protein production, or perhaps discover additional functions of alternative CNL gene transcripts (Ferrier-Cana et al. 2005).

What regulates the changes in alternative R gene transcript profiles during the resistance response is a second important open question. The trigger may not be the same in every gene-for-gene interaction. Conceivably, in some cases it could be a direct effect of the pathogen effector virulence function to alter or negatively impact the host transcript processing machinery. In other cases, it could be a result of the general host stress response. Finally, a cell that is programmed to undergo self-destruction may simply do away with the niceties of quality control. These open questions will undoubtedly be answered in the near future, as in vivo protein imaging and transcript profiling tools become more sensitive and sophisticated. The R gene-mediated response has the advantage that the function of R genes can be easily quantified in terms of disease resistance and that R genes are not essential for plant viability in the absence of pathogens. The study of R gene alternative splicing therefore promises increased functional insights into the important process of alternative splicing in plants.

References

Aarts N, Metz M, Holub E, Staskawicz BJ, Daniels MJ, Parker JE (1998) Different requirements for *EDS1* and *NDR1* by disease resistance genes define at least two R gene-mediated signaling pathways in *Arabidopsis*. Proc Natl Acad Sci USA 95:10306–10311

Abramovitch RB, Anderson JC, Martin GB (2006) Bacterial elicitation and evasion of plant innate immunity. Nat Rev Mol Cell Biol 7:601–611

Ade J, DeYoung BJ, Golstein C, Innes RW (2007) Indirect activation of a plant nucleotide binding site-leucine-rich repeat protein by a bacterial protease. Proc Natl Acad Sci USA 104:2531–2536

Anderson PA, Lawrence GJ, Morrish BC, Ayliffe MA, Finnegan EJ, Ellis JG (1997) Inactivation of the flax rust resistance gene *M* associated with loss of a repeated unit within the leucine-rich repeat coding region. Plant Cell 9:641–651

Arciga-Reyes L, Wootton L, Kieffer M, Davies B (2006) UPF1 is required for nonsense-mediated mRNA decay (NMD) and RNAi in Arabidopsis. Plant J 47:480–489

Austin MJ, Muskett P, Kahn K, Feys BJ, Jones JDG, Parker JE (2002) Regulatory role of *SGT1* in early R gene-mediated plant defenses. Science 295:2077–2080

Ayliffe MA, Frost DV, Finnegan EJ, Lawrence GJ, Anderson PA, Ellis JG (1999) Analysis of alternative transcripts of the flax *L6* rust resistance gene. Plant J 17:287–292

Ayliffe MA, Steinau M, Park RF, Rooke L, Pacheco MG Hulbert SH, Trick HN, Pryor AJ (2004) Aberrant mRNA processing of the maize *Rp1-D* rust resistance gene in wheat and barley. Mol Plant-Microbe Interact 17:853–864

Azevedo C, Sadanandom A, Kitagawa K, Freialdenhoven A, Shirasu K, Schulze-Lefert P (2002) The RAR1 interactor SGT1, an essential component of R gene-triggered disease resistance. Science 295:2073–2076

Belkhadir Y, Subramaniam R, Dangl JL (2004) Plant disease resistance protein signaling: NBS-LRR proteins and their partners. Curr Opin Plant Biol 7:391–399

Bendahmane A, Farnham G, Moffett P, Baulcombe DC (2002) Constitutive gain-of-function mutants in a nucleotide binding site-leucine rich repeat protein encoded at the *Rx* locus of potato. Plant J 32:195–204

Bent AF, Kunkel BN, Dahlbeck D, Brown KL, Schmidt R, Giraudat J, Leung J, Staskawicz BJ (1994) *RPS2* of *Arabidopsis thaliana*: A leucine-rich repeat class of plant disease resistance genes. Science 265:1856–1860

Bieri S, Mauch S, Shen QH, Peart J, Devoto A, Casais C, Ceron F, Schulze S, Steinbiß HH, Shirasu K, Schulze-Lefert P (2004) RAR1 positively controls steady state levels of barley MLA resistance proteins and enables sufficient MLA6 accumulation for effective resistance. Plant Cell 16:3480–3495

Bittner-Eddy PD, Beynon JL (2001) The Arabidopsis downy mildew resistance gene, *RPP13-Nd*, functions independently of *NDR1* and *EDS1* and does not require the accumulation of salicylic acid. Mol Plant-Microbe Interact 14:416–421

Borhan MH, Holub EB, Beynon JL, Rozwadowski K, Rimmer SR (2004) The Arabidopsis TIR-NB-LRR gene *RAC1* confers resistance to *Albugo candida* (white rust) and is dependant on *EDS1* but not *PAD4*. Mol Plant-Microbe Interact 17:711–719

Botella MA, Parker JE, Frost LN, Bittner-Eddy PD, Beynon JL, Daniels MJ, Holub EB, Jones JDG (1998) Three genes of the Arabidopsis *RPP1* complex resistance locus recognize distinct *Peronospora parasitica* avirulence determinants. Plant Cell 10:1847–1860

Boyes DC, Nam J, and Dangl JL (1998) The *Arabidopsis thaliana RPM1* disease resistance gene product is a peripheral plasma membrane protein that is degraded coincident with the hypersensitive response. Proc Natl Acad Sci USA 95:15849–15854

Burch-Smith TM, Schiff M, Caplan JL, Tsao J, Czymmek K, Dinesh-Kumar SP (2007) A novel role for the TIR domain in association with pathogen-derived elicitors. PLoS Biol 5:e68

Caldo RA, Nettleton D, Wise RP (2004) Interaction-dependent gene expression in *Mla*-specified response to barley powdery mildew. Plant Cell 16:2514–2528

Caldo RA, Nettleton D, Peng J, Wise RP (2006) Stage-specific suppression of basal defense discriminates barley plants containing fast- and delayed-acting *Mla* powdery mildew resistance alleles. Mol Plant-Microbe Interact 19:939–947

Century KS, Holub EB, Staskawicz BJ (1995) *NDR1*, a locus of *Arabidopsis thaliana* that is required for disease resistance to both a bacterial and a fungal pathogen. Proc Natl Acad Sci USA 92:6597–6601

Century KS, Shapiro AD, Repetti PP, Dahlbeck D, Holub E, Staskawicz BJ (1997) *NDR1*, a pathogen-induced component required for *Arabidopsis* disease resistance. Science 278:1963–1965

Chandra-Shekara AC, Navarre D, Kachroo A, Kang H-G, Klessig D, Kachroo P (2004) Signaling requirements and role of salicylic acid in *HRT*- and *rrt*-mediated resistance to turnip crinkle virus in Arabidopsis. Plant J 40:647–659

Chisholm ST, Coaker G, Day B, Staskawicz BJ (2006) Host-microbe interactions: shaping the evolution of the plant immune response. Cell 124:803–814

Deslandes L, Olivier J, Peeters N, Feng DX, Khounlotham M, Boucher C, Somssich L, Genin S, Marco Y (2003) Physical interaction between RRS1-R, a protein conferring resistance to bacterial wilt, and PopP2, a type III effector targeted to the plant nucleus. Proc Natl Acad Sci USA 100:8024–8029

Dinesh-Kumar SP, Baker BJ (2000) Alternatively spliced *N* resistance gene transcripts: their possible role in tobacco mosaic virus resistance. Proc Natl Acad Sci USA 97:1908–1913

Dodds PN, Lawrence GJ, Catanzariti AM, Teh T, Wang CIA, Ayliffe MA, Kobe B, Ellis JG (2006) Direct protein interaction underlies gene-for-gene specificity and coevolution of the flax resistance genes and flax rust avirulence genes. Proc Natl Acad Sci USA 103:8888–8893

Eulgem T, Tsuchiya T, Wang X-J, Beasley B, Cuzick A, Tör M, Zhu T, McDowell JM, Holub E, Dangl JL (2007) EDM2 is required for *RPP7*-dependent disease resistance in Arabidopsis and affects *RPP7* transcript levels. Plant J 49:829–839

Falk A, Feys BJ, Frost LN, Jones JDG, Daniels MJ, Parker JE (1999) *EDS1*, an essential component of *R* gene-mediated disease resistance in *Arabidopsis* has homology to eukaryotic lipases. Proc Natl Acad Sci USA 96:3292–3297

Ferrier-Cana E, Macadré C, Sévignac M, David P, Langin T, Geffroy V (2005) Distinct post-transcriptional modifications result into seven alternative transcripts of the CC-NBS-LRR gene *JA1tr* of *Phaseolus vulgaris*. Theor Appl Genet 110:895–905

Flor HH (1971) Current status of the gene-for-gene concept. Annu Rev Phytopathol 9:275–296

Gassmann W, Hinsch ME, Staskawicz BJ (1999) The *Arabidopsis RPS4* bacterial-resistance gene is a member of the TIR-NBS-LRR family of disease-resistance genes. Plant J 20:265–277

Goodman RN, Novacky AJ (1994) The Hypersensitive Reaction in Plants to Pathogens: A Resistance Phenomenon. (St Paul, MN: APS Press)

Grant MR, Godiard L, Straube E, Ashfield T, Lewald J, Sattler A, Innes RW, Dangl JL (1995) Structure of the *Arabidopsis RPM1* gene enabling dual specificity disease resistance. Science 269:843–846

Grant SR, Fisher EJ, Chang JH, Mole BM, Dangl JL (2006) Subterfuge and manipulation: Type III effector proteins of phytopathogenic bacteria. Annu Rev Microbiol 60:425–449

Greenberg JT, Yao N (2004) The role and regulation of programmed cell death in plant-pathogen interactions. Cell Microbiol 6:201–211

Gupta S, Zink D, Korn B, Vingron M, Haas SA (2004) Genome wide identification and classification of alternative splicing based on EST data. Bioinformatics 20:2579–2585

Halterman D, Zhou FS, Wei FS, Wise RP, Schulze-Lefert P (2001) The MLA6 coiled-coil, NBS-LRR protein confers *AvrMla6*-dependent resistance specificity to *Blumeria graminis* f. sp *hordei* in barley and wheat. Plant J 25:335–348

Halterman DA, Wise RP (2004) A single-amino acid substitution in the sixth leucine-rich repeat of barley MLA6 and MLA13 alleviates dependence on RAR1 for disease resistance signaling. Plant J 38:215–226

Halterman DA, Wise RP (2006) Upstream open reading frames of the barley *Mla13* powdery mildew resistance gene function co-operatively to down-regulate translation. Mol Plant Pathol 7:167–176

Halterman DA, Wei FS, Wise RP (2003) Powdery mildew-induced *Mla* mRNAs are alternatively spliced and contain multiple upstream open reading frames. Plant Physiol 131:558–567

Hori K, Watanabe Y (2005) UPF3 suppresses aberrant spliced mRNA in Arabidopsis. Plant J 43:530–540

Hubert DA, Tornero P, Belkhadir Y, Krishna P, Takahashi A, Shirasu K, Dangl JL (2003) Cytosolic HSP90 associates with and modulates the *Arabidopsis* RPM1 disease resistance protein. EMBO J 22:5679–5689

Hwang C-F, Bhakta AV, Truesdell GM, Pudlo WM, Williamson VM (2000) Evidence for a role of the N terminus and leucine-rich repeat region of the *Mi* gene product in regulation of localized cell death. Plant Cell 12:1319–1329

Jones JDG, Dangl JL (2006) The plant immune system. Nature 444:323–329

Kornblihtt AR (2005) Promoter usage and alternative splicing. Curr Opin Cell Biol 17:262–268

Lawrence GJ, Finnegan EJ, Ayliffe MA, Ellis JG (1995) The *L6* gene for flax rust resistance is related to the Arabidopsis bacterial resistance gene *RPS2* and the tobacco viral resistance gene *N*. Plant Cell 7:1195–1206

Levy M, Edelbaum O, Sela I (2004) Tobacco mosaic virus regulates the expression of its own resistance gene *N*. Plant Physiol 135:2392–2397

Li L, Howe GA (2001) Alternative splicing of prosystemin pre-mRNA produces two isoforms that are active as signals in the wound response pathway. Plant Mol Biol 46:409–419

Liu Y, Schiff M, Serino G, Deng XW, Dinesh-Kumar SP (2002) Role of SCF ubiquitin-ligase and the COP9 signalosome in the *N* gene-mediated resistance response to tobacco mosaic virus. Plant Cell 14:1483–1496

Liu Y, Burch-Smith T, Schiff M, Feng S, Dinesh-Kumar SP (2004) Molecular chaperone Hsp90 associates with resistance protein N and its signaling proteins SGT1 and Rar1 to modulate an innate immune response in plants. J Biol Chem 279:2101–2108

Lu R, Malcuit I, Moffett P, Ruiz MT, Peart J, Wu AJ, Rathjen JP, Bendahmane A, Day L, Baulcombe DC (2003) High throughput virus-induced gene silencing implicates heat shock protein 90 in plant disease resistance. EMBO J 22:5690–5699

Maquat LE (2004) Nonsense-mediated mRNA decay: splicing, translation and mRNP dynamics. Nat Rev Mol Cell Biol 5:89–99

Martin GB, Bogdanove AJ, Sessa G (2003) Understanding the functions of plant disease resistance proteins. Annu Rev Plant Biol 54:23–61

McDowell JM, Simon SA (2006) Recent insights into R gene evolution. Mol Plant Pathol 7:437–448

McDowell JM, Cuzick A, Can C, Beynon J, Dangl JL, Holub EB (2000) Downy mildew (*Peronospora parasitica*) resistance genes in Arabidopsis vary in functional requirements for *NDR1*, *EDS1*, *NPR1* and salicylic acid accumulation. Plant J 22:523–529

McDowell JM, Dhandaydham M, Long TA, Aarts MGM, Goff S, Holub EB, Dangl JL (1998) Intragenic recombination and diversifying selection contribute to the evolution of downy mildew resistance at the *RPP8* locus of Arabidopsis. Plant Cell 10:1861–1874

Meyers BC, Morgante M, Michelmore RW (2002) TIR-X and TIR-NBS proteins: two new families related to disease resistance TIR-NBS-LRR proteins encoded in *Arabidopsis* and other plant genomes. Plant J 32:77–92

Meyers BC, Kozik A, Griego A, Kuang HH, Michelmore RW (2003) Genome-wide analysis of NBS-LRR-encoding genes in Arabidopsis. Plant Cell 15:809–834

Meyers BC, Dickerman AW, Michelmore RW, Sivaramakrishnan S, Sobral BW, Young ND (1999) Plant disease resistance genes encode members of an ancient and diverse protein family within the nucleotide-binding superfamily. Plant J 20:317–332

Moffett P, Farnham G, Peart J, Baulcombe DC (2002) Interaction between domains of a plant NBS-LRR protein in disease resistance-related cell death. EMBO J 21:4511–4519

Muskett PR, Kahn K, Austin MJ, Moisan LJ, Sadanandom A, Shirasu K, Jones JDG, Parker JE (2002) Arabidopsis *RAR1* exerts rate-limiting control of R gene-mediated defenses against multiple pathogens. Plant Cell 14:979–992

Ner-Gaon H, Halachmi R, Savaldi-Goldstein S, Rubin E, Ophir R, Fluhr R (2004) Intron retention is a major phenomenon in alternative splicing in *Arabidopsis*. Plant J 39:877–885

Nimchuk Z, Eulgem T, Holt BE, Dangl JL (2003) Recognition and response in the plant immune system. Annu Rev Genet 37:579–609

Noël L, Moores TL, van der Biezen EA, Parniske M, Daniels MJ, Parker JE, Jones JDG (1999) Pronounced intraspecific haplotype divergence at the RPP5 complex disease resistance locus of Arabidopsis. Plant Cell 11:2099–2111

O'Neill L (2000) The Toll/interleukin-1 receptor domain: a molecular switch for inflammation and host defence. Biochem Soc Trans 28:557–563

Parker JE, Holub EB, Frost LN, Falk A, Gunn ND, Daniels MJ (1996) Characterization of *eds1*, a mutation in Arabidopsis suppressing resistance to *Peronospora parasitica* specified by several different *RPP* genes. Plant Cell 8:2033–2046

Parker JE, Coleman MJ, Szabò V, Frost LN, Schmidt R, van der Biezen E, Moores T, Dean C, Daniels M, Jones JDG (1997) The Arabidopsis downy mildew resistance gene *RPP5* shares similarity to the Toll and interleukin-1 receptors with *N* and *L6*. Plant Cell 9:879–894

Reddy ASN (2004) Plant serine/arginine-rich proteins and their role in pre-mRNA splicing. Trends Plant Sci 9:541–547

Schornack S, Ballvora A, Gürlebeck D, Peart J, Baulcombe D, Ganal M, Baker B, Bonas U, Lahaye T (2004) The tomato resistance protein Bs4 is a predicted non-nuclear TIR-NB-LRR protein that mediates defense responses to severely truncated derivatives of AvrBs4 and overexpressed AvrBs3. Plant J 37:46–60

Shen Q-H, Saijo Y, Mauch S, Biskup C, Bieri S, Keller B, Seki H, Ülker B, Somssich IE, Schulze-Lefert P (2007) Nuclear activity of MLA immune receptors links isolate-specific and basal disease-resistance responses. Science 315:1098–1103

Shen QH, Zhou FS, Bieri S, Haizel T, Shirasu K, Schulze-Lefert P (2003) Recognition specificity and RAR1/SGT1 dependence in barley *Mla* disease resistance genes to the powdery mildew fungus. Plant Cell 15:732–744

Shigeoka S, Ishikawa T, Tamoi M, Miyagawa Y, Takeda T, Yabuta Y, Yoshimura K (2002) Regulation and function of ascorbate peroxidase isoenzymes. J Exp Bot 53:1305–1319

Shirano Y, Kachroo P, Shah J, Klessig DF (2002) A gain-of-function mutation in an Arabidopsis Toll interleukin-1 receptor-nucleotide binding site-leucine-rich repeat type R gene triggers defense responses and results in enhanced disease resistance. Plant Cell 14:3149–3162

Takahashi A, Casais C, Ichimura K, Shirasu K (2003) HSP90 interacts with RAR1 and SGT1 and is essential for RPS2-mediated disease resistance in *Arabidopsis*. Proc Natl Acad Sci USA 100:11777–11782

Tameling WIL, Elzinga SDJ, Darmin PS, Vossen JH, Takken FLW, Haring MA, Cornelissen BJC (2002) The tomato R gene products I-2 and Mi-1 are functional ATP binding proteins with ATPase activity. Plant Cell 14:2929–2939

Tameling WIL, Vossen JH, Albrecht M, Lengauer T, Berden JA, Haring MA, Cornelissen BJC, Takken FLW (2006) Mutations in the NB-ARC domain of I-2 that impair ATP hydrolysis cause autoactivation. Plant Physiol 140:1233–1245

Tao Y, Xie Z, Chen W, Glazebrook J, Chang HS, Han B, Zhu T, Zou G, Katagiri F (2003) Quantitative nature of Arabidopsis responses during compatible and incompatible interactions with the bacterial pathogen *Pseudomonas syringae*. Plant Cell 15:317–330

Tauszig-Delamasure S, Bilak H, Capovilla M, Hoffmann JA, Imler JL (2002) Drosophila MyD88 is required for the response to fungal and Gram-positive bacterial infections. Nat Immunol 3:91–97

Tör M, Gordon P, Cuzick A, Eulgem T, Sinapidou E, Mert-Türk F, Can C, Dangl JL, Holub EB (2002) Arabidopsis SGT1b is required for defense signaling conferred by several downy mildew resistance genes. Plant Cell 14:993–1003

Tornero P, Merritt P, Sadanandom A, Shirasu K, Innes RW, Dangl JL (2002) *RAR1* and *NDR1* contribute quantitatively to disease resistance in Arabidopsis, and their relative contributions are dependent on the *R* gene assayed. Plant Cell 14:1005–1015

Wang BB, Brendel V (2006) Genomewide comparative analysis of alternative splicing in plants. Proc Natl Acad Sci USA 103:7175–7180

Warren RF, Merritt PM, Holub E, Innes RW (1999) Identification of three putative signal transduction genes involved in *R* gene-specified disease resistance in Arabidopsis. Genetics 152:401–412

Warren RF, Henk A, Mowery P, Holub E, Innes RW (1998) A mutation within the leucine-rich repeat domain of the Arabidopsis disease resistance gene *RPS5* partially suppresses multiple bacterial and downy mildew resistance genes. Plant Cell 10:1439–1452

Wei F, Wing R, Wise RP (2002) Genome dynamics and evolution of the *Mla* (powdery mildew) resistance locus in barley. Plant Cell 14:1903–1917

Whitham S, Dinesh-Kumar SP, Choi D, Hehl R, Corr C, Baker B (1994) The product of the tobacco mosaic virus resistance gene *N*: Similarity to Toll and the interleukin-1 receptor. Cell 78:1101–1115

Zhang XC, Gassmann W (2003) *RPS4*-mediated disease resistance requires the combined presence of *RPS4* transcripts with full-length and truncated open reading frames. Plant Cell 15:2333–2342

Zhang XC, Gassmann W (2007) Alternative splicing and mRNA levels of the disease resistance gene *RPS4* are induced during defense responses. Plant Physiol 145:1577–1587

Zhang Y, Dorey S, Swiderski M, Jones JDG (2004) Expression of *RPS4* in tobacco induces an AvrRps4-independent HR that requires EDS1, SGT1 and HSP90. Plant J 40:213–224

Zhang YL, Goritschnig S, Dong XN, Li X (2003) A gain-of-function mutation in a plant disease resistance gene leads to constitutive activation of downstream signal transduction pathways in *suppressor of npr1-1, constitutive 1*. Plant Cell 15:2636–2646

Zhou FS, Kurth JC, Wei FS, Elliott C, Valé G, Yahiaoui N, Keller B, Somerville S, Wise R, Schulze-Lefert P (2001) Cell-autonomous expression of barley *Mla1* confers race-specific resistance to the powdery mildew fungus via a *Rar1*-independent signaling pathway. Plant Cell 13:337–350

Nuclear RNA Export and Its Importance in Abiotic Stress Responses of Plants

V. Chinnusamy, Z. Gong, and J.-K. Zhu(✉)

Contents

Introduction 236
   Nuclear mRNA Export 237
   Nuclear Pore Complex 237
   Nuclear RNA Export Factors 238
   Transcription-Coupled mRNP Formation 240
   mRNA Processing Coupled with mRNP Formation 241
   Export of mRNP Through NPCs 242
Nuclear Export of miRNAs and siRNAs in Plants 243
Stress-Specific mRNA Export Pathways 245
Role of mRNA Export in Hormonal and Biotic Stress Responses of Plants 246
Role of mRNA Export in Abiotic Stress Response of Plants 248
Conclusion and Perspectives 251
References 251

Abstract Transduction of developmental and environmental cues into the nucleus to induce transcription and the export of RNAs to the cytoplasm through the nuclear pore complex (NPC) play pivotal roles in regulation of gene expression. The process of bulk export of mRNAs from nucleus to cytoplasm is highly conserved across eukaryotes. Assembly of export-competent mRNA ribonucleoprotein (mRNP) is coupled with both transcription and mRNA processing. The export-competent mRNP consists of mRNAs and a dozen nucleocytoplasmic shuttling nuclear proteins, including RNA export factors (Mex67-Mtr2 heterodimer, Npl3), poly(A)-binding proteins, DEAD-box protein 5 (Dbp5), and nucleoporins (NUPs) in yeast. Mobile NUPs help docking of mRNP to the NPC nuclear basket. A partially

---

J.-K. Zhu
Institute for Integrative Genome Biology and Department of Botany and Plant Sciences,
University of California, Riverside, CA 92521, USA
e-mail: Jian-kang.zhu@ucr.edu

unfolded mRNP complex appears to be pulled through the NPC by using energy from Dbp5-catalyzed ATP hydrolysis. Dbp5 probably catalyzes the release of mRNA from mRNP in the cytoplasm. In contrast to bulk export of mRNAs by a Mex67-Mtr2/Npl3-dependent pathway, a specific subset of mRNA export under stress and export of microRNAs are mediated through the karyopherin (importin β) family of proteins in a Ran-GTPase-dependent pathway. Our knowledge of mRNA export mechanisms in flowering plants is in its infancy. Some proteins of the NUP107–160 complex, NUPs and DEAD-box proteins (DBPs), have been studied in flowering plants. *Arabidopsis* NUP160/SAR1 plays a critical role in mRNA export, regulation of flowering, and hormone and abiotic stress responses, whereas NUP96/SAR3/MOS3 is required for mRNA export to modulate hormonal and biotic stress responses. DEAD-box proteins have been implicated in mRNA export and abiotic stress response of yeast and higher plants. *Arabidopsis* DBP CRYOPHYTE/LOS4 plays an important role in mRNA export, abiotic stress response, germination, and plant development. Further studies on various components of nuclear mRNA export in plants during nonstress and stress conditions will be necessary to understand the link between mRNA export and stress-responsive gene expression.

Introduction

Abiotic stresses such as drought, salinity, and temperature extremes pose severe limitations to plant growth and development, thus limiting agricultural productivity. Drought is a major production constraint for agriculture in 45% of the world's geographical areas, where 38% of the human population lives (Bot et al. 2000). This problem will further increase as per capita fresh water availability is expected to decrease drastically in the near future. About 20% of irrigated agricultural land is adversely affected by salinity (Flowers and Yeo 1995). Approximately 93% of the continental area of Earth experiences, at least some of the time, temperatures below +15°C, which is cold stress for many crops. Further, high-temperature stress is a major problem, affecting about 40% of the irrigated areas of wheat alone (Fischer and Byerlee 1991). Global climate changes may increase drought and temperature stresses in many parts of the world by the end of this century. Hence, concerted efforts are being made worldwide to understand the mechanisms of abiotic stress tolerance in plants and to employ these mechanisms for genetic improvement of crop plants. Significant progress has been made in understanding the molecular genetic basis of abiotic stress tolerance in plants.

The perception and transduction of abiotic stresses to switch on genes involved in adaptive responses are critical to the survival and reproduction of plants exposed to adverse environments. Plants have evolved multiple stress response pathways, some of which are specific, but others may be common for various abiotic stresses (Chinnusamy et al. 2004). Transcriptional regulation of gene expression is relatively well understood (Zhu 2002; Chinnusamy et al. 2005, 2006; Yamaguchi-Shinozaki and Shinozaki 2006), but only limited progress has been made in unraveling posttranscriptional regulation of gene expression under various abiotic stresses (Sunkar and Zhu 2004; Borsani et al. 2005; Sunkar et al. 2006) (see the chapter by G. S. Ali and A. S. N. Reddy, this volume).

Eukaryotes have a well-defined nucleus separated from the cytoplasm by a nuclear envelope, the hallmark of eukaryotic organisms. Therefore, nucleocytoplasmic trafficking—entry of developmental/environmental signals for gene expression and export of expressed gene products from the nucleus through nuclear pores—plays a fundamental role in gene expression in eukaryotic organisms. The processes of nucleocytoplasmic trafficking are fairly well understood in animals and yeast, but only a beginning has been made in higher plants (Cullen 2003; Cole and Scarcelli 2006). In eukaryotes, gene expression is regulated posttranscriptionally by pre-mRNA processing, mRNA stability, RNA export from the nucleus, and translation. Thus nuclear export of mRNA and microRNAs (miRNAs)/short interfering RNAs (siRNAs) is an integral part of gene regulation in response to abiotic stresses. This chapter discusses the processes of nuclear export of mRNAs and miRNAs/siRNAs in nonstress and stress environments, mainly based on studies from yeast and higher plants.

Nuclear mRNA Export

In eukaryotes, the nucleus is surrounded by a double-layered membrane called the nuclear envelope. Therefore, transduction of environmental/development signals into the nucleus to regulate gene transcription and export of RNA molecules from the nucleus depends highly on the macromolecular traffic through the nuclear envelope. The composition of the outer membrane of the nuclear envelope is similar to that of the endoplasmic reticulum (ER), whereas the inner membrane of the nuclear envelope contains a distinct protein composition. The space between the two layers/lumen is called the perinuclear space. The nucleocytoplasmic transport of macromolecules occurs through nuclear pores present on the nuclear envelope, except during cell division, when dissolution of the nuclear envelope occurs. The aqueous channel of the nuclear pore is about ~9 nm in diameter at rest but during active transport expands up to ~25 nm. Larger molecules of >40 kDa are selectively transported through nuclear pores with the help of transport receptors or nuclear export factors. Although the nuclear pore permits the diffusion of molecules <40 kDa in size, even small proteins and RNAs are actively transported. Transport through the nuclear pore is energy dependent (Cullen 2003; Cole and Scarcelli 2006).

Nuclear Pore Complex

The nuclear pore is composed of several proteins forming a nuclear pore complex (NPC). Proteins that constitute the NPC are collectively called nucleoporins

(NUPs). In yeast, the molecular mass of the NPC is ~50 MDa, and the core is composed of about 30 NUPs, multiple copies that form an octagonal symmetry. Most NUPs are stationary, and some shuttle between the nucleus and cytoplasm (Rout et al. 2000). The NPC is octagonally symmetric around its cylindrical axis. Peripheral filaments emanating from the core into the nucleoplasm conjoin distally to form a basketlike structure that extends up to 100 nm into the nucleoplasm, whereas peripheral filaments emanating from the core into the cytoplasm spread outward, and eight fibrils extend ~50 nm into the cytoplasm. This architecture and many of the NUPs are conserved in all eukaryotes (Rout et al. 2000).

NUPs bind to a cargo (proteins and transported RNAs) or nuclear transporter that in turn binds to a cargo. Transport of the cargo through nuclear pores requires binding of the cargo to soluble NUPs and nuclear transport receptors or nuclear export factors (NEFs) (Weis 2002; Pemberton and Paschal 2005). Nuclear transport receptors/NEFs transiently interact with NUPs in the NPC and transport the cargo across the nuclear envelope. About one-third of NUPs provide binding sites for transport receptors and thus play crucial roles in the transport of cargo-receptor complexes. The receptor/export factor binding sites on these NUPs have multiple phenylalanineglycine (FG) repeat motifs (FxFG, GLFG, or FG) flanked by polar residues (Suntharalingam et al. 2003). Different FG motifs may facilitate interaction of NUPs with distinct subsets of transport receptors, as evident from yeast mutants that affect the translocation of particular transport receptors (Blevins et al. 2003). Interaction between FG repeats of NUPs with export factors helps in docking of cargo and its transport through NPCs. Binding of the export factors to FG motifs (FXFG, GLFG, and FG) of NUPs is necessary for mRNA export (Strasser et al. 2000). Understanding the interaction between the transport complex and specific NUPs will shed further light on the regulation of nucleocytoplasmic transport.

Nuclear RNA Export Factors

Nuclear transport receptors can be classified into the karyopherin/importin-β family of proteins and non-karyopherin NEFs. Export of proteins, tRNA, U-rich snRNAs, 5S RNAs, and miRNAs/siRNAs is mediated through importin-β family export receptors. The non-karyopherin NEFs mediate export of mRNAs (Cullen 2003; Cole and Scarcelli 2006). Karyopherin binding to a nuclear cargo requires the GTP-bound form of the Ran GTPase, whereas cytoplasmic hydrolysis of Ran-GTP to Ran-GDP induces cargo release. However, NEFs function independently of Ran-GTPase. Ran-dependent karyopherins play a limited role in mRNA export but a vital role in miRNA export (Kimura et al. 2004; Cole and Scarcelli 2006). Some of the NEFs involved in mRNA export in yeast and their homologs in other eukaryotes are listed in Table 1. Under nonstress conditions, yeast MEX67p (a homolog of vertebrate Tap/NXF1), an export factor unrelated to the importin-β family, mediates bulk export of mRNAs (Segref et al. 1997). MEX67p binds to

Table 1 Nuclear export factors involved in mRNA export in yeast

Yeast gene for nuclear export factors	Metazoan homolog	Remarks	References
MEX67p (mRNA export factor of 67 kDa)	Human Tap/Nuclear Export Factor 1 (NXF1)	Yeast mutant deficient in Mex67p accumulates mRNA in the nucleus. MEX67p interacts with FG of NUPs in NPC.	Segref et al. 1997
Mtr2p (mRNA transport)	Human NXT1/p15	Heterodimer formation between Mtr2p and Mex67p is required for the export of mRNA in yeast.	Strasser et al. 2000
Yra1p (yeast RNA annealing protein 1)	Mice Aly/ human REF (RNA export factor)	Yra1p is a member of the REF (RNA and export factor binding proteins) family. It binds to mRNA and to Mex67p. Yra2p can substitute for Yra1p function. Mutation in Yra1p results in mRNA accumulation in the nucleus.	Strasser and Hurt 2000; Zenklusen et al. 2001
Npl3 (nuclear protein localization 3)		mRNA export factor that functions in proper packaging of the mRNP; loss-of-function temperature-sensitive *NPL3* alleles accumulate poly(A)$^+$ RNA in the nucleus at the nonpermissive temperature.	Lee et al. 1996

both mRNAs and NUPs. The yeast MEX67p-Mtr2p heterodimer interacts with the mRNA ribonucleoprotein (mRNP) complex for export through NPCs (Vinciguerra and Stutz 2004; Cole and Scarcelli 2006).

Export of mRNAs requires processing, packaging by RNA-binding proteins, recognition by export factors, and translocation through the NPC into the cytoplasm. mRNA export does not depend on a specific motif in the cargo, but sequestration of mRNA into the mRNP complex is necessary for export (Cullen 2003; Dimaano and Ullman 2004). However, in some specific cases, such as karyopherin-dependent stress mRNA export, an adenylate uridylate-rich element (ARE) in the 3′ UTR of mRNAs is necessary for recognition by karyopherins. Formation of the mRNP complex is coupled with transcription and processing of pre-mRNA.

Transcription-Coupled mRNP Formation

Pre-mRNA transcribed by RNA polymerase (Pol) II is processed to add a 5′ monomethyl cap soon after transcription initiation, whereas splicing of introns and 3′ cleavage and polyadenylation occur immediately after transcription in the nucleus. mRNA processing occurs cotranscriptionally, and RNA Pol II coordinates these activities. The carboxy-terminal domain (CTD) of Pol II plays a crucial role in these processes in a manner dependent on the state of CTD phosphorylation (Hirose and Manley 2000). mRNPs are formed in the nucleus by packing of mRNAs into heterogenous nuclear RNPs (hnRNPs). A subset of these hnRNP proteins is retained in the nucleus, whereas others accompany the mRNA into the cytoplasm, where they dissociate to release mRNA. Then these RNPs move back into the nucleus for further rounds of export. In *Saccharomyces cerevisiae*, Npl3 (also termed Mtr13/Mts1/Nab1/Nop3), a shuttling NEF, contains two RNA-recognition motifs. Npl3p is a major mRNA-binding protein. The role of Npl3p in mRNA export was revealed by nuclear accumulation of mRNA in *npl3* mutants (Lee et al. 1996). Furthermore, Npl3p, along with other nuclear proteins, packages pre-mRNA into an export-competent RNP (Shen et al. 1998). In the cytoplasm, Npl3 releases mRNA and is transported back into the nucleus by the importin Mtr10 (Senger et al. 1998). In *S. cerevisiae*, Npl3p is recruited into the transcription complex with RNA Pol II. Mutations in both Npl3 and TATA-binding protein block mRNA export. Chromatin immunoprecipitation assays in yeast showed that Npl3 is recruited to mRNA during transcription at an early stage, whereas another mRNA export factor, Yra1p (=Aly/REF in metazoans), is recruited cotranscriptionally at a later step (Lei et al. 2001). This finding is consistent with the role of Yra1p, because it appears to tag the completely processed mRNP for nuclear export (Strässer and Hurt 2000). Npl3 purified from yeast showed 17 methylated arginines and 10 Arg-Gly-Gly tripeptides exclusively dimethylated. Arginine methylation of Npl3 appears to facilitate export directly by weakening contacts with nuclear proteins (McBride et al. 2005).

Yra1p, Sub2p, THO (suppressor of the transcriptional defect of Hpr1 by overexpression) complex, Tex1, and Hpr1p (hyperrecombination) form a protein complex named TREX (transcription-export) complex in yeast (Strasser et al. 2002). THO complex consists of four proteins, namely, Tho2, Hpr1, Mft1 (mitochondrial fusion targeting 1), and Thp2 (THO2-HPR1 phenotype). The TREX complex is specifically recruited to genes during transcription and plays a conserved role in coupling transcription to bulk mRNA export (Strasser et al. 2002). Hpr1p genetically and physically interacts with Yra1p and Sub2p (an RNA helicase). Hpr1p is necessary for efficient targeting of Yra1p and Sub2p to genes undergoing active transcription (Zenklusen et al. 2002). Cotranscriptional recruitment to the mRNA export receptor Mex67p contributes to nuclear pore anchoring of activated genes (Dieppois et al. 2006). Yra1 interacts directly with the mRNA export factor Mex67, which localizes primarily to nuclear pores. This finding suggests that Yra1 can function to bridge the mRNP formation with the actual translocation machinery (Strässer and Hurt 2000). Thus cotranscriptional recruitment of RNA export factors into mRNP is a conserved mechanism of bulk mRNA export.

An RNA helicase DEAD-box protein, Dbp5, has been implicated in transcription-coupled formation of export competent mRNPs. Yeast Dbp5 associates with mRNA early during transcription and accompanies it to the cytoplasm (Zhao et al. 2002). Yeast Dbp5 interacts with TFIIH, which may help in loading Dbp5 onto nascent mRNP (Estruch and Cole 2003). Further evidence for the involvement of TFIIH in mRNA export came from the finding that *Schizosaccharomyces pombe ptr8+* mutation and *S. cerevisiae ssl2* mutation cause nuclear accumulation of poly(A)$^+$ RNA. *PTR8* and *SSL2* genes encode a component of TFIIH homologous to human XPB, a protein involved in nucleotide excision repair and transcription. Expression of human XPB in these yeast mutants rescues them from mRNA export defects. Moreover, Ptr8p functionally interacts with Tho2p, a component of the TREX complex involved in mRNA export (Mizuki et al. 2007). Thus Dbp helps in formation of mRNPs for export.

mRNA Processing Coupled with mRNP Formation

In addition to transcription-dependent mRNP formation, mRNA processing such as 5′ capping, splicing, and polyadenylation also helps in formation of export-competent mRNPs. Splicing appears to promote mRNA export but is not necessary for mRNA export. The exon junction complex (EJC) is a multisubunit complex deposited by the spliceosome 20 to 24 nucleotides 5′ to the site of intron removal (Le Hir et al. 2000). The EJC helps binding of factors involved in mRNA export and nonsense-mediated mRNA decay. EJC consists of SRM160, RNPS1, Y14, Magoh, and Aly (=Yra1p) proteins. As discussed above, Aly (=Yra1p), a member of the REF binding proteins, plays a crucial role in mRNA export (Table 1). EJC formed during splicing provides a binding site for the Tap-Nxt (=MEX67p-Mtr2p in yeast) heterodimer, and thus splicing enhances mRNA export (Le Hir et al. 2001). The recruitment of Yra1p to EJC appears to be mediated by Sub2p, a member of the DEAD-box family of RNA helicases, which play an important role in spliceosome assembly. Sub2p mutation blocks poly(A)$^+$ mRNA export in yeast. Sub2p and Yra1p directly interact both in vitro and in vivo, and Sub2p helps in recruitment of Yra1p to the mRNP (Strässer and Hurt, 2001). Since Yra1p can interact with Mex67p (Strasser et al. 2000), Yra1p in turn recruits the Mex67p-Mtr2p heterodimer to form an export-competent mRNP.

mRNA export depends upon proper 3′ end processing such as poly(A) tail formation (see the chapter by A.G. Hunt, this volume). Yeast mutants defective in 3′ mRNA processing are also impaired in mRNA export. Furthermore, deletion of the *cis*-acting sequences required to couple 3′ processing and termination results in production of transcripts that fail to exit the nucleus. This evidence suggests that cleavage, termination, and export are coupled (Hammell et al. 2002). Yra1p is preferentially recruited to mRNPs of intron-containing genes in a splicing-dependent pathway, whereas Yra1p recruitment into mRNPs of all genes depends on 3′-end formation, regardless of intron status (Lei and Silver 2002). Furthermore, yeast poly(A)-binding protein Pab1 has been shown to shuttle between the nucleus and the cytoplasm and plays a role in mRNA export. Inhibition of nuclear import of Pab1 results in a kinetic delay in the export of mRNA. In addition to bulk mRNA export by the Mex67p-Mtr2p pathway, Pap1 also plays a role in the karyopherin chromosome region maintenance 1 (CRM1)-dependent pathway of nuclear export. A *pab1* deletion strain is rescued by a mutation in the 5'–3' exoribonuclease *RRP6*, a component of the nuclear exosome. Thus nuclear Pab1 may be required for efficient mRNA export and quality control of mRNA in the nucleus (Brune et al. 2005).

These results explain why splicing enhances the nuclear export of mRNA but is not necessary for mRNA export. This observation may be due to the fact that cotranscriptional splicing- and 3'-end processing–coupled mRNP formation–recruits the same nuclear factor, Yra1p. Thus Yra1p is recruited into mRNP by any one of the processes of mRNA synthesis and helps in recruitment of the Mex67p-Mtr2p heterodimer to form export-competent mRNP.

Export of mRNP Through NPCs

As discussed above, assembly of export-competent mRNP occurs cotranscriptionally and is coupled to mRNA processing. Export-competent mRNP consists of mRNA and a dozen nuclear proteins, including REFs (Mex67-Mtr2 heterodimer, Npl3), poly(A)-binding proteins (Pab1 and Pap2), Dead box protein Dbp5, and mobile NUPs. Mutations in yeast Mex67, Gle1, and Dbp5 result in nuclear accumulation of poly(A)$^+$ mRNAs (Segref et al. 1997; Tseng et al. 1998). Studies of Balbiani ring mRNP granule export revealed that at least partial unfolding of mRNPs is necessary for their transport through NPCs, and the 5' end of the mRNP enters the channel first (Daneholt 2001). mRNPs are possibly pulled through NPCs by using energy from ATP hydrolysis. Dbp5 probably acts as the ATPase and unwinds mRNP during transport through the NPC. Mutations that impair ATPase activity of Dbp5 also impair the function of Dbp5 in vivo (Schmitt et al. 1999). Gle1 stimulates ATPase activity of Dbp5 in an IP6-dependent manner; this activation may facilitate the remodeling of mRNP protein composition during directional transport and provide energy to power transport cycles (Alcazar-Roman et al. 2006; Weirich et al. 2006). In yeast, a mutation in Gle1 affects the overall structural integrity of NPCs and impairs mRNA export (Murphy et al. 1996). A Gle1 homolog in humans, hGle1, is also a nuclear-cytoplasmic shuttling protein involved in mRNA export (Watkins et al. 1998). Proteomic analysis of vertebrate NPCs and the nucleocytoplasmic shuttling nature of Gle1 suggested that the association of hGle1 with NPCs is transient. The N-terminal region of hGle1 interacts with the nucleoporin hNup155. NPC localization of hGle1p requires hNup155p and hNup42p/hCG1/NPL1 (Rayala et al. 2004; Kendirgi et al. 2005). Another NUP-interacting export factor, Gle2p/Rae1p, was shown to play a crucial role in mRNA export (Murphy et al. 1996) through its interaction with yeast Nup116p, which shares similarity with human hNUP98 (Pritchard et al. 1999).

Yeast Nup145p is similar to yeast Nup116p and hNup98. Depletion of Nup145p in vivo leads to rapid accumulation of mRNA in the nucleus. Genetic analysis in yeast revealed that Nup145p, Nup116p, and Nup100p play redundant functions in nucleocytoplasmic transport (Fabre et al. 1994). Nup160 and Nup133 are localized on the basket side of the NPC. Nucleoporins Nup160, Nup133, Nup107, and Nup96 exist as a complex and are collectively termed the Nup160 complex. Among these, Nup160 and Nup133 interact with Nup98 and play a role in mRNA export (Vasu et al. 2001). Nup153p is a highly mobile protein recruited to the mRNP export complex during transcription (Griffis et al. 2004).

The mechanism of translocation of mRNP through NPC is much less known. Interactions between nuclear export factors and FG repeats on NUPs drive the translocation of receptor-cargo complexes through nuclear pores. Tap, the metazoan homolog of yeast nuclear export factor MEX67, contains UBA-like and NTF2-like folds that can associate directly with FG repeats. Mutations in the Tap-UBA domain abolished the interaction of Tap with nucleoporins Nup98, p62, and RanBP2, whereas mutations in the NTF2-like domain of Tap impaired Nxt1 (=yeast Mtr2) binding. Although both of these mutations impaired mRNA export, Tap interaction with the NPC in vivo or its nucleocytoplasmic shuttling was not affected. Thus Tap requires both the UBA- and NTF2-like domains to mediate the export of RNA cargo, but it shuttles through the NPC independently of these domains when free of RNA cargo (Levesque et al. 2006).

Cytoplasmic filaments of NPC consist of Nup159, Gle1, Nup82, and Nup42/Rip1. Mutations in these genes affect mRNA export (Cole and Scarcelli 2006). Dbp5 travels along with the mRNP complex through NPCs. Once mRNP reaches the cytoplasmic side of the NPC, Dbp5 interacts with the cytoplasmic filament Nup159 of the NPC. The N-terminal region of Nup159 is necessary for Dbp5 binding, and deletion of this domain significantly reduces the Dbp5 at the nuclear periphery. In addition to Dbp5, Nup159 also interacts with nucleoporins Nup82 and NSP1 (nucleoskeletal-like protein 1). Nsp1p contains multiple repeats of the amino acids FXFG. Nup159 mutant yeast is impaired in mRNA export (Del Priore et al. 1997). Mutations in highly conserved residues of the N-terminal region, which interact with Dbp5, render Nup159-Dbp5 interaction temperature sensitive and also result in temperature-sensitive mRNA export. These findings suggest that the Nup159 N-terminal domain functions in mRNA export as a binding platform, tethering shuttling Dbp5 molecules at the nuclear periphery (cytoplasmic side), and the Dbp5 may release mRNA from mRNP at the cytoplasmic face of the NPC (Weirich et al. 2004).

Nuclear Export of miRNAs and siRNAs in Plants

Small RNAs, which do not code for proteins but negatively regulate gene expression, are classified into two types—miRNAs and siRNAs—on the basis of their biogenesis. miRNAs are synthesized from single-stranded primary miRNA (pri-miRNA) transcripts, which are transcribed from miRNA genes (*MIR* genes) by RNA Pol II. In plants, the pri-miRNA transcript forms one or more stem-loop secondary structures, which is(are) cleaved by a ribonuclease III-like enzyme called Dicer-like 1 (DCL1) protein to produce an miRNA-miRNA* duplex in the nucleus. DCL1 protein interacts with a nuclear dsRNA-binding protein, HYPONASTIC LEAVES 1 (HYL1) (Kurihara et al. 2006), to precisely cleave pri-miRNA into the miRNA-miRNA* duplex (stem region of the hairpin) of ~21-nt length. A nuclear methyltransferase protein, HUA ENHANCER 1 (HEN1), methylates the three terminal nucleotides on their 2′-OH group in the miRNA-miRNA* duplex, which is then exported from the nucleus into the cytoplasm by HASTY (HST), a member of the importin β family of nucleocytoplasmic transporters, via the Ran-GDP-Ran-GTP transport system (Yi et al. 2003; Lund et al. 2004; Park et al. 2005). The miRNA-miRNA* duplex is then unwound into a single-stranded mature miRNA by an unknown helicase. The miRNA (~21-nt length) then enters the RNA-induced silencing complex (Bartel 2004; Kidner and Martienssen 2005; Jones-Rhoades et al. 2006; Mallory and Vaucheret 2006).

Endogenous siRNAs are synthesized from long double-stranded RNAs (dsRNAs) of endogenous origin such as (a) miRNA-directed cleavage products of noncoding single-stranded RNAs (ssRNAs)/transgene mRNAs, which are then converted into dsRNAs by RNA-dependent RNA polymerases (RDRs), generating *trans*-acting siRNAs (ta-siRNAs); (b) dsRNAs formed from the mRNAs encoded by natural *cis*-antisense gene pairs (nat-siRNAs; Borsani et al. 2005); and (c) dsRNAs generated from heterochromatin and DNA repeats (Mallory and Vaucheret 2006). RDRs and DCL-like proteins process the dsRNAs formed from different sources. Biogenesis of different classes of siRNAs is carried out by specific RDR-DCL protein combinations in the nucleus (Jones-Rhoades et al. 2006; Mallory and Vaucheret 2006).

Small RNAs are incorporated into an RNA-induced silencing complex (RISC) or a RNAi-induced transcriptional silencing (RITS) complex, which contain Argonaute (AGO) family proteins. miRNA and siRNAs regulate gene expression by (a) cleavage of target mRNAs that are complementary to the miRNA or siRNA, (b) miRNA- or siRNA-mediated translational repression, and (c) transcriptional silencing mediated mainly by heterochromatic siRNAs and, in some cases, by miRNA (Bartel 2004; Bao et al. 2004; Chan et al. 2005). Repression of gene expression by mRNA cleavage and translational repression necessitates the export of miRNAs or siRNAs from the nucleus to the cytoplasm. *Arabidopsis* HST is an ortholog of mammalian exportin 5 (Exp5) and yeast Msn5p. Msn5p exports phosphorylated proteins and imports replication protein A, involved in DNA replication and repair, whereas EXP5 exports pre-microRNAs (pre-miRNAs), tRNAs, and short hairpin RNAs into the cytoplasm (Yi et al. 2005). However, loss-of-function mutations in *HST* reduced the export of most but not all miRNAs but did not affect the export of tRNAs and endogenous siRNAs or transgene silencing in *Arabidopsis*. These results suggest that HST plays a crucial role in miRNA export, but other nuclear export pathways are also involved in export of small RNAs in *Arabidopsis* (Park et al. 2005).

Stress-Specific mRNA Export Pathways

In yeast, in situ hybridization showed that abiotic stresses (heat shock and ethanol) blocked the nuclear export of most poly(A)$^+$ RNA, whereas the heat shock protein 70 (HSP70) mRNAs were exported through functional NPCs under stress. These results suggest a separate nuclear traffic for normal mRNAs and HSP mRNAs under heat stress (Saavedra et al. 1996). In yeast, bulk export of mRNA is independent of Ran but dependent on nuclear export factors MEX67, MTR2, and NPL3. In yeast, an FG repeat containing 42-kDa nucleoporin, NUP42/Rip1p, is associated with NPCs and is necessary for mRNA export under heat stress. NUP42 is dispensable for growth and mRNA export under normal conditions, because *HSP70* mRNAs transcribed from the *GAL* promoter::SSA4 under non-heat-stress conditions was efficiently exported to the cytoplasm in the *nup42* mutant. Yeast Nup42p interacts with the REFs Gle1p (GLFG LEthal), Gle2p, and Nup100p (Saavedra et al. 1997); Gle1p and Gle2p are required for specific mRNA export under stress.

Heat shock (42°C) to yeast cells results in rapid dissociation of mRNA export factor Gle2p from the nuclear envelope into the cytoplasm, which is accompanied by inhibition of bulk export of mRNAs. Acclimation treatment (37°C for 1 h) helped to stabilize association of Gle2p to NPC at 42°C and thus permitted export of bulk poly(A)$^+$ mRNA. Yeast deletion mutants *gle2Δ* and *nup42Δ* could not induce the acclimation response and thus failed to adapt the export of bulk poly(A)$^+$ mRNA to heat shock. These results suggest that Gle2p and Nup42p are necessary for mRNA export, especially under heat-shock conditions (Izawa et al. 2004), whereas nucleoporins Nup159p/Rat7p, Nup120p/Rat2p, and Nup145p/Rat10p are required for mRNA export in both normal and heat stress pathways (Saavedra et al. 1997).

Gle1 is a conserved mRNA export factor from yeast to humans. hGle1 interacts with hNup155, which is necessary to target hGle1 to NPCs. In addition to hNup155, hCG1/NPL1 (=yNup42) and hGle1B are required for Gle1 targeting to NPCs. siRNA-directed suppression of hCG1 resulted in decreased levels of hCG1, which in turn resulted in hGle1 accumulation in cytoplasmic foci and also inhibited export of the Hsp70 mRNA in HeLa cells. Thus the Gle1-yNup42 (=hGle1-hCG1) mRNA export mechanism is highly conserved and plays a pivotal role in specific mRNA export under heat stress (Kendirgi et al. 2005).

Gle1 and Nup42p involved in mRNA export are in the Nup82p subcomplex, on the cytoplasmic face of the NPC in yeast. The Nup82p subcomplex contains Nup82p, Rat7p/Nup159p, Nsp1p, Gle1p/Rss1p, and Rip1p/Nup42p. Nup159p and Gle1p contain binding sites for Dbp5p. In the yeast Nup42p deletion mutant *nup42Δ*, both Gle1p and Dbp5p dissociate from NPCs after heat shock at 42°C. Efficient export of HSP mRNA after heat shock depends upon a novel 6-amino acid motif of Dbp5p. This motif is not required for mRNA export under normal growth conditions or ethanol shock (Rollenhagen et al. 2004).

In addition to Nup42p-Gle1p-Dbp5p, yeast RNA Pol II subunit Rpb4p also appears to mediate mRNA export under abiotic stresses in yeast. Yeast Rpb4p is required for transcription, mRNA export, and cell survival under heat stress conditions. Class II mutants of RBP4 (*Rbp4-2*), which affect only the export role but are normal in transcription function, could not acquire thermotolerance even under acclimation treatment. Rpb4p appears to perform different roles in export of bulk mRNA at 37°C and heat shock mRNAs at 42°C (Farago et al. 2003). Rpb4p and Rpb7p form a dissociable heterodimeric complex, which can bind to mRNA and shuttle between the nucleus and cytoplasm as a heterodimer. Shuttling of Rpb4p and Rpb7p depends on ongoing transcription under normal conditions, but during the severe stresses of heat shock, ethanol, and starvation, the two proteins shuttle via a transcription-independent pathway (Selitrennik et al. 2006).

The karyopherin CRM1 is involved in export of proteins, HIV mRNA, U small nuclear RNAs, and all rRNAs in humans. Although bulk mRNA export is independent of karyopherins, a specific subset of mRNA appears to be exported through CRM1 under stress. Human RNA binding protein (HuR) binds to an adenylate uridylate-rich element (ARE) in the 3′ UTRs. The interaction of HuR, pp32 and APRIL, the leucine-rich nuclear export signal (NES)-containing ligands, with CRM1 enhances the export of mRNAs of certain early response genes (ERGs) (Gallouzi and Steitz, 2001). In addition, the nuclear export of human *interferon-$\alpha 1$* (*IFN-$\alpha 1$*) mRNA, an ERG mRNA induced upon viral infection, has been shown to be exported through CRM1, and ARE in the 3′ UTR of *IFN-$\alpha 1$* mRNA is not essential for this (Kimura et al. 2004). Thus the CRM1 pathway appears to mediate both ARE-dependent and -independent export of specific mRNAs under stress.

Role of mRNA Export in Hormonal and Biotic Stress Responses of Plants

Relatively little information exists on the NUPs and NEFs required for mRNA export in plants. Bioinformatic analyses have identified several conserved NUPs from rice and *Arabidopsis* that are homologous to yeast and vertebrate NUPs (Neumann et al. 2006). Since the nuclear mRNA export pathway is highly conserved across species (fungi, insects, vertebrate) and many proteins similar to NUPs and NEFs have been found in higher plants, the mRNA export process in higher plants appears to be similar to that of other eukaryotes. The following section briefly discusses the NUPs characterized in plants and their role in mRNA export (Table 2).

The animal NUP107–160 complex is functionally equivalent to the yeast NUP84 complex, which is involved in mRNA export (Bai et al., 2004). Most subunits of the NUP107–160 complex (NUP160, NUP133, NUP107, NUP96, NUP85, NUP43, and two proteins similar to Sec13) have identifiable homologs in the *Arabidopsis* proteome, which suggests that the NUP107–160 complex is also conserved in *Arabidopsis* (Parry et al. 2006). *Arabidopsis snc1* (suppressor of npr1-1, constitutive 1), a gain-of-function mutant in a Toll Interleukin1 receptor-nucleotide binding-Leu-rich repeat-type resistance gene (*R*-gene), shows constitutive expression of disease resistance genes. A suppressor screen for *snc1* led to the isolation of *mos3* (modifier of snc1,3). mos3 is susceptible to bacterial diseases,

Table 2 Higher plant homologs of yeast/vertebrate NUPs involved in mRNA export

Plant Protein	Function	Remarks	References
Arabidopsis suppressor of auxin resistance 1 (SAR1)/ AtNUP160	Nucleoporin, part of Nup107–160 subcomplex	Similar to human Nup160; sar1 mutants accumulate poly(A) RNA within the nucleus	Parry et al. 2006
Arabidopsis NUP160	Nucleoporin, part of Nup107–160 subcomplex	atnup160–1 is defective in poly(A) mRNA export, reduces induction of CBF3-LUC under cold stress, and is hypersensitive to chilling and freezing stress	Dong et al. 2006
Suppressor of auxin resistance 3 (SAR3) of Arabidopsis	Nucleoporin, part of Nup107–160 subcomplex	Similar to human Nup96; sar3 mutants accumulate poly(A) RNA within the nucleus	Parry et al. 2006
Arabidopsis modifier of SNC1,3 (MOS3/SAR3)	Nucleoporin, part of Nup107–160 subcomplex	Required for disease resistance; similar to human Nup96, which is required for reversal of inhibition of mRNA nuclear export by a viral protein	Zhang and Li 2005
Lotus NUP133	Nucleoporin, part of Nup107–160 subcomplex	NUP133 participates in host-plant recognition of symbiotic microbes; Nup133 in yeast and vertebrate plays crucial role in mRNA export	Kanamori et al. 2006

which suggests that MOS3 is required for basal resistance to pathogens. *MOS3* encodes a putative NUP, which is localized on the nuclear envelope. *MOS3* shows high sequence similarity to human Nup96, which is involved in mRNA export. Thus nucleocytoplasmic trafficking appears to play a crucial role in biotic stress resistance (Zhang and Li 2005).

Direct evidence for the role of Nup96 in mRNA export came from studies of *sar* (suppressor of auxin resistance) mutants. A screen to identify suppressors of the *axr1* (auxin-resistant 1) mutant resulted in identification of *sar1* and *sar3* mutants in *Arabidopsis*. *sar1* and *sar3* mutations affect the nuclear localization of the transcriptional repressor AXR3/INDOLE ACETIC ACID17 and thus suppress the phenotype conferred by *axr1*. *SAR1* and *SAR3* encode proteins with similarities to vertebrate NUP160 and NUP96, respectively. SAR1 and SAR3 localize to the nuclear membrane. *sar1* and *sar3* mutant plants accumulate poly(A) RNA within the nucleus. These results suggest that SAR1 and SAR3 nucleoporins are required for mRNA export and modulate plant responses to the hormone auxin (Parry et al. 2006). Another member of the NUP107–160 complex, NUP133, characterized from *Lotus*, also showed a participation in host-plant recognition of symbiotic microbes (Kanamori et al. 2006).

Role of mRNA Export in Abiotic Stress Response of Plants

Dead box proteins (DBPs) with RNA helicase activity are involved in RNA metabolism such as transcription, RNA processing, RNA decay, and nucleocytoplasmic transport. As discussed previously, in yeast and vertebrates, Dbp5 plays a crucial role in export competent mRNP formation in nucleus, transport of mRNP through the NPC, and release of mRNA from mRNP in the cytosol. The role of DBP in mRNA export and abiotic stress response of flowering plants came from the analysis of the mutant *los4* (low expression of osmotically responsive genes 4) of *Arabidopsis*.

Screening for mutants with altered expression of the *RD29A-LUC* reporter gene under abiotic stresses led to the isolation of the mutant *los4-1*. *los4-1* shows reduced *RD29A-LUC* expression in response to cold but not ABA or high salt. In *los4-1*, *CBF3* expression is reduced and *CBF1* and *CBF2* expression is delayed, and thus the mutant exhibits hypersensitivity to chilling temperatures. Map-based cloning of *LOS4* showed that it encodes a DEAD-box RNA helicase (Gong et al 2002). Later, the *cryophyte/los4–2* mutant (allelic to *los4–1*) was isolated; the mutant showed superinduction of *CBF2* under cold stress and enhanced cold tolerance. The CRYOPHYTE/LOS4-GFP protein is enriched in the nuclear rim. Consistent with the cold-sensitive phenotype of *los4–1*, *los4–1* showed inhibited mRNA export under both normal and cold stress conditions. In contrast, the cold-tolerant but heat-sensitive *cryophyte/los4-2* showed normal mRNA export under cold stress but was defective in mRNA export from the nucleus at warm temperatures (Gong et al. 2005). The results from *los4* mutants demonstrated that LOS4 DEAD-box RNA helicase plays an important role in mRNA export, which is necessary for abiotic stress response. LOS4 helicase is also involved in many physiological processes such as germination (ABA hypersensitivity of *los4–2*) and plant development (*los4* mutant flowers earlier), in addition to its role in low-temperature responses (Gong et al. 2005).

In addition to their presence in *Arabidopsis*, abiotic stress-responsive DEAD-box-related helicases have been reported from pea. Pea DNA helicases 45 (*PDH45*) and *PDH47* are upregulated by various abiotic stresses. Tissue-specific differences in expression of *PDH47* were observed, because ABA upregulated *PDH47* expression in roots but not in shoots. PDH47 was localized to both the cytosol and nucleus. Similar to the Dbp5 of yeast, PDH47 also showed ATPase activity (Vashisht et al. 2005). Transgenic tobacco plants overexpressing PDH45 showed enhanced tolerance to salt stress (Sanan-Mishra et al. 2005). Thus DEAD-box RNA helicases play a pivotal role in mRNA export and abiotic stress tolerance. Further studies will be necessary to understand the link between DBP-mediated mRNA export and abiotic stress-responsive gene expression.

The role of NUPs in abiotic stress response was demonstrated recently. Use of a *CBF3* promoter-driven LUC reporter gene screen isolated *atnup160-1* mutant, which was hypersensitive to chilling stress and also defective in acquired freezing tolerance. Map-based cloning of At*NUP160* revealed it to encode a putative NUP, homologous to the mammal Nup160. Microarray analysis revealed that *atnup160* mutation impaired the expression of *CBFs* and a number of other genes involved in plant cold tolerance. *atnup160* flowers early and shows retarded seedling growth, especially at low temperatures. The AtNUP160-GFP fusion protein is localized at the nuclear rim. *atnup160–1* mutant plants are also impaired in mRNA export. AtNUP160 is ubiquitously expressed in all tissues and not regulated by cold stress. Thus AtNUP160 is critical for mRNA export, cold-responsive gene expression, cold tolerance, as well as plant development at normal temperatures (Dong et al. 2006). AtNUP160/SAR1 plays a pivotal role in nucleocytoplasmic transport of RNAs and regulates plant development (flowering), abiotic stress, and hormonal responses in plants (Dong et al. 2006; Parry et al. 2006).

Nucleocytoplasmic shuttling of NUPs and transport receptors is important for mRNA export. Once mRNA is delivered into the cytosol, the proteins (NUPs and export factors) are imported back into the nucleus. Importin β proteins interact with importin α and a cytoplasmic cargo containing the nuclear localization signal; then this complex is imported into the nucleus through the NPC. For example, nuclear mRNA export factor Npl3p is imported back by importin Mtr10 (Senger et al. 1998). Moreover, the importin β-domain protein CRM1-dependent pathway plays a crucial role in export of specific mRNAs under stress in humans. We used a *P RD29A:LUC* genetic screen to identify a SAD2/Importin β-domain protein involved in nucleocytoplasmic trafficking. *sad2-1* knockout mutation enhanced the expression of *RD29A:LUC* reporter gene and ABA- and stress-responsive genes under cold, salt, polyethylene glycol (PEG), and ABA treatments. However, the expression of SAD2 is not regulated by stress or ABA. *sad2–1* also exhibited ABA hypersensitivity in seed germination and seedling growth. SAD2 is predominantly localized in the nucleus. These results suggest that SAD2 may play a crucial role in nuclear import of a negative regulator of cold, salt, PEG, and ABA responses (Verslues et al. 2006).

Arabidopsis homologs for some of the yeast proteins involved in nuclear export of mRNA have been identified by sequence similarity. Our studies of *AtNUP160*, *LOS4*, and *SAD2* showed that these genes involved in mRNA export are critical for abiotic stress responses, although their transcript levels are not regulated by abiotic stresses. We examined the expression pattern of some of the *Arabidopsis* genes potentially involved in nuclear export of mRNA by using the Genevestigator response viewer (Zimmermann et al. 2004). Interestingly, some of the genes showed more than twofold induction under stress (Table 3). Putative *Arabidopsis* nucleoporin At2g45000, a homolog of yeast NSP1, appears to be upregulated by drought, salt, cold, and heat stresses. Some of the genes showed more than twofold induction under hypoxia. Further molecular genetic analysis of these genes will help unravel their role in abiotic stress responses of plants.

Table 3 Arabidopsis homologs of yeast NUPs/nuclear proteins involved in mRNA export

Yeast	Arabidopsis	Fold change in expression of target genes under stress (https://www.genevestigator.ethz.ch)						
		Drought	Salt	Cold	Heat	ABA	Oxidative	Maximum increase*
NSP1	At2g45000[1]	2.50	2.00	1.50	1.75	1.00	0.75	2.50, drought
Nup42/Rip1	At1g75340[1]	0.93	0.92	0.93	1.12	0.72	0.87	2.02, hypoxia
Nup82	At5g05680[1]	0.99	0.68	0.91	0.94	0.58	1.06	2.63, *A. tumefaciens*
Nup84	At3g14120[1]	1.03	0.77	0.72	1.30	0.66	1.04	1.68, *A. tumefaciens*
Nup100/	At1g59660/At1g	1.25	1.23	1.29	1.09	0.69	1.67	13.22, hypoxia
Nup116	10390[1]	1.15	0.89	1.14	1.31	0.98	1.32	
Nup120	At1g33410/AtNUP160/SAR1	0.93	0.80	0.72	1.16	0.99	0.92	1.45, low nitrate
Nup133	At2g05120[1,a]	0.88	0.93	0.82	1.26	0.98	0.89	2.10, anoxia
Nup145	At1g80680 [1]/MOS3/SAR3	1.00	0.73	0.85	1.45	0.83	1.07	1.45, heat
Gle1	At1g13120[1]	0.93	0.84	0.96	1.16	0.80	0.86	1.68, BL
Gle2	At1g80670[1]	1.07	0.84	0.91	1.37	0.79	1.70	7.85, hypoxia
Yra1	At5g59950	1.16	1.15	1.61	1.64	1.00	1.36	1.97, hypoxia
Dbp5	At3g53110/LOS4	0.95	1.01	0.91	1.15	0.88	0.99	1.59, anoxia
	At1g14850/NUP155	0.97	0.84	0.91	0.93	1.14	0.98	1.50, cold 24h/7days
Mago	At1g02140[b]	0.93	0.74	0.93	1.06	0.76	0.97	1.43, heat; 1.39 salicylic acid
Y14	At1g51510[b]	0.94	0.69	1.32	0.95	0.82	0.87	1.53, cold 24h/7days

[1]*Arabidopsis* homologs of yeast proteins identified by Neumann et al. 2006
*Maximum fold increase under stress/hormone treatment
[a]*Arabidopsis* homologs of Lotus *NUP133*
[b]Park and Muench 2007

Conclusion and Perspectives

Our knowledge of mRNA export mechanisms in flowering plants is in its infancy. Being poikilothermic and sessile, plants need a high order of gene regulation to respond to environmental cues, including abiotic stresses. Transduction of developmental and environmental cues into the nucleus to induce transcription and export of mRNA and regulatory small RNAs (miRNAs/siRNAs) to the cytoplasm through the NPC plays a pivotal role in regulation of gene expression. The process of bulk export of mRNA from the nucleus to the cytoplasm takes place through a distinct pathway independent of Ran-regulated karyopherins. The assembly of export-competent mRNP is coupled to transcription and 3′ mRNA processing. Export-competent mRNP consists of mRNA and a dozen nuclear proteins, including RNA-export factors (Mex67-Mtr2), Npl3, poly(A)-binding proteins, mobile NUPs, and the DEAD-box protein Dbp5. Dbp5 appears to power the transport through NPCs and catalyzes the release of mRNA from mRNP in the cytoplasm. It is known that in yeast and humans under stress conditions, a specific subset of mRNA is exported through specific pathways other than the bulk mRNA export pathway. However, in plants such pathways are yet to be discovered.

Bioinformatic analysis of *Arabidopsis* and rice genome sequences led to the in silico identification of nucleoporin and NEF homologs of yeast/vertebrates. Although the mRNA export pathway appears to be conserved in vertebrates, insects, and yeast, whether it is completely conserved in flowering plants is still unknown. Current evidence suggests that some variations in RNA export pathways exist in plants. For example, the mammalian and yeast nuclear export receptor EXP5 exports pre-microRNAs (pre-miRNAs) as well as tRNAs into the cytoplasm. However, the EXP5 ortholog of *Arabidopsis*, HST, is involved in nuclear export of miRNA but not tRNAs. In plants, NUP107–160 complex proteins have been shown to participate in mRNA export and regulate germination, flowering, hormone response, host-plant symbiotic interaction, and biotic and abiotic stress tolerance. Furthermore, the DBP RNA helicases have been implicated in mRNA export and abiotic stress tolerance of plants. Further studies on various components of the nuclear mRNA export machinery in plants and their role in abiotic stress response will be necessary to better understand the link between mRNA export and abiotic stress-responsive gene expression.

Acknowledgements Our work has been supported by grants from the US Department of Agriculture, National Science Foundation, and National Institutes of Health to J.-K. Zhu.

References

Alcazar-Roman AR, Tran EJ, Guo S, Wente SR (2006) Inositol hexakisphosphate and Gle1 activate the DEAD-box protein Dbp5 for nuclear mRNA export. Nat Cell Biol 8:711–716

Bai SW, Rouquette J, Umeda M, Faigle W, Loew D, Sazer S, Doye V (2004) The fission yeast Nup107–120 complex functionally interacts with the small GTPase Ran/Spi1 and is required for mRNA export, nuclear pore distribution, and proper cell division. Mol Cell Biol 24:6379–6392

Bao N, Lye KW, Barton MK (2004) MicroRNA binding sites in Arabidopsis class III HD-ZIP mRNAs are required for methylation of the template chromosome. Dev Cell 7:653–662

Bartel DP (2004) MicroRNAs: genomics, biogenesis, mechanism, and function. Cell 116:281–297

Blevins MB, Smith AM, Phillips EM, Powers MA (2003) Complex formation among the RNA export proteins Nup98, Rae1/Gle2, and TAP. J Biol Chem 278:20979–20988

Borsani O, Zhu J, Verslues PE, Sunkar R, Zhu JK (2005) Endogenous siRNAs derived from a pair of natural *cis*-antisense transcripts regulate salt tolerance in Arabidopsis. Cell 123:1279–1291

Bot A.J, Nachtergaele FO, Young A (2000) Land resource potential and constraints at regional and country levels. In: World Soil Resources Reports 90, Land and Water Development Division, FAO, Rome, pp 17–24

Brune C, Munchel SE, Fischer N, Podtelejnikov AV, Weis K (2005) Yeast poly(A)-binding protein Pab1 shuttles between the nucleus and the cytoplasm and functions in mRNA export. RNA 11:517–531

Chan SW, Henderson IR, Jacobsen SE (2005) Gardening the genome: DNA methylation in *Arabidopsis thaliana*. Nat Rev Genet 6:351–360

Chinnusamy V, Jagendorf A, Zhu JK (2005) Understanding and improving salt tolerance in plants. Crop Sci 45:437–448

Chinnusamy V, Schumaker K, Zhu JK (2004) Molecular genetic perspectives on cross-talk and specificity in abiotic stress signalling in plants. J Exp Bot 55:225–236

Chinnusamy V, Zhu J, Zhu JK (2006) Gene regulation during cold acclimation in plants. Physiol Plant 126:52–61

Cole CN, Scarcelli JJ (2006) Transport of messenger RNA from the nucleus to the cytoplasm. Curr Opin Cell Biol 18:299–306

Cullen BR (2003) Nuclear RNA export. J Cell Sci 116:587–597

Danehalt B (2001) Assembly and transport of a premessenger RNP particle. Proc Natl Acad Sci USA 98:7012–7017

Del Priore V, Heath C, Snay C, MacMillan A, Gorsch L, Dagher S, Cole C (1997) A structure/function analysis of Rat7p/Nup159p, an essential nucleoporin of *Saccharomyces cerevisiae*. J Cell Sci 110:2987–2999

Dieppois G, Iglesias N, Stutz F (2006) Cotranscriptional recruitment to the mRNA export receptor Mex67p contributes to nuclear pore anchoring of activated genes. Mol Cell Biol 26:7858–7870

Dimaano C, Ullman KS (2004) Nucleocytoplasmic transport: Integrating mRNA production and turnover with export through the nuclear pore. Mol Cell Biol 24:3069–3076

Dong CH, Hu X, Tang W, Zheng X, Kim YS, Lee BH, Zhu JK (2006) A putative Arabidopsis nucleoporin AtNUP160 is critical for RNA export and required for plant tolerance to cold stress. Mol Cell Biol 26:9533–9543

Estruch F, Cole CN (2003) An early function during transcription for the yeast mRNA export factor Dbp5p/Rat8p suggested by its genetic and physical interactions with transcription factor IIH components. Mol Biol Cell 14:1664–1676

Fabre E, Boelens WC, Wimmer C, Mattaj IW, Hurt EC (1994) Nup145p is required for nuclear export of mRNA and binds homopolymeric RNA in vitro via a novel conserved motif. Cell 78:275–289

Farago M, Nahari, T, Hammel C, Cole CN, Choder M (2003) Rpb4p, a subunit of RNA Polymerase II, mediates mRNA export during stress. Mol Biol Cell 14:2744–2755

Fischer RA, Byerlee DR (1991) Trends of wheat production in the warmer areas: major issues and economic considerations. In: D.A. Saunders (Ed), Wheat for Nontraditional, Warm Areas, pp. 3–27. CIMMYT, Mexico City.

Flowers TJ, Yeo AR (1995) Breeding for salinity resistance in crop plants. Where next? Aust J Plant Physiol 22:875–884

Gallouzi IE, Steitz JA (2001) Delineation of mRNA export pathways by the use of cell-permeable peptides. Science 294:1895–1901

Gong Z, Dong CH, Lee H, Zhu J, Xiong L, Gong D, Stevenson B, Zhu JK (2005) A DEAD box RNA helicase is essential for mRNA export and important for development and stress responses in Arabidopsis. Plant Cell 17:256–267

Gong Z, Lee H, Xiong L, Jagendorf A, Stevenson B, Zhu JK (2002) RNA helicase-like protein as an early regulator of transcription factors for plant chilling and freezing tolerance. Proc Natl Acad Sci USA 99:11507–11512

Griffis ER, Craige B, Dimaano C, Ullman KS, Powers MA (2004) Distinct functional domains within nucleoporins Nup153 and Nup98 mediate transcription dependent mobility. Mol Biol Cell 15:1991–2002

Hammell CM, Gross S, Zenklusen D, Heath CV, Stutz F, Moore C, and Cole CN (2002) Coupling of termination, 3′ processing, and mRNA export. Mol Cell Biol 22:6441–6457

Hirose Y, Manley JL (2000) RNA polymerase II and the integration of nuclear events. Genes Dev 14:1415–1429

Izawa S, Takemura R, Inoue Y (2004) Gle2p is essential to induce adaptation of the export of bulk poly(A)+ mRNA to heat shock in *Saccharomyces cerevisiae*. J Biol Chem 279:35469–35478

Jones-Rhoades MW, Bartel DP, Bartel B (2006) MicroRNAs and their regulatory roles in plants. Annu Rev Plant Biol 57:19–53

Kanamori N, Madsen LH, Radutoiu S, Frantescu M, Quistgaard EM, Miwa H, Downie JA, James EK, Felle HH, Haaning LL, Jensen TH, Sato S, Nakamura Y, Tabata S, Sandal N, Stougaard J (2006) A nucleoporin is required for induction of Ca^{2+} spiking in legume nodule development and essential for rhizobial and fungal symbiosis. Proc Natl Acad Sci USA 103:359–364

Kendirgi F, Rexer DJ, Alcazar-Roman AR, Onishko HM, Wente SR (2005) Interaction between the shuttling mRNA export factor Gle1 and the nucleoporin hCG1: A conserved mechanism in the export of HSP70 mRNA. Mol Biol Cell 16:4304–4315

Kidner CA, Martienssen RA (2005) The developmental role of microRNA in plants. Curr Opin Plant Biol 8:38–44

Kimura T, Hashimoto I, Nagase T, Fujisawa JI (2004) CRM1-dependent, but not ARE-mediated, nuclear export of IFN-α1 mRNA. J Cell Sci 117:2259–2270

Kurihara Y, Takashi Y, Watanabe Y (2006) The interaction between DCL1 and HYL1 is important for efficient and precise processing of pri-miRNA in plant microRNA biogenesis. RNA 12:206–212

Le Hir H, Gatfield D, Izaurralde E, Moore MJ (2001) The exon-exon junction complex provides a binding platform for factors involved in mRNA export and nonsense-mediated mRNA decay. EMBO J 20:4987–4997

Le Hir H, Izaurralde E, Maquat LE, Moore MJ (2000) The spliceosome deposits multiple proteins 20–24 nucleotides upstream of mRNA exon-exon junctions. EMBO J 19:6860–6869

Lee MS, Henry M, Silver PA (1996) A protein that shuttles between the nucleus and the cytoplasm is an important mediator of RNA export. Genes Dev 10:1233–1246

Lei E, Silver PA (2002) Intron status and 3′-end formation control cotranscriptional export of mRNA. Genes Dev 16:2761–2766

Lei EP, Krebber H, Silver PA (2001) Messenger RNAs are recruited for nuclear export during transcription. Genes Dev 15:1771–1782

Levesque L, Bor YC, Matzat LH, Jin L, Berberoglu S, Rekosh D, Hammarskjold ML, Paschal BM (2006) Mutations in Tap uncouple RNA export activity from translocation through the nuclear pore complex. Mol Biol Cell 17:931–943

Lund E, Guttinger S, Calado A, Dahlberg JE, Kutay U (2004) Nuclear export of MicroRNA precursors. Science 303:95–98

Mallory AC, Vaucheret H (2006) Functions of microRNAs and related small RNAs in plants. Nat Genet 38:S31–S36

McBride AE, Cook JT, Stemmler EA, Rutledge KL, McGrath KA, Rubens JA (2005) Arginine methylation of yeast mRNA-binding protein Npl3 directly affects its function, nuclear export, and intranuclear protein interactions. J Biol Chem 280:30888–30898

Mizuki F, Namiki T, Sato H, Furukawa H, Matsusaka T, Ohshima Y, Ishibashi R, Andoh T, Tani T (2007) Participation of XPB/Ptr8p, a component of TFIIH, in nucleocytoplasmic transport of mRNA in fission yeast. Genes Cells 12:35–47

Murphy R, Watkins JL, Wente SR (1996) GLE2, a *Saccharomyces cerevisiae* homologue of the *Schizosaccharomyces pombe* export factor RAE1, is required for nuclear pore complex structure and function. Mol Biol Cell 7:1921–1937

Neumann N, Jeffares DC, Poole AM (2006) Outsourcing the nucleus: nuclear pore complex genes are no longer encoded in nucleomorph genomes. Evol Bioinformatics 2:389–400

Park MY, Wu G, Gonzalez-Sulser A, Vaucheret H, Poethig RS (2005) Nuclear processing and export of microRNAs in Arabidopsis. Proc Natl Acad Sci USA 102:3691–3696

Park NI, Muench DG (2007) Biochemical and cellular characterization of the plant ortholog of PYM, a protein that interacts with the exon junction complex core proteins Mago and Y14. Planta 225:625–639

Parry G, Ward S, Cernac, A, Dharmasiri S, Estelle M (2006) The Arabidopsis SUPPRESSOR OF AUXIN RESISTANCE proteins are nucleoporins with an important role in hormone signaling and development. Plant Cell 18:1590–603

Pemberton LF, Paschal BM (2005) Mechanisms of receptor-mediated nuclear import and nuclear export. Traffic 6:187–198

Pritchard CE, Fornerod M, Kasper LH, van Deursen JM (1999) RAE1 is a shuttling mRNA export factor that binds to a GLEBS-like NUP98 motif at the nuclear pore complex through multiple domains. J. Cell Biol 145:237–254

Rayala HJ, Kendirgi F, Barry DM, Majerus PA, Wente SR (2004) The mRNA export factor human Gle1 interacts with the nuclear pore complex protein Nup155. Mol Cell Proteomics 3:145–155

Rollenhagen C, Hodge CA, Cole CN (2004) The nuclear pore complex and the DEAD box protein Rat8p/Dbp5p have nonessential features which appear to facilitate mRNA export following heat shock. Mol Cell Biol 24:4869–4879

Rout MP, Aitchison, JD, Suprapto A, Hjertaas K, Zhao Y, Chait BT (2000) The yeast nuclear pore complex: composition, architecture, and transport mechanism. J Cell Biol 148:635–651

Saavedra C, Tung, KS, Amberg DC, Hopper AK, and Cole CN (1996) Regulation of mRNA export in response to stress in *Saccharomyces cerevisiae*. Genes Dev 10:1608–1620

Saavedra CA, Hammell CM, Heath CV, and Cole CN (1997) Yeast heat-shock mRNAs are exported through a distinct pathway defined by Rip1p. Genes Dev 11:2845–2856

Sanan-Mishra N, Pham, XH, Sopory SK, Tuteja N (2005) Pea DNA helicase 45 overexpression in tobacco confers high salinity tolerance without affecting yield. Proc Natl Acad Sci USA 102:509–514

Schmitt C, von Kobbe C, Bachi A, Pante N, Rodrigues JP, Boscheron C, Rigaut G, Wilm M, Seraphin B, Carmo-Fonseca M, Izaurralde E (1999) Dbp5, a DEAD-box protein required for mRNA export, is recruited to the cytoplasmic fibrils of nuclear pore complex via a conserved interaction with CAN/Nup159p. EMBO J 18:4332–4347

Segref A, Sharma K, Doye V, Hellwig A, Huber J, Luhrmann R, Hurt E (1997) Mex67p, a novel factor for nuclear mRNA export, binds to both poly(A) RNA and nuclear pores. EMBO J 16:3256–3271

Selitrennik M, Duek L, Lotan R, Choder M (2006) Nucleocytoplasmic shuttling of the Rpb4p and Rpb7p subunits of *Saccharomyces cerevisiae* RNA Polymerase II by two pathways. Eukaryot Cell 5:2092–2103

Senger B, Simos G, Bischoff FR, Podtelejnikov A, Mann M, Hurt E (1998) Mtr10p functions as a nuclear import receptor for the mRNA-binding protein Npl3p. EMBO J 17:2196–2207

Shen EC, Henry MF, Weiss VH, Valentini SR, Silver PA, Lee MS (1998) Arginine methylation facilitates the nuclear export of hnRNP proteins. Genes Dev 12:679–691

Strasser K, Hurt E (2000) Yra1p, a conserved nuclear RNA-binding protein, interacts directly with Mex67p and is required for mRNA export. EMBO J 19:410–420

Strasser K, Hurt E (2001) Splicing factor Sub2p is required for nuclear mRNA export through its interaction with Yra1p. Nature 413:648–652

Strasser K, Bassler J, Hurt E (2000) Binding of the Mex67p/Mtr2p heterodimer to FXFG, GLFG, and FG repeat nucleoporins is essential for nuclear mRNA export. J Cell Biol 150:695–706

Strasser K, Masuda S, Mason P, Pfannstiel J, Oppizzi M, Rodriguez-Navarro S, Rondon AG, Aguilera A, Struhl K, Reed R, and Hurt E (2002) TREX is a conserved complex coupling transcription with messenger RNA export. Nature 417:304–308

Sunkar R, Zhu JK (2004) Novel and stress-regulated microRNAs and other small RNAs from Arabidopsis. Plant Cell 16:2001–2019

Sunkar R, Kapoor A, Zhu JK (2006) Posttranscriptional induction of two Cu/Zn superoxide dismutase genes in Arabidopsis is mediated by downregulation of miR398 and important for oxidative stress tolerance. Plant Cell 18:2051–2065

Suntharalingam M, Wente SR (2003) Peering through the pore: nuclear pore complex structure, assembly, and function. Dev Cell 4:775–789

Tseng SS, Weaver PL, Liu Y, Hitomi M, Tartakoff AM, Chang TH (1998) Dbp5p, a cytosolic RNA helicase, is required for poly(A)+ RNA export. EMBO J 17:2651–2662

Vashisht AA, Pradhan A, Tuteja R, and Tuteja N (2005) Cold- and salinity stress-induced bipolar pea DNA helicase 47 is involved in protein synthesis and stimulated by phosphorylation with protein kinase C. Plant J 44:76–87

Vasu S, Shah S, Orjalo A, Park M, Fischer WH, Forbes DJ (2001) Novel vertebrate nucleoporins Nup133 and Nup160 play a role in mRNA export. J Cell Biol 155:339–354

Verslues PE, Guo Y, Dong CH, Ma W, Zhu JK (2006) Mutation of SAD2, an importin -domain protein in Arabidopsis, alters abscisic acid sensitivity. Plant J 47:776–787

Vinciguerra P, Stutz F (2004) mRNA export: an assembly line from genes to nuclear pores. Curr Opin Cell Biol 16:285–292

Watkins JL, Murphy R, Emtage JL, Wente SR (1998) The human homologue of *Saccharomyces cerevisiae* Gle1p is required for poly(A) RNA export. Proc Natl Acad Sci USA 95:6779–6784

Weirich CS, Erzberger JP, Berger JM, Weis K (2004) The N-terminal domain of Nup159 forms a beta-propeller that functions in mRNA export by tethering the helicase Dbp5 to the nuclear pore. Mol Cell 16:749–760

Weirich CS, Erzberger JP, Flick JS, Berger JM, Thorner J, Weis K (2006) Activation of the DExD/H-box protein Dbp5 by the nuclear-pore protein Gle1 and its coactivator InsP6 is required for mRNA export. Nat Cell Biol 8:668–676

Weis K (2002) Nucleocytoplasmic transport: cargo trafficking across the border. Curr Opin Cell Biol 14:328–335

Wiegand HL, Coburn G.A, Zeng Y, Kang Y, Bogerd HP, Cullen BR (2002) Formation of Tap/NXT1 heterodimers activates Tap-dependent nuclear mRNA export by enhancing recruitment to nuclear pore complexes. Mol Cell Biol 22:245–256

Yamaguchi-Shinozaki K, Shinozaki K (2006) Transcriptional regulatory networks in cellular responses and tolerance to dehydration and cold stresses. Annu Rev Plant Biol 57:781–803

Yi R, Qin Y, Macara IG, Cullen BR (2003) Exportin-5 mediates the nuclear export of pre-microRNAs and short hairpin RNAs. Genes Dev 17:3011–3016

Zenklusen D, Vinciguerra P, Strahm Y, Stutz F (2001) The yeast hnRNP-like proteins Yra1p and Yra2p participate in mRNA export through interaction with Mex67p. Mol Cell Biol 21:4219–4232

Zenklusen D, Vinciguerra P, Wyss JC, Stutz (2002) Stable mRNP formation and export require cotranscriptional recruitment of the mRNA export factors Yra1p and Sub2p by Hpr1p. Mol Cell Biol 22:8241–8253

Zhang Y, Li X (2005) A putative nucleoporin 96 Is required for both basal defense and constitutive resistance responses mediated by suppressor of npr1-1, constitutive 1. Plant Cell 17:1306–1316

Zhao J, Jin SB, Bjorkroth B, Wieslander L, Daneholt B (2002) The mRNA export factor Dbp5 is associated with Balbiani ring mRNP from gene to cytoplasm. EMBO J 21:1177–1187

Zhu JK (2002) Salt and drought stress signal transduction in plants. Annu Rev Plant Biol 53:247–273

Zimmermann P, Hirsch-Hoffmann M, Hennig L, Gruissem W (2004) GENEVESTIGATOR. Arabidopsis microarray database and analysis toolbox. Plant Physiol 136:2621–2632

Regulation of Alternative Splicing of Pre-mRNAs by Stresses

G.S. Ali, A.S.N. Reddy(✉)

Contents

Introduction	258
Cis and *Trans* Elements That Affect Splicing	258
Analysis of Effects of Stresses on Alternative Splicing with Bioinformatics Approaches	259
Role of Alternative Splicing in Biotic Stresses	260
Role of Alternative Splicing in Abiotic Stresses	261
Temperature Stress	261
Salt, Drought, and Osmotic Stress	265
Wound Stress	267
Radiation and Light Stress	267
Signaling Pathways That Affect Stress-Regulated Alternative Splicing	269
Subcellular Distribution of SR Proteins in Response to Stresses and Its Implications for Alternative Splicing	270
Conclusions and Future Directions	271
References	272

Abstract A substantial fraction (~30%) of plant genes is alternatively spliced, but how alternative splicing is regulated remains unknown. Many plant genes undergo alternative splicing in response to a variety of stresses. Large-scale computational analyses and experimental approaches focused on select genes are beginning to reveal that alternative splicing constitutes an integral part of gene regulation in stress responses. Based on the studies discussed in this chapter, it appears that alternative splicing generates transcriptome/proteome complexity that is likely to be important for stress adaptation. However, the signaling pathways that relay stress conditions to splicing machinery and if and how the alternative spliced products confer adaptive advantages to plants are poorly understood.

A.S.N. Reddy
Department of Biology and Program in Molecular Plant Biology, Colorado State University, Fort Collins, CO 80523, USA
e-mail: reddy@colostate.edu

Introduction

Precursor messenger RNA (pre-mRNA) splicing is critical for the expression of intron-containing genes in most eukaryotes including plants. Approximately eighty percent of plant genes contain one or more introns, which are constitutively removed by the spliceosome (Reddy 2007). 'In addition to constitutive splicing, regulated or alternative splicing produces multiple mRNAs from the same gene by using alternative splice sites (Graveley 2001; Reddy 2001, 2007; Maniatis and Tasic 2002; Ast 2004; Reddy 2007). These variant mRNAs can code for proteins with different functions and subcellular localization, essentially increasing the proteome diversity of organisms (Blencowe 2006). In addition, splicing variants may not code for different proteins but play a role at the posttranscriptional level by regulating RNA stability and turnover. For example, several alternatively spliced mRNA isoforms harbor premature termination codons (PTC) and are targeted for degradation through the nonsense-mediated decay (NMD) pathway (Lejeune and Maquat 2005), providing another layer to the regulation of gene expression. Our understanding of alternative splicing in plants has lagged far behind advancements in animal systems. Nevertheless, several recent large-scale, primarily computational, analyses in model and crop plants have revealed the widespread occurrence of alternatively spliced products in plants (Haas et al. 2003; Zhu et al. 2003; Iida et al. 2004; Ner-Gaon et al. 2004; Campbell et al. 2006; Iida and Go 2006; Wang and Brendel 2006; Chen et al. 2007) (see the chapter by B. J. Haas, this volume). These alternatively spliced genes represent a wide range of functional groups, suggesting that alternative splicing affects almost all aspects of plant development and physiology. In this article we discuss potential roles of alternative splicing in plant stress responses. We limit our discussion to mostly abiotic stresses as the role of alternative splicing in biotic stresses is covered in the chapter by W. Gassmann in this volume. We also provide several possible scenarios as to how stresses regulate alternative splicing and what functional significance alternative splicing products might have, especially in providing adaptive advantages to plants under stress conditions. Although our focus is primarily on the role of some selected alternatively spliced genes in stress responses, we also review some of the recent computational analyses that have uncovered potential alternatively spliced genes in stress adaptation.

Cis and *Trans* Elements That Affect Splicing

Accurate splicing requires a combination of *cis*-acting RNA sequences present in pre-mRNA and a large number of *trans*-acting spliceosomal elements. The *cis*-acting RNA sequences are the 5' and 3' splice sites, the branchpoint, and the polypyrimidine tract (see the chapters by M. A. Schuler and C. G. Simpson and J. W. S. Brown, this volume). The *trans*-acting elements consist of five major spliceosomal particles: U1, U2, U4, U5, and U6 small nuclear ribonucleoprotein particles (snRNP) (Jurica and Moore 2003; Nilsen 2003). In metazoans, the recruitment of U1 snRNP to the 5' splice site is facilitated by members of the serine/arginine-rich (SR) family splicing factors (Eperon et al. 1993; Wu and Maniatis 1993; Kohtz et al. 1994; Jamison et al. 1995; Zahler and Roth 1995; Xiao and Manley 1997; Cao et al. 1998; Graveley 2000). These proteins constitute a highly conserved family of structurally and functionally related non-snRNP proteins with multiple roles in pre-mRNA splicing and other aspects of RNA metabolism (Graveley 2000; Reddy 2001, 2004; Sanford et al. 2003; Huang and Steitz 2005). A detailed discussion on plant SR proteins and their functions is covered in the chapter by A. Barta et al. in this volume. SR proteins function as essential splicing factors in constitutive pre-mRNA splicing and also regulate alternative splicing by influencing splice site selection in a concentration-dependent manner (Manley and Tacke 1996; Graveley 2000). They have a modular domain structure consisting of one or two N-terminal RNA recognition motifs (RRMs) and a C-terminal arginine/serine-rich (RS) domain. The RRM, which confers RNA-binding specificity, binds to specific regulatory sequences in pre-mRNA, and the RS domain mediates protein-protein and protein-RNA interactions in the splicing machinery (Graveley 2000; Shen et al. 2004). In higher eukaryotes, sequences around the splice sites are less conserved, and, therefore, for correct and efficient definition of splice sites additional regulatory sequences adjacent to the splice sites, collectively called splicing enhancers/repressors, are required (Graveley 2000; Reddy 2004). SR proteins are thought to bind to these sequences and promote the recruitment of U1 snRNP to the correct 5′ splice sites and other snRNPs to assemble the spliceosome (Eperon et al. 1993; Kohtz et al. 1994; Jamison et al. 1995; Zahler and Roth 1995; Graveley 2000) (see the chapter by A. Barta et al., this volume). In addition, because of deviation from the consensus splice sites, SR proteins also function in selecting alternative weak splice sites and thus probably are one of the major contributors to increasing transcriptome complexity, protein diversity, and phenotypes (Graveley 2000).

A majority of *SR* genes display alternative splicing specific to a given stress, suggesting that these isoforms might be playing a role in stress responses (Lazar and Goodman 2000; Palusa et al. 2007; Tanabe et al. 2007). Except for only a few studies involving overexpression of functional isoforms, for example, *atRSZ33* and *SR30* that exhibit pleiotropic effects (Lopato et al. 1999; Kalyna et al. 2003) or mutant lines such as *SR45* (Ali et al. 2007), virtually nothing is known about the function of these isoforms. Future studies will undoubtedly reveal the right combination of alternatively spliced SR isoforms and RNA sequences that are required for a specific stress response.

Analysis of Effects of Stresses on Alternative Splicing with Bioinformatics Approaches

Both in metazoans and in plants, computational and experimental analyses have demonstrated that intron-containing genes undergo extensive alternative splicing (Lopato et al. 1999; Lazar and Goodman 2000; Kalyna et al. 2003; Iida et al. 2004; Wang and Brendel 2004; Ner-Gaon et al. 2007; Palusa et al. 2007). Detailed computational analyses based on alignment of ESTs/cDNAs against the human genome and experimental analyses using RT-PCR and more recently splicing-sensitive microarrays that can differentiate between alternative splicing isoforms have estimated that over 70% of human genes undergo alternative splicing (Modrek and Lee 2002; Johnson et al. 2003; Blencowe 2006). In plants such detailed analyses are limited primarily because of relatively smaller numbers of ESTs and also because of the lack of high-throughput tools such as splicing-sensitive arrays (Blencowe 2006; Reddy 2007). Nevertheless, with available plant ESTs/cDNAs, it is now estimated that at least 30% of plant genes are alternatively spliced (Haas et al. 2003; Zhu et al. 2003; Iida et al. 2004; Ner-Gaon et al. 2004; Campbell et al. 2006; Iida and Go 2006; Wang and Brendel 2006; Chen et al. 2007). However, this estimate is likely to grow because rare alternative spliced isoforms that appear under special circumstances or in specific tissues may simply not be represented in the available EST libraries. Annotations of the alternatively spliced genes revealed that they represent diverse biological and molecular processes, suggesting a fundamental role for alternative splicing in affecting the development and physiology of plants.

A GO (Gene Ontology) analysis of the alternatively spliced gene list reported by Wang and Brendel (2006) and Chen et al (2007) showed that a majority of stress-related genes undergo alternative splicing. This is significantly higher than the 30% figure of alternatively spliced genes in the genomes of Arabidopsis and rice. A further breakdown of the list shows that almost all major biotic and abiotic stresses are represented in the alternative spliced category. Transcription factors are known to affect the expression of genes under specific stress conditions. Interestingly, several transcription factors themselves are subjected to splicing regulation, which in turn may regulate downstream genes and coordinate interactions between major networks at major nodes. Iida et al (2005) have recently reported that of the 1,968 total Arabidopsis transcription factors, 110 undergo alternative splicing. Several of these transcription factors are associated with stresses. The exact mechanism by which these alternative isoforms are regulated and how they affect downstream targets remains to be studied. Recent computational analysis have identified several exonic and intronic splicing enhancer sequences in Arabidopsis (Sheth et al. 2006; Pertea et al. 2007). It is suggested that these sequences together with SR proteins likely play important roles in regulating alternative splicing in response to various stresses. This suggestion is supported by the observation that the tissue-specific alternative splicing of ascorbate peroxidase is conferred by an intronic splicing regulatory element (Yoshimura et al. 2002). The occurrence of alternative splicing in transcription factors alludes to another level of relationship between transcription and splicing. These studies are also consistent with the proposal that various stages of genetic information transmission display extensive coupling at various levels.

Role of Alternative Splicing in Biotic Stresses

Plants devote considerable resources to defending against pathogens by reprogramming the gene expression profile of thousands of genes (see, e.g., Wang et al. 2006). Consistent with these modifications, several published reports on individual genes and large-scale analysis have revealed that alternative splicing plays an important role in plant disease resistance. It has been shown that several *R* genes undergo alternative splicing. A detailed account of the alternative splicing of *R* genes and their functional significance in disease resistance is covered in the chapter by W. Gassmann in this volume. Apart from *R* genes, other genes involved in the disease resistance pathways such as *EDS1* also undergo alternative splicing (Iida et al. 2004). The functional relevance of many of these alternatively spliced isoforms of *R* genes in disease resistance remains unknown except in a few cases, and should be one of the priority areas for determining the role of alternative splicing in agronomically important traits in the future.

Role of Alternative Splicing in Abiotic Stresses

Similar to biotic stresses, abiotic stresses also make plants divert considerable metabolic resources toward adapting and surviving in adverse conditions by modulating gene expression (Zhu 2002; Yamaguchi-Shinozaki and Shinozaki 2006). As a consequence, plants under abiotic stress conditions usually display reduced growth and reproductive success. Therefore, to deal with abiotic stresses and to ensure their reproductive success, plants tightly regulate their stress-related gene expression at multiple levels. Investigations of regulation at the transcriptional level have resulted in identification of several key transcription factors and other potential signaling components (Yamaguchi-Shinozaki and Shinozaki 2006). In addition to transcriptional regulation, over the past several years several studies have also shown the occurrence of alternative splicing in several genes under various abiotic stresses. Here we review most of these studies that have established a relationship between alternative splicing and an abiotic stress. Since SR proteins are widely believed be the major regulators of alternative splicing, we start our discussion with the regulation of alternative splicing of SR genes where known. Then we discuss specific examples of genes in each stress, and at the end we provide a speculative unifying model for the regulation of stress-dependent alternative splicing. Here we define abiotic stress as any stress condition that does not involve plant pathogens or symbiotic organisms.

Temperature Stress

Studies using DNA microarrays or focused studies with several genes have revealed an elaborate network of genes that respond to cold and provide tolerance to plants (Chinnusamy et al. 2006; Yamaguchi-Shinozaki and Shinozaki 2006) (see the chapter by V. Chinnusamy et al., this volume). To adapt to extreme or suboptimal temperatures, plants use several mechanisms, which include induction of cold and other stress-related genes, modification in the composition and level of membrane lipids, and modulation of sugar and proline metabolism. The alternative splicing of genes with different molecular functions is affected by low temperature. These include splicing-related proteins, transcription factors, and protein involved in lipid and sugar metabolism. Here we review several of these genes that code for SR splicing factors (Palusa et al., 2007), a ribokinase, a C3H2C3 RING-finger protein (Mastrangelo et al. 2005), β-hydroxyacyl ACP dehydratase (Tai et al. 2007), Citrus Low Temperature (CLT) protein (Jia et al. 2004), Alternative Oxidase (AOX) (Fung et al. 2006), the cereal DREB2 (Egawa et al. 2006), and a Potato Invertase (Bournay et al. 1996).

Some studies have shown that the alternative splicing pattern of several different *SR* genes is changed in Arabidopsis treated with cold or heat (Lazar and Goodman 2000; Palusa et al. 2007). In most of these studies, although the functional significance of the alternatively spliced isoforms is not studied, the prediction of alternative spliced products suggests a change in the ratio of transcripts and proteins with different domain structure, which in turn may affect the splicing of downstream genes. For example, the alternative splicing patterns of *SR34/SR1, SR34b, RS40 RS31*, and *SR33* were changed by cold, whereas those of *SR30, SR1, SR34b, RS31a, RS40, RSZ32, RSZ33, SR33*, and *SCL30a* were changed by heat. These data show that several SR genes are affected by only one stress, whereas there is a set that are changed by both heat and cold, sometimes in an opposite manner. For example isoform 2 of *SR33* was increased by heat but reduced by cold (Palusa et al. 2007). Taken together these results indicate that changes in temperature affect the alternative splicing pattern of several different SR genes simultaneously, suggesting that these different isoforms act in combination to specifically alter the level of splicing factor isoforms and/or alternative splicing of downstream cold-regulated genes.

The wheat *DREB2* gene produces three isoforms named as *Wdreb2α, Wdreb2β*, and *Wdreb2γ*, which are remarkably similar to the barley HvDRF1.1, HvDRF1.2, and HvDRF1.3 isoforms (Xue and Loveridge 2004; Egawa et al. 2006). These isoforms differ from one another in the inclusion/exclusion of exon 2 and 3; *Wdreb2γ* consists of exons 1 and 4 and codes for a 344-amino acid full-length protein, whereas *Wdreb2β* consists of exons 1, 2, and 4, and codes for a 60-amino acid short peptide. A change in the alternative splicing pattern of these isoforms in response to abiotic stresses, including cold, was analyzed independently in wheat and barley. In response to cold, all three transcripts of the *Wdreb2* increased with prolonged cold treatment. A comparative kinetic analysis indicated that later during the cold treatment isoforms 1 and 3 also appeared. The maize *ZmDREB2A* homolog produced two isoforms, with the longer transcript coding for the functional protein, which was preferentially induced by cold and repressed by heat (Qin et al. 2007). Although no definitive conclusion can be drawn from these results, the kinetics of alternative splicing patterns indicate that prolonged cold treatment promotes inclusion of exons 2 and 4, probably leading to downregulation of the functional isoforms. Whether the *Hv*DRF isoforms will behave in a similar fashion in response to cold is not known. With their high sequence and gene structure similarity to *Wdreb2*, it is tempting to suggest that they may be regulated in a similar fashion. A survey of the literature and the databases of alternative spliced Arabidopsis genes (Iida et al. 2005; Wang and Brendel 2006) suggests that the homologs of *Wdreb2* in Arabidopsis do not undergo alternative splicing.

However, several other members of the ERF/AP2 family transcription factors are predicted to undergo alternative splicing (Iida et al. 2005).

The composition and level of lipids play pivotal roles in the adaptation of plants to cold (Falcone et al. 2004). In black spruce, intron 1 of the gene for β-hydroxyacyl ACP dehydratase, which is a fatty acid biosynthetic enzyme, is alternatively spliced in a cold-dependent manner (Tai et al. 2007). Under normal conditions, most of the transcripts retain intron 1, which results in a severely truncated and probably non-functional peptide. Temperatures below freezing during the winter weeks promoted the splicing of intron 1 in the buds. Similarly, imbibing of seeds at 4°C also promoted intron 1 splicing in a reversible manner. These analyses indicate that in response to cold, black spruce regulates the production of functional β-hydroxyacyl ACP dehydratase by alternative splicing. Interestingly, the gene for acetyl CoA carboxylase (ACC) enzyme, which is involved in lipid metabolism, also displays alternative splicing and alternative promoter initiation both in metazoans and plants (Podkowinski et al. 2003; Barber et al. 2005). This suggests that the regulation of lipid metabolism by alternative splicing is conserved and probably evolved before the divergence of animals and plants. However, differences in the gene structure and the alternative splicing pattern suggest a lack of phylogenetic relationship between the two lineages, and it is likely that alternative splicing in response to cold evolved independently in plants and animals. Whether the alternative splicing of *ACC* is regulated by stresses is not known.

Plants produce reactive oxygen species (ROS) under most stress conditions. ROS are harmful, and therefore cells have evolved mechanisms to detoxify them. The tomato *LeAOX* gene produces three isoforms by alternative splicing of the last intron (Fung et al. 2006). These three isoforms code for proteins with differential activities. It was shown that in addition to a complex alternative splicing pattern in tissues, cold inhibited the splicing of the last intron, resulting in changing the ratio of various isoforms of *AOX*. Interestingly, in the same study this change in cold-dependent splicing pattern was correlated with a change in the splicing pattern of the *9G8* SR splicing factor gene. The longer isoform of *9G8*, presumably containing extra introns, was reduced by cold storage. Although no sequence information was determined for the longer isoform, it is possible that it codes for a nonfunctional isoform. The Arabidopsis homolog of the *Le9G8* gene, however, does not undergo alternative splicing (Palusa et al. 2007), suggesting its species-specific splicing regulation by cold. Since SR proteins are well known to affect alternative splicing of other genes, the apparent change in the splicing of *9G8* may be responsible for changing the splicing of *AOX* (Fung et al. 2006).

Sugar metabolism plays an essential role in cold stresses, and genes involved in sugar metabolism are differentially regulated under cold (Mastrangelo et al. 2005). In durum wheat two early cold-regulated (*e-cor*) genes, coding for a ribokinase (*7H8*) and a RING-finger protein (*6G2*), were shown to undergo alternative splicing (Mastrangelo et al. 2005). Alternative transcripts of these genes were produced by alternative splicing of introns either in the coding region (*7H8*) or in the 3' UTR (*6G2*). The ratio of transcripts that retained alternative introns increased with cold treatment. For *7H8*, all alternative isoforms resulted from a complex combination of several introns, all of which resulted in the introduction of a premature termination codon (PTC), which would predictably result in nonfunctional truncated proteins and a net decrease in the functional protein. In contrast, the alternative splicing of *6G2* does not influence the protein sequence, but a net increase in the alternative spliced products may also increase the net level of total protein or the alternative 3' UTR may be regulating the stability of the transcript. Similar results in terms of cold-dependent intron retention were observed with the barley and bread wheat homologs of these genes (Mastrangelo et al. 2005). Interestingly, however, cold-dependent intron retention of the Arabidopsis homolog of the *7H8* gene was not observed, suggesting differential species-specific splicing activities between dicots and monocots in response to cold.

Freezing temperatures (2 to $-1°C$) promoted the usage of a proximal 3' splice site in the trifoliate orange *CLT* gene (Jia et al. 2004). Although the alternative splicing event, which was mapped to the 3' UTR, did not alter the amino acid sequence, the inclusion of additional nucleotides in the 3' UTR may have a role in regulating the stability of the *CLT* transcript. A 9-bp exon, which codes for three amino acids in a highly conserved motif in potato invertase, was omitted by cold stress (Bournay et al. 1996). The omission of these amino acids could alter the function of invertase and sugar metabolism, which is likely to play a role in cold adaptation.

In addition to cold, several studies have shown changes in the alternative splicing pattern of genes associated with heat stress. For example, *HSP70* (Christensen et al. 1992; Hopf et al. 1992), *HSP81*, and waxy (Larkin and Park 1999), which are known to be induced by elevated temperatures, also undergo alternative splicing. In several of these cases, the change in the alternative splicing apparently leads to either an enhancement or reduction of the functional isoforms.

From the above examples we can derive the following principles. (a) Certain splicing events are conserved across different plant species, whereas others are species specific. This suggests that evolution of alternative splicing may be shaped by temperature stress in an ecological-specific manner according to the prevalent environmental conditions and, therefore, is likely to perform an important role in the ecological distribution of plants. (b) The above studies also suggest that temperature stress does not affect one particular splicing type but rather affects several different types in different genes. (c) Also evident from these studies is that temperature stress does not always lead to an increase in the functional isoform of a gene. It may be that certain genes play positive and others play negative roles and, therefore, are likely affected accordingly in such a way that those playing positive roles are enhanced and those playing negative roles are downregulated.

Although none of these studies has addressed the molecular mechanisms that regulate these alternative splicing events, by inference from studies with metazoan systems it is likely due to a combination of several *cis* and *trans* elements. *Cis* elements in plants that respond to temperature stress are mostly unknown, whereas the role of *trans* elements is most likely performed by SR proteins in combination with other spliceosomal proteins. A signaling pathway that starts with the perception of a stress leads to either de novo expression of a set of SR proteins or a change in their phosphorylation status. Depending upon whether an SR protein acts as a splicing enhancer or repressor, it would lead to activation or inactivation of a set of splicing factors, which in turn would lead to a specific splicing outcome most probably by acting in a specific combination. It is also possible that the phosphorylation state of SR and other splicing factors causes a change in their subcellular localization, which in turn provides a quick means of regulating their activity in splicing by spatial confinement. Also, another likely scenario could be that a change in temperature can rearrange the 3D conformation of an RNA *cis*-splicing element or the thermal-dependent stability of protein-protein, RNA-RNA, or RNA-protein interactions, which would eventually result in a particular alternative splicing outcome. Similarly, transcription rate has been shown to affect alternative splicing (de la Mata et al. 2003). Since under stress conditions several genes are induced, it is conceivable that because of their increased transcription rate an exon or intron is skipped, resulting in unproductive spliced variants.

Salt, Drought, and Osmotic Stress

Similar to temperature stress, a variety of genes are affected by salt and drought stresses also. It is worth mentioning here that the alternative splicing of several genes were affected by more than one stress, consistent with the observation that several of these stresses affect a common set of genes at the transcriptional level, too. Here we discuss these genes with an emphasis on the effect of alternative splicing on a change in the structure of the encoded protein or on potential regulation of gene expression. Most of these studies have only addressed whether the alternative splicing pattern of a gene is changed under a stress condition. A mechanistic understanding as to how these alternative splicing events are affected by a stress is unknown. These genes include several *SR* genes (Palusa et al. 2007), *AOX* (Kong et al. 2003), *Dreb2* (Egawa et al. 2006), and *SOS4* (Shi et al. 2002).

In contrast to many SR genes affected by temperature, the splicing patterns of only three Arabidopsis SR genes, *SR34b*, *SR33*, and *SR30*, were changed by higher salt levels (Palusa et al. 2007; Tanabe et al. 2007). Isoform 2 of *SR33* was reduced whereas isoform 3 was increased by salt stress. In the case of *SR34b* isoform 3 was reduced. Similarly, isoforms 3 and 4 of *SR30* were reduced by high-salt stress (Tanabe et al. 2007). In the absence of mechanistic analysis, we cannot draw any definitive conclusion about the physiological relevance of these changes, but it is conceivable that the regulation of different SR isoforms affects salt-stress related genes. Apart from SR genes, the splicing pattern of several other genes is also affected by salt stress. The rice gene *OsIM*, which codes for Alternative Oxidase (AOX), produces three isoforms by alternative splicing (Kong et al. 2003). Only one of them, *OsIM1*, codes for a full-length functional protein. Under salt stress conditions the ratio of *OsIm1* to *OsIm2* increased steadily in the salt-tolerant variety, whereas in the salt-sensitive variety this ratio peaked at 24 h but returned back to normal level. Since *OsIM1* codes for a functional protein, its increase probably resulted in increased functional proteins. Since *OsIM2* retains intron 6, these data suggest that salt stress promotes the splicing of this intron. These analyses also indicate that the alternative splicing of this gene is genetically controlled and probably regulates a plant's adaptation to salt stress. The increased AOX likely helps in scavenging the excess ROS that is produced under high-salt stress conditions. The alternative isoforms of *Wdreb2*, *Wdrebα*, and *Wdreb2β* increased under salt stress, whereas *Wdreb2γ* remained relatively steady (Egawa et al. 2006). Since, as mentioned above, *Wdreb2α* and *Wdreb2β* code for functional and nonfunctional proteins, respectively, their simultaneous induction under salt stress does not provide useful information about their functional relevance. Another gene that is associated with salt stress and that undergoes alternative splicing but not in a stress-dependent manner is *the salt overly sensitive (SOS4)* gene (Shi et al. 2002). *SOS4* produces two transcripts by alternative splicing in intron 1, resulting in proteins with a 34-amino acid difference in their N-termini. The functional significance of these isoforms remains unknown, as both of them are equally induced by salt stress and the overexpression of either isoform was effective in rescuing the salt overly sensitive phenotype of *sos4* mutant plants. A connection between SR-like proteins and salt tolerance was provided by Forment et al (2002). These authors showed that expression of the RS domain of Arabidopsis SR-like proteins was able to provide salt tolerance to yeast cells and transgenic Arabidopsis plants (Forment et al. 2002).

Similar to salt and cold, drought also induces the alternative splicing of several genes. Drought induced the retention of the fourth intron in *6G2* gene (Mastrangelo et al. 2005). Since the retention of this intron was also promoted by cold, one can reason that these two stresses regulate *6G2* through a common mechanism, likely through an SR protein. Dehydrins are a group of hydrophilic proteins with typical YSK2 domains that are enriched in tyrosine, serine, and lysine, respectively (Xiao and Nassuth 2006). These domains are involved in interaction with other cellular components and in some cases may be involved in scavenging free radicals. Dehydrins are differentially induced by cold, drought, or ABA and have cryoprotective and antifreeze activities. Their involvement in stress responses is substantiated by their induction by low temperature, salt, and drought, and by improvement of stress tolerance when they are overexpressed in Arabidopsis (Puhakainen et al. 2004). Two *Vitis* dehydrin genes (*DHN1a* and *DHN1b*), which contain a single intron, produced an unspliced and a spliced isoform (Xiao and Nassuth 2006). The spliced isoform codes for a full-length dehydrin with a single Y- and S-domain and two K domains (YSK$_2$). In *Vitis raparia* the unspliced isoform was induced by cold. The unspliced product likely codes for two ORFs, each coding for truncated proteins; isoform 1 lacks one full and one partial K domain in the last part of the protein (YS), and isoform 2 starts near the beginning of the first K-domain and therefore consists of only one full and one partial K-domains with no Y- and S-domain (K2). Drought induced both spliced and unspliced *DHN1* but with different kinetics; the spliced product progressively increased from day 1 and unspliced products from day 4 of the onset of drought conditions in two different species of *Vitis*. ABA also induced the production of both isoforms although with different kinetics. Since the unspliced product may produce protein that lacks important domains, it may therefore have impaired interaction with other components. It is interesting to know that one of these isoforms retains its K2 domain but lacks the S domain, which is proposed to bind calcium. Their appearance later in the stress response may be a way to downregulate the activity of elevated cytosolic calcium that occurs in cold and drought stresses. Similar to salt stress, the alternative isoforms *Wdreb2α* and *Wdreb2β* of the wheat *Dreb2* gene displayed increased induction under drought conditions whereas *Wdreb2γ* remained constant (Egawa et al. 2006). Similar change by drought and cold stresses suggests that a common factor that converges on the splicing of this gene under different stress conditions is involved in mediating these stress responses. Their barley homologs, *HvDRF1.1*, *HvDRF1.2*, and *HvDRF1.3*, are also upregulated by drought; however, no significant difference in their alternative splicing pattern was observed (Xue and Loveridge 2004). In contrast, their *ZmDREB2A* homologs produced two isoforms, with the functional isoform preferentially induced more by drought (Qin et al. 2007).

Wound Stress

There are not many reports that address a relationship between wounding and alternative splicing in plants. The alternative splicing of the tomato prosystemin was reported by Li and Howe (2001). Sytemin is a short polypeptide that plays an essential role in the signaling of tomato wound responses. The gene for prosystemin produces two transcripts by the alternative 3′ splice site selection in the third intron, which results in proteins in which Arg57 in isoform A is replaced by Gly-Thr. The relative ratio of these two transcripts was, however, not changed by wounding, suggesting that the constitutive presence of the two isoforms may be needed for wound response (Li and Howe 2001). Since wounding is also caused by insect attack, and several genes are known to be induced by herbivorous insects, it would be interesting to find a relationship between insect feeding and alternative splicing. Because of substantial overlap between wounding and other stresses such as freezing injury (Yamaguchi-Shinozaki and Shinozaki 2006), the alternative splicing pattern of several of the genes affected by cold likely also changes in response to general wound response.

Radiation and Light Stress

Like all living organisms, plants are also prone to DNA and other damage from harmful radiations and sunlight. They have therefore evolved several damage control and reversal mechanisms. Most studies have been conducted in nonplant systems and have uncovered several genes that play important roles in defense against radiations, particularly UV light. In metazoans alternative splicing has been shown to play a role in adapting to UV light stress. For example, the murine double-minute protein MDM2 undergoes alternative splicing after UV irradiation, leading to an alternate isoform that downregulates the functional MDM2, leading to the activation of p53 transcription factor, which plays an important role in cell stress responses (Chandler et al. 2006).

In plants several reports have begun to shed light on the role of alternative splicing in radiation and light stress. Intense light stress changed the splicing pattern of *SR30* and of a putative Arabidopsis homolog of transformer-2-like, atSR45a (Tanabe et al. 2007). The rice *OSMUS81* gene, which codes for an endonuclease and plays a role in DNA damage repair, produces two transcripts, a full-length functional isoform and a shorter isoform resulting from the splicing of a cryptic intron in exon 14 leading to the deletion of the second Helix-hairpin-Helix (HhH) (Mimida et al. 2007). This deletion probably impairs its interaction with other components of the DNA repair complex. The ratio of the *OSMUS81α* isoform increased relative to the *OSMUS81β* isoform under gamma and UV-C radiations. This indicates that under radiation stress alternative splicing leads to an increase in the functional isoform of *OSMUS81* to meet its increased demand for DNA repair. Similarly, other plant DNA damage-associated genes undergo alternative splicing, but the functional relevance of radiation damage and the modulation of alternative splicing is not known (Hirouchi et al. 2003). The *AtRAD1* (Vonarx et al. 2002), *AtXRCC3*, and *AtRAD51C* genes (Osakabe et al. 2002) have been reported to undergo alternative splicing. The *AtRAD1* gene produces four alternative isoforms by the usage of alternative 5' splice sites (*AtRAD1–2*), 3' splice site (*AtRAD1–3*), or both (*AtRAD1–4*), each coding for proteins that lack a region required for endonuclease complex formation and hence are nonfunctional. The *AtXRCC3* and *AtRAD51C* both produce two alternatively spliced isoforms. The alternative splicing event in *AtXRCC3* does not change the amino acid sequence of the encoded protein isoforms and therefore may play a role in transcript stability or translation. In contrast, the alternative spliced isoforms of *AtRAD51C* result from the usage of an alternative 5' splice site in the sixth exon. The usage of the alternative proximal 5' splice site results in a truncated protein, which lacks a Walker B motif involved in nucleotide binding. Both these genes are induced by gamma radiation. Since the alternative splicing ratio was not changed by radiation, a relationship between radiation and splicing cannot be established. Interestingly, several mammalian genes of the *recA/RAD51* family also undergo alternative splicing (Cartwright et al. 1998), indicating that alternative splicing of these genes may have an ancient origin playing a role in radiation stress adaptation. How the alternative splicing pattern of these genes is related to radiation damage is not clear and will require further exploration. In this context it is interesting to note that at least one metazoan SR protein was recently associated with DNA stability and may provide a connection between these essential splicing factors and DNA damage control (Li and Manley 2005). Their study in plants will undoubtedly provide important clues about the regulation of the DNA damage caused by radiation.

Signaling Pathways That Affect Stress-Regulated Alternative Splicing

In plants signaling pathways that connect a stress stimulus to a change in an alternative splicing event are largely unknown. For any meaningful understanding of alternative splicing it would be important to elucidate these pathways. Except for a few biochemical analysis with SR proteins (Golovkin and Reddy 1999; Savaldi-Goldstein et al. 2000), this area of alternative splicing in plants remains unexplored (see the chapter by Fluhr and the other chapter by G. S. Ali and A. S. N. Reddy, this volume). A Clk/Sty protein kinase, AFC2, has been shown to interact with and phosphorylate several SR proteins (Golovkin and Reddy 1999). Interestingly, one of these kinases, the tobacco LAMMER type PK12 kinase, which interacts with and phosphorylates atSRp34/SR1, has been shown to be induced by ethylene, suggesting that those abiotic stresses that involve ethylene in their pathways may be communicating to splicing machinery through this kinase. Beyond the finding that these kinases were localized to the nucleus (Savaldi-Goldstein et al. 2000), nothing is known about the regulation of their subcellular localization under stresses. A stress-dependent translocation across the nuclear membrane would strongly suggest their involvement in stresses. Additional evidence for the involvement of SR phosphorylation was provided by showing that salt tolerance conferred by the expression of an RS domain in yeast was dependent upon the yeast SR Sky1p kinase (Forment et al. 2002). Interestingly, a GFP-tagged AFC2 localized to speckles in a manner very similar to other SR proteins, suggesting that phosphorylation of SR proteins may be occurring in the speckles (unpublished data). The MAP kinases MPK6 and/or MPK3, which are induced by many stresses, also phosphorylate several SR proteins (Feilner et al. 2005), providing a possible link between stresses and the regulation of splicing. Several members of the mitogen-activated protein kinase (MAPK) cascade, which are involved in stresses and development of plants, themselves undergo alternative splicing (Xiong and Yang 2003; Castells et al. 2006). At least in one case alternatively spliced isoforms were shown to display differential activity and interaction with other signaling components. Since several plant stresses are mediated through similar kinases (Xiong and Yang 2003), it is likely that these kinases provide a fine-tuning mechanism for providing specificity to stress responses. A hypothetical model showing the potential role of kinases and SR proteins is shown in Fig. 1.

Another potential and interesting signaling connection between stresses and splicing is the transcription factors. Most of the studies reported in plants and discussed above have only focused on the change in the alternative splicing pattern of target genes. Although no study to uncover the signaling components that link a stress to splicing machinery have been undertaken, there is at least one example where involvement of some sort of signaling can be envisioned. The wheat *Dreb2* gene has a well-known function in stress signaling. Its alternative splicing likely provides a rapid response to stresses and provides a quick adaptive advantage to plants. Since cold and most of the other stresses are known to induce an increase in

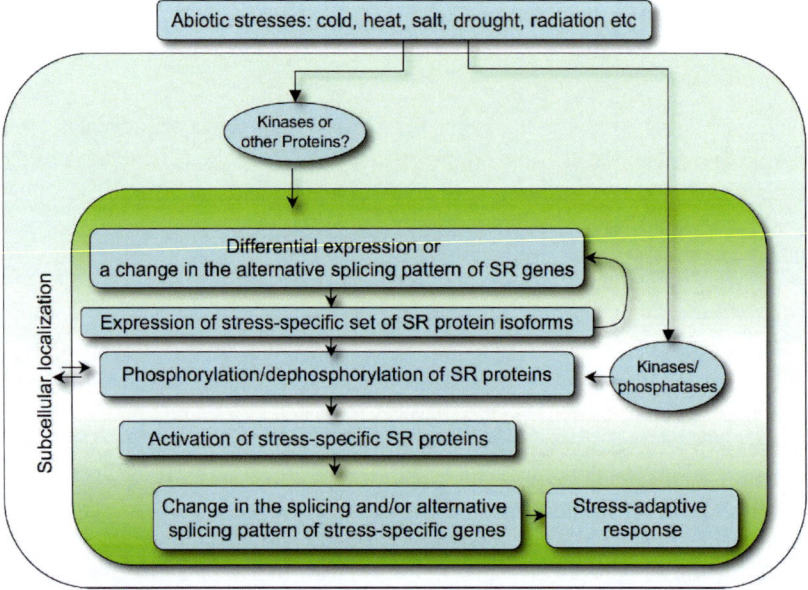

Fig. 1 A hypothetical model for the role of alternative splicing and SR proteins in stress responses. In response to a particular stress, SR genes undergo alternative splicing, producing a set of specific SR protein isoforms, which are activated by phosphorylation/dephosphorylation and/or translocation across the nuclear membrane. This combination of SR proteins then participate in splicing and/or alternative splicing of stress-regulated genes (SRs and others), eventually resulting in adaptation to stresses.

cytosolic calcium, a calcium-dependent pathway cannot be ruled out. ABA is a general stress hormone that mediates signaling in several abiotic stress responses. Recently, it was shown that only one of the four isoforms of *FCA*, which undergoes alternative splicing and produces four transcripts, binds ABA and affects flowering (Razem et al. 2006). However, a connection between flowering and any other stress response was not shown, although it is known that stresses usually promote flowering.

Subcellular Distribution of SR Proteins in Response to Stresses and Its Implications for Alternative Splicing

Several studies in metazoans indicate that the subcellular localization and concentrations of splicing-related proteins vary under different stress conditions, tissues, and growth stages, for example, hSlu (Shomron et al. 2005) and SF2 (Chiodi et al. 2004). More interestingly, in several of these studies a change in the subcellular localization in response to heat, UV, and other stresses was correlated with a change in the alternative splicing pattern of reporter and/or endogenous genes (Shomron et al. 2005). Considering similarity in the cellular localization pattern of SR proteins to their animal counterparts, one can reason that the alternative splicing pattern of plant genes is likely to be regulated by the regulation of subcellular localization of plant SR proteins in a similar fashion. In plants, using fluorescent-protein tagged SR proteins, several recent studies have demonstrated that the subcellular localization of SR protein is altered by heat and cold (Ali et al. 2003; Tillemans et al. 2005; Ali and Reddy 2006) (see the other chapter by G. S. Ali and A. S. N. Reddy, this volume). For example, Ali et al. (2003) showed that exposing Arabidopsis plants expressing GFP-tagged SR45 to elevated temperatures led to relocalization to enlarged speckles, whereas cold relocalized it to a diffuse pattern. It was further demonstrated that the heat-dependent relocalization of SR45 was dependent upon phosphorylation (Ali et al. 2003). Although no direct correlation between change in the subcellular localization and alternative splicing was assayed in these studies, heat and cold were shown to alter the alternative splicing pattern of several SR genes (Palusa et al. 2007). Taken together these studies suggest that temperature stress may be regulating alternative splicing by changing the subnuclear distribution of SR proteins.

Conclusions and Future Directions

The above discussion clearly shows that alternative splicing of a variety of genes representing different molecular functions is changed under abiotic stresses. In any given stress, several types of splicing are observed, indicating that a complex system of alternative splicing operates at different levels. Several genes are affected by more than one stress, suggesting that alternative splicing may be another important node in the cross talk among these stresses. Alternative splicing of several SR proteins has been shown to change under various stress conditions. These changes in turn may change the concentration of different SR proteins and their isoforms in a stress-specific manner, which in turn changes the alternative splicing of downstream target genes. Thus it seems that several SR proteins and their isoforms affect alternative splicing in a combinatorial fashion (Fig. 1). These analyses clearly underscore a need for dissecting the mechanisms that regulate alternative splicing and coordinate with other levels of regulations such as at the transcription and posttranslational levels, for example, production of nonfunctional or perhaps dominant-negative protein isoforms. The next key questions to address are (a) to functionally characterize these splicing events and (b) to derive stress-specific principles in terms of *cis*-regulatory sequences and the combinatorial set of SR and other splicing proteins that affect a specific splicing outcome. These goals will be accomplished by a combination of several techniques of cell biology, molecular biology, and large-scale analysis such as splicing-sensitive microarrays. (c) The goal of relating subcellular level events such as relocalization of splicing factors to enlarged speckles to a splicing outcome and eventually an organism level response will be a major challenge. (d) Another goal is to dissect the signaling pathways that relay stresses to splicing machinery. Addressing these questions will require devising novel experimental tools based on fluorescent proteins combined with higher-level genomewide computational analyses. Another major challenge will be to start taking steps in translating the fundamental knowledge gained through studying alternative splicing for improving agricultural crops.

Acknowledgements This work was supported by a grant from NSF.

References

Ali GS, Reddy AS (2006) ATP, phosphorylation and transcription regulate the mobility of plant splicing factors. J Cell Sci 119:3527–3538
Ali GS, Golovkin M, Reddy AS (2003) Nuclear localization and in vivo dynamics of a plant-specific serine/arginine-rich protein. Plant J 36:883–893
Ali GS, Palusa SG, Golovkin M, Prasad J, Manley JL et al (2007) Regulation of plant developmental processes by a novel splicing factor. PLoS ONE 2: e471
Ast G (2004) How did alternative splicing evolve? Nat Rev Genet 5:773–782
Barber MC, Price NT, Travers MT (2005) Structure and regulation of acetyl-CoA carboxylase genes of metazoa. Biochim Biophys Acta 1733:1-28
Blencowe BJ (2006) Alternative splicing: new insights from global analyses. Cell 126:37–47
Bournay AS, Hedley PE, Maddison A, Waugh R, Machray GC (1996) Exon skipping induced by cold stress in a potato invertase gene transcript. Nucleic Acids Res 24:2347–2351
Campbell MA, Haas BJ, Hamilton JP, Mount SM, Buell CR (2006) Comprehensive analysis of alternative splicing in rice and comparative analyses with Arabidopsis. BMC Genomics 7:327
Cao H, Li X, Dong X (1998) Generation of broad-spectrum disease resistance by overexpression of an essential regulatory gene in systemic acquired resistance. Proc Natl Acad Sci USA 95:6531–6536
Cartwright R, Dunn AM, Simpson PJ, Tambini CE, Thacker J (1998) Isolation of novel human and mouse genes of the recA/RAD51 recombination-repair gene family. Nucleic Acids Res 26:1653–1659
Castells E, Puigdomenech P, Casacuberta JM (2006) Regulation of the kinase activity of the MIK GCK-like MAP4K by alternative splicing. Plant Mol Biol 61:747–756
Chandler DS, Singh RK, Caldwell LC, Bitler JL, Lozano G (2006) Genotoxic stress induces coordinately regulated alternative splicing of the p53 modulators MDM2 and MDM4. Cancer Res 66:9502–9508
Chen FC, Wang SS, Chaw SM, Huang YT, Chuang TJ (2007) Plant Gene and Alternatively Spliced Variant Annotator. A plant genome annotation pipeline for rice gene and alternatively spliced variant identification with cross-species EST conservation from seven plant species. Plant Physiol 143:1086–1095
Chinnusamy V, Zhu J, Zhu JK (2006) Gene regulation during cold acclimation in plants. Physiol Plant 126:52–61
Chiodi I, Corioni M, Giordano M, Valgardsdottir R, Ghigna C et al (2004) RNA recognition motif 2 directs the recruitment of SF2/ASF to nuclear stress bodies. Nucleic Acids Res 32:4127–4136
Christensen AH, Sharrock RA, Quail PH (1992) Maize polyubiquitin genes: structure, thermal perturbation of expression and transcript splicing, and promoter activity following transfer to protoplasts by electroporation. Plant Mol Biol 18:675–689
de la Mata M, Alonso CR, Kadener S, Fededa JP, Blaustein M et al (2003) A slow RNA polymerase II affects alternative splicing in vivo. Mol Cell 12:525–532
Egawa C, Kobayashi F, Ishibashi M, Nakamura T, Nakamura C et al (2006) Differential regulation of transcript accumulation and alternative splicing of a DREB2 homolog under abiotic stress conditions in common wheat. Genes Genet Syst 81:77–91
Eperon IC, Ireland DC, Smith RA, Mayeda A, Krainer AR (1993) Pathways for selection of 5′ splice sites by U1 snRNPs and SF2/ASF. EMBO J 12:3607–3617

Falcone DL, Ogas JP, Somerville CR (2004) Regulation of membrane fatty acid composition by temperature in mutants of Arabidopsis with alterations in membrane lipid composition. BMC Plant Biol 4:17

Feilner T, Hultschig C, Lee J, Meyer S, Immink RG et al (2005) High throughput identification of potential Arabidopsis mitogen-activated protein kinases substrates. Mol Cell Proteomics 4:1558–1568

Forment J, Naranjo MA, Roldan M, Serrano R, Vicente O (2002) Expression of Arabidopsis SR-like splicing proteins confers salt tolerance to yeast and transgenic plants. Plant J 30:511–519

Fung RW, Wang CY, Smith DL, Gross KC, Tao Y et al (2006) Characterization of alternative oxidase (AOX) gene expression in response to methyl salicylate and methyl jasmonate pre-treatment and low temperature in tomatoes. J Plant Physiol 163:1049–1060

Golovkin M, Reddy ASN (1999) An SC35-like protein and a novel serine/arginine-rich protein interact with Arabidopsis U1-70K protein. J Biol Chem 274:36428–36438

Graveley BR (2000) Sorting out the complexity of SR protein functions. RNA 6:1197–1211

Graveley BR (2001) Alternative splicing: increasing diversity in the proteomic world. Trends Genet 17:100–107

Haas BJ, Delcher AL, Mount SM, Wortman JR, Smith RK Jr. et al (2003) Improving the Arabidopsis genome annotation using maximal transcript alignment assemblies. Nucleic Acids Res 31:5654–5666

Hirouchi T, Nakajima S, Najrana T, Tanaka M, Matsunaga T et al (2003) A gene for a Class II DNA photolyase from *Oryza sativa*: cloning of the cDNA by dilution-amplification. Mol Genet Genomics 269:508–516

Hopf N, Plesofsky-Vig N, Brambl R (1992) The heat shock response of pollen and other tissues of maize. Plant Mol Biol 19:623–630

Huang Y, Steitz JA (2005) SRprises along a messenger's journey. Mol Cell 17:613–615

Iida K, Go M (2006) Survey of conserved alternative splicing events of mRNAs encoding SR proteins in land plants. Mol Biol Evol 23:1085–1094

Iida K, Seki M, Sakurai T, Satou M, Akiyama K et al (2004) Genome-wide analysis of alternative pre-mRNA splicing in *Arabidopsis thaliana* based on full-length cDNA sequences. Nucleic Acids Res 32:5096–5103

Iida K, Seki M, Sakurai T, Satou M, Akiyama K et al (2005) RARTF: Database and tools for complete sets of arabidopsis transcription factors. DNA Res 12:247–256

Jamison SF, Pasman Z, Wang J, Will C, Luhrmann R et al (1995) U1 snRNP-ASF/SF2 interaction and 5′ splice site recognition: characterization of required elements. Nucleic Acids Res 23:3260–3267

Jia Y, del Rio HS, Robbins AL, Louzada ES (2004) Cloning and sequence analysis of a low temperature-induced gene from trifoliate orange with unusual pre-mRNA processing. Plant Cell Rep 23:159–166

Johnson JM, Castle J, Garrett-Engele P, Kan Z, Loerch PM et al (2003) Genome-wide survey of human alternative pre-mRNA splicing with exon junction microarrays. Science 302:2141–2144

Jurica MS, Moore MJ (2003) Pre-mRNA splicing: awash in a sea of proteins. Mol Cell 12:5–14

Kalyna M, Lopato S, Barta A (2003) Ectopic expression of atRSZ33 reveals its function in splicing and causes pleiotropic changes in development. Mol Biol Cell 14:3565–3577

Kohtz JD, Jamison SF, Will CL, Zuo P, Luhrmann R et al (1994) Protein-protein interactions and 5′-splice-site recognition in mammalian mRNA precursors. Nature 368:119–124

Kong J, Gong JM, Zhang ZG, Zhang JS, Chen SY (2003) A new AOX homologous gene OsIM1 from rice (*Oryza sativa* L.) with an alternative splicing mechanism under salt stress. Theor Appl Genet 107:326–331

Larkin PD, Park WD (1999) Transcript accumulation and utilization of alternate and non-consensus splice sites in rice granule-bound starch synthase are temperature-sensitive and controlled by a single-nucleotide polymorphism. Plant Mol Biol 40:719–727

Lazar G, Goodman HM (2000) The Arabidopsis splicing factor SR1 is regulated by alternative splicing. Plant Mol Biol 42:571–581

Lejeune F, Maquat LE (2005) Mechanistic links between nonsense-mediated mRNA decay and pre-mRNA splicing in mammalian cells. Curr Opin Cell Biol 17:309–315

Li L, Howe GA (2001) Alternative splicing of prosystemin pre-mRNA produces two isoforms that are active as signals in the wound response pathway. Plant Mol Biol 46:409–419

Li X, Manley JL (2005) Inactivation of the SR protein splicing factor ASF/SF2 results in genomic instability. Cell 122:365–378

Lopato S, Kalyna M, Dorner S, Kobayashi R, Krainer AR et al (1999) atSRp30, one of two SF2/ASF-like proteins from *Arabidopsis thaliana*, regulates splicing of specific plant genes. Genes Dev 13:987–1001

Maniatis T, Tasic B (2002) Alternative pre-mRNA splicing and proteome expansion in metazoans. Nature 418:236–243

Manley JL, Tacke R (1996) SR proteins and splicing control. Genes Dev 10:1569–1579

Mastrangelo AM, Belloni S, Barilli S, Ruperti B, Di Fonzo N et al (2005) Low temperature promotes intron retention in two e-cor genes of durum wheat. Planta 221:705–715

Mimida N, Kitamoto H, Osakabe K, Nakashima M, Ito Y et al (2007) Two alternatively spliced transcripts generated from OsMUS81, a rice homologue of yeast MUS81, are upregulated by DNA-damaging treatments. Plant Cell Physiol 48:648–654

Modrek B, Lee C (2002) A genomic view of alternative splicing. Nat Genet 30:13–19

Ner-Gaon H, Leviatan N, Rubin E, Fluhr R (2007) Comparative cross-species alternative splicing in plants. Plant Physiol 144:1632–1641

Ner-Gaon H, Halachmi R, Savaldi-Goldstein S, Rubin E, Ophir R et al (2004) Intron retention is a major phenomenon in alternative splicing in Arabidopsis. Plant J 39:877–885

Nilsen TW (2003) The spliceosome: the most complex macromolecular machine in the cell? Bioessays 25:1147–1149

Osakabe K, Yoshioka T, Ichikawa H, Toki S (2002) Molecular cloning and characterization of RAD51-like genes from *Arabidopsis thaliana*. Plant Mol Biol 50:71–81

Palusa SG, Ali GS, Reddy ASN (2007) Alternative splicing of pre-mRNA of Arabidopsis serine/arginine-rich proteins:regulation by hormones and stresses. Plant J 49:1091–1107

Pertea M, Mount SM, Salzberg SL (2007) A computational survey of candidate exonic splicing enhancer motifs in the model plant *Arabidopsis thaliana*. BMC Bioinformatics 8:159

Podkowinski J, Jelenska J, Sirikhachornkit A, Zuther E, Haselkorn R et al (2003) Expression of cytosolic and plastid acetyl-coenzyme A carboxylase genes in young wheat plants. Plant Physiol 131:763–772

Puhakainen T, Hess MW, Makela P, Svensson J, Heino P et al (2004) Overexpression of multiple dehydrin genes enhances tolerance to freezing stress in Arabidopsis. Plant Mol Biol 54:743–753

Qin F, Kakimoto M, Sakuma Y, Maruyama K, Osakabe Y et al (2007) Regulation and functional analysis of ZmDREB2A in response to drought and heat stresses in *Zea mays* L. Plant J 50:54–69

Razem FA, El-Kereamy A, Abrams SR, Hill RD (2006) The RNA-binding protein FCA is an abscisic acid receptor. Nature 439:290–294

Reddy ASN (2001) Nuclear pre-mRNA processing in plants. CRC Crit Rev Plant Sci 20:523–572

Reddy ASN (2004) Plant serine/arginine-rich proteins and their role in pre-mRNA splicing. Trends Plant Sci 9:541–547

Reddy ASN (2007) Alternative splicing of pre-messenger RNAs in plants in the Genomic Era. Annu Rev Plant Biol 58:267–294

Sanford JR, Longman D, Caceres JF (2003) Multiple roles of the SR protein family in splicing regulation. In: Jeanteur P, editor. Regulation of alternative splicing. New York: Springer. pp. 33–58

Savaldi-Goldstein S, Sessa G, Fluhr R (2000) The ethylene-inducible PK12 kinase mediates the phosphorylation of SR splicing factors. Plant J 21:91–96

Shen H, Kan JL, Green MR (2004) Arginine-serine-rich domains bound at splicing enhancers contact the branchpoint to promote prespliceosome assembly. Mol Cell 13:367–376

Sheth N, Roca X, Hastings ML, Roeder T, Krainer AR et al (2006) Comprehensive splice-site analysis using comparative genomics. Nucleic Acids Res 34:3955–3967

Shi H, Xiong L, Stevenson B, Lu T, Zhu JK (2002) The Arabidopsis salt overly sensitive 4 mutants uncover a critical role for vitamin B6 in plant salt tolerance. Plant Cell 14:575–588

Shomron N, Alberstein M, Reznik M, Ast G (2005) Stress alters the subcellular distribution of hSlu7 and thus modulates alternative splicing. J Cell Sci 118:1151–1159

Tai HH, Williams M, Iyengar A, Yeates J, Beardmore T (2007) Regulation of the beta-hydroxyacyl ACP dehydratase gene of *Picea mariana* by alternative splicing. Plant Cell Rep 26:105–113

Tanabe N, Yoshimura K, Kimura A, Yabuta Y, Shigeoka S (2007) Differential expression of alternatively spliced mRNAs of Arabidopsis SR protein homologs, atSR30 and atSR45a, in response to environmental stress. Plant Cell Physiol 48:1036–1049

Tillemans V, Dispa L, Remacle C, Collinge M, Motte P (2005) Functional distribution and dynamics of Arabidopsis SR splicing factors in living plant cells. Plant J 41:567–582

Vonarx EJ, Howlett NG, Schiestl RH, Kunz BA (2002) Detection of *Arabidopsis thaliana* AtRAD1 cDNA variants and assessment of function by expression in a yeast rad1 mutant. Gene 296:1-9

Wang BB, Brendel V (2004) The ASRG database: identification and survey of *Arabidopsis thaliana* genes involved in pre-mRNA splicing. Genome Biol 5: R102

Wang BB, Brendel V (2006) Genomewide comparative analysis of alternative splicing in plants. Proc Natl Acad Sci USA 103:7175–7180

Wang D, Amornsiripanitch N, Dong X (2006) A genomic approach to identify regulatory nodes in the transcriptional network of systemic acquired resistance in plants. PLoS Pathog 2: e123

Wu JY, Maniatis T (1993) Specific interactions between proteins implicated in splice site selection and regulated alternative splicing. Cell 75:1061–1070

Xiao H, Nassuth A (2006) Stress- and development-induced expression of spliced and unspliced transcripts from two highly similar dehydrin 1 genes in *V. riparia* and *V. vinifera*. Plant Cell Rep 25:968–977

Xiao SH, Manley JL (1997) Phosphorylation of the ASF/SF2 RS domain affects both protein-protein and protein-RNA interactions and is necessary for splicing. Genes Dev 11:334–344

Xiong L, Yang Y (2003) Disease resistance and abiotic stress tolerance in rice are inversely modulated by an abscisic acid-inducible mitogen-activated protein kinase. Plant Cell 15:745–759

Xue GP, Loveridge CW (2004) HvDRF1 is involved in abscisic acid-mediated gene regulation in barley and produces two forms of AP2 transcriptional activators, interacting preferably with a CT-rich element. Plant J 37:326–339

Yamaguchi-Shinozaki K, Shinozaki K (2006) Transcriptional regulatory networks in cellular responses and tolerance to dehydration and cold stresses. Annu Rev Plant Biol 57:781–803

Yoshimura K, Yabuta Y, Ishikawa T, Shigeoka S (2002) Identification of a *cis* element for tissue-specific alternative splicing of chloroplast ascorbate peroxidase pre-mRNA in higher plants. J Biol Chem 277:40623–40632

Zahler AM, Roth MB (1995) Distinct functions of SR proteins in recruitment of U1 small nuclear ribonucleoprotein to alternative 5′ splice sites. Proc Natl Acad Sci USA 92:2642–2646

Zhu JK (2002) Salt and drought stress signal transduction in plants. Annu Rev Plant Biol 53:247–273

Zhu W, Schlueter SD, Brendel V (2003) Refined annotation of the Arabidopsis genome by complete expressed sequence tag mapping. Plant Physiol 132:469–484

Intron-Mediated Regulation of Gene Expression

A.B. Rose

Contents

What Good Are Introns?	278
Indirect Gene Regulation by Introns	278
Direct Gene Regulation by Introns	279
Enhancers and Promoters Within Introns	280
Enhancers	280
Alternative Promoters	281
Intron-Mediated Enhancement	282
Monocots and Dicots	282
Sequences Involved in IME	283
The Role of Splicing in IME	284
Intron Position	285
The Biochemistry of IME	286
A Model for IME	286
Conclusions	287
References	288

Abstract Introns can significantly affect gene expression in plants and many other eukaryotes in a variety of ways. Several types of gene regulation, both positive and negative, that involve plant introns are reviewed in this chapter. Some introns contain enhancer elements or alternative promoters, while many others elevate mRNA accumulation by a different process that has been named intron-mediated enhancement (IME). The introns involved in IME must be within transcribed sequences near the start of a gene and in their natural orientation to increase expression. The intron sequences involved are still poorly defined, and the mechanism of IME remains mysterious. A model of IME is presented in which introns increase transcript elongation.

A.B. Rose
Molecular and Cellular Biology, University of California, Davis, CA 95616, USA
email: abrose@ucdavis.edu

Abbreviations IME: Intron-mediated enhancement; NMD: Nonsense-mediated-mRNA decay; EJC: Exon junction complex; GUS: β-Glucuronidase; UTR: Untranslated region; PolII: RNA polymerase II; CTD: Carboxy-terminal domain; mRNA: Messenger RNA

What Good Are Introns?

Introns have been somewhat puzzling ever since their discovery in 1977. What possible benefit could be derived from interrupting most eukaryotic genes with extraneous nucleotides that must be precisely and completely removed from a primary transcript before it can serve as a functional mRNA? Answers to this question often focus on long-term advantages that introns provide. For example, introns may facilitate the evolution of new proteins through exon shuffling, and alternative splicing increases the coding capacity of a genome. While certainly advantageous, these returns seem somewhat modest compared to the high costs of maintaining introns and the complicated machinery needed to remove them. It appears wasteful on a scale not normally tolerated by evolution to expend significant energy to incorporate nucleotides into introns, only to remove and discard them almost immediately. In addition, introns present risks because inappropriate intron retention or even a single nucleotide inaccuracy in splicing is likely to render that mRNA useless, and a mutation in any essential component of the spliceosome could fatally cripple the expression of most genes.

The persistence of introns in all known eukaryotes despite these costs implies that introns provide some immediate and important benefits. One often overlooked but vital role for introns is that they can have profound effects on gene expression. Removing the introns from a gene by replacing genomic sequences with the corresponding cDNA (leaving the promoter and terminator intact) often causes expression to plummet. Conversely, inserting an intron into a normally intronless gene, including those of bacterial origin, can significantly increase the expression of that gene. Not all introns boost expression, however. Some have no effect, and others have negative regulatory functions. The purpose of this chapter is to briefly review several types of gene regulation by introns in plants. Readers are referred to other articles for reviews of splicing (Reddy 2001) and gene expression in plants (Belostotsky and Rose 2005) and the effect of introns on gene expression in other eukaryotes (Le Hir et al. 2003).

Indirect Gene Regulation by Introns

An indirect way in which introns affect gene expression in some organisms is through their role in nonsense-mediated mRNA decay (NMD), the process in which premature stop codons trigger the rapid degradation of the mRNA containing them.

Various signals can cause a stop codon to be recognized as premature, including the presence of an intron more than 50 nt downstream of the termination codon in mammals (Maquat 2004b). Introns seem to be at least one NMD signal in plants (Kertesz et al. 2006), although the role of introns in NMD may be different in plants than in mammals (Maquat 2004a). Most of the plant genes in which NMD has been documented are intronless, and NMD of the rice *waxy* gene depends on splicing of an intron that is upstream (not downstream) of a stop codon (Isshiki et al. 2001).

For NMD to be activated by the relative locations of introns and stop codons, there must be a connection between nuclear splicing and cytoplasmic translation. This communication is provided by the exon junction complex (EJC), a collection of proteins that are deposited on the mRNA during splicing, marking the former sites of introns (Behm-Ansmant and Izaurralde 2006). The EJC can also facilitate export of the mRNA to the cytoplasm (Dimaano and Ullman 2004) and promote an association of the mRNA with ribosomes (Nott et al. 2004). Either mechanism can increase translational efficiency-the amount of protein made per unit of mRNA-from an intron-containing gene compared to an intronless one. Homologs of most of the known components of the EJC can be found in plants (Belostotsky and Rose 2005), and it is possible that the EJC is responsible for the observed additional boost to expression that some introns provide beyond that seen at the level of mRNA accumulation. In Arabidopsis, four of six introns tested approximately doubled translational efficiency, and this effect is independent of intron location in the gene (Rose 2004). Similarly, introns from the Arabidopsis *COX5c-1* and *COX5c-2* genes (Curi et al. 2005), the potato *ubi7* gene (Garbarino et al. 1995), and the maize *Adh1* intron (Mascarenhas et al. 1990) cause a greater increase in enzyme activity than can be explained by mRNA levels alone.

Direct Gene Regulation by Introns

Introns have been observed to stimulate gene expression in a wide range of diverse organisms, including mammals, nematodes, insects, fungi, and plants. Plants in which this phenomenon has been demonstrated include algae (*Chlamydomonas*, *Volvox*), moss (*Physcomitrella*), gymnosperms (pine), and many monocots (maize, rice, wheat, oat, barley, banana) and dicots (Arabidopsis, tobacco, tomato, potato, petunia, bean, cotton). This broad distribution suggests that introns play a fundamental and evolutionarily conserved role in regulating eukaryotic gene expression.

There are several examples of introns whose effect on expression is larger than that of the promoter from the same gene. Many genes with an intact promoter are essentially not expressed at all without an intron. A striking demonstration of the relative importance of introns and promoters can be found in a study in which introns were swapped between genes. The three Arabidopsis genes that encode vegetative profilins (*PRF1*, *2*, and *3*) are expressed throughout the plant, while expression of the two reproductive profilins *PRF4* and *5* is limited to the stigma and anthers. Not only is the first *PRF2* intron required for full expression of a fusion between the *PRF2* promoter and the β-glucuronidase (*GUS*) gene, it also converts the expression of a *PRF5:GUS* fusion from a reproductive to a vegetative pattern (Jeong et al. 2006). This indicates that the *PRF2* intron has more influence than the *PRF5* promoter in determining the tissue-specific pattern of expression and has a greater impact than either promoter in controlling the amount of expression.

This example illustrates the experimental approach typically used to study the role of introns in regulating genes. The expression of at least two constructs, one with an intron and the other without, is monitored after their introduction into plant cells either transiently or as a transgene. Usually the gene is a fusion between a plant promoter and a bacterial reporter such as *GUS*, but some are entirely native genes from which the introns have been removed. The level at which introns affect gene expression is not always investigated, although when examined introns are usually found to cause an increase in mRNA accumulation.

Introns can elevate mRNA levels by at least two general methods. One is by increasing the synthesis of mRNA through the action of a transcriptional enhancer element or an alternative promoter that is located within the intron. However, many introns must increase expression by a less familiar mechanism because the characteristics of the enhancement are inconsistent with either enhancers or promoters. This second category has been termed intron-mediated enhancement (IME). Diagnostic tests to differentiate between enhancers/promoters and IME include whether or not an intron can enhance expression from outside of transcribed sequences or in either orientation, whether the intron stimulates expression of a gene that has a minimal or no promoter, and whether or not the intron contains any specific sequence elements (other than those needed for splicing) that are required for the enhancing activity or to which specific factors bind. Examples in which at least some of these tests have been performed are detailed below. There are many more published cases in which genes are more highly expressed with an intron than without but the underlying mechanism remains uncharacterized.

Enhancers and Promoters Within Introns

Enhancers

Several plant introns can increase the expression of a gene from upstream of transcribed sequences and in either orientation and therefore probably contain enhancer elements. For example, the first intron from an Arabidopsis gene encoding elongation factor eEF-1β increases expression in transgenic tobacco leaves when the intron is upstream of the start of transcription and in either orientation, although it is less effective when backwards (Gidekel et al. 1996). Similarly, the 5′-UTR of the Arabidopsis *ACT1* gene contains an intron that stimulates anther-specific expression of an *ACT1:GUS* fusion approximately 10-fold when in either orientation upstream of the promoter (Vitale et al. 2003). Two algal introns that elevate expression independent of orientation are the *RBCS2* first intron from *Chlamydomonas reinhardtii* and the fifth intron from the *regA* gene in *Volvox carteri* (Lumbreras et al. 1998; Stark et al. 2001).

A particularly well-characterized example of an enhancer-containing intron is the large (2,999 bp) second intron of the Arabidopsis *agamous* (*AG*) gene. This intron can function in both orientations to drive expression of a reporter from a minimal promoter (Deyholos and Sieburth 2000). The intron can be further divided into a 5' and a 3' enhancer, both of which stimulate expression, although the 3' enhancer has been more thoroughly characterized (Busch et al. 1999). Binding sites within the intron for proteins known to regulate *AG* expression have been defined by small mutations, gel shift assays, and chromatin immunoprecipitation (Sridhar et al. 2006), and consensus binding sequences for other proteins have been identified in this intron but await experimental validation.

The *AG* second intron is also important for restricting *AG* expression to specific tissues within the flower. *AG* is a class C floral organ identity gene, meaning that its activity is limited to the inner two whorls of the developing flower, where it directs the formation of carpels and stamens. Thus the tissue-specific expression of *AG* must be tightly regulated, and sequences in the second intron are needed to prevent inappropriate expression in the outer flower whorls and vegetative organs (Sieburth and Meyerowitz 1997). The combination of positive and negative effectors makes this intron a very complex regulatory element.

Several other introns are like *AG* intron 2 in that they are required to restrict gene expression in some way, and may be considered negative enhancer elements. Removal of the 1328 nt first intron of the *Seedstick* (*STK*) gene causes *STK:GUS* expression to expand from just the ovules and septum, where the native *STK* gene is expressed, to include all parts of the flower (Kooiker et al. 2005). The 3493 nt first intron from the *Flowering Locus C* (*FLC*) gene of Arabidopsis is required for maintaining the reduction in *FLC* expression brought about by vernalization, as well as the repression of *FLC* expression by the *FCA* gene product (Sheldon et al. 2002). Similarly, deletions within the first intron of the wheat *VRN-1* gene are associated with reduced repression of this gene in response to vernalization (Fu et al. 2005). The *AG, STK, FLC*, and *VRN-1* introns are all large, which may provide sufficient room for numerous controlling elements or permit the establishment of a local chromatin conformation required for appropriate expression.

Alternative Promoters

A few introns are capable of driving at least weak expression of a promoterless gene and therefore contain promoters that could be partly responsible for increasing expression. These include the first intron from the *Ostub16* and *OsCDPK2* genes in rice (Morello et al., 2002, 2006), the introns from *PpAct1* and *PpAct5* genes in the moss *Physcomitrella patens* (Weise et al. 2006), and the first intron from the sesame *FAD2* gene assayed in Arabidopsis (Kim et al. 2006). The conclusion that the rice introns mentioned contain alternative promoters is supported by evidence that some transcripts originate within intron sequences. These are spliced by using the normal 3' splice site of the intron to create transcripts that differ at their 5' end from the mRNA originating upstream of the intron. The functional significance of the intronic promoter is unclear, as both forms of mRNA encode identical proteins and transcripts starting in the intron are far less abundant.

Intron-Mediated Enhancement

Not long after introns were first shown to elevate expression in mammalian cells, Callis et al. (1987) published a seminal study describing the effect of introns on gene expression in maize. This work is remarkable for its thoroughness and the number of features of IME for which this paper is the first published record in plants. These include the observations that several different coding sequences (including bacterial genes) respond to introns, that IME is not limited to a specific promoter, that several different introns have the ability to stimulate expression of the same gene, that first introns tend to enhance more than later introns, that introns only increase expression when within transcribed sequences and in the proper orientation, that introns fail to enhance from within transcribed sequences at the 3' end of a gene, and that the effect of introns is visible as an increase in mRNA accumulation. These characteristics have become the defining features of IME and made it clear that the introns studied were operating by a mechanism unrelated to conventional enhancer elements.

Monocots and Dicots

The earliest experiments on IME in plants were done in maize, in which several introns were shown to stimulate expression up to 100-fold (Callis et al. 1987; Vasil et al. 1989). In contrast, some of the first dicot genes to be tested are expressed at similar or possibly higher levels without their introns (Chee et al. 1986; Kuhlemeier et al. 1988). This led to the widespread belief that IME is more robust in monocots than dicots. Since then, claims of dicot introns that increase expression 1,000-fold or more have been published (Curie et al. 1993; Zhang et al. 1994), as have many reports of introns whose stimulation is less than 10-fold in monocots. What appeared to be a trend may simply reflect differences in the enhancing ability of the introns that happened to be chosen first for study. As more introns are analyzed, the difference in the magnitude of IME in monocots and dicots is becoming less obvious.

The approach of testing the same intron in both monocot and dicot species has given mixed results (Last et al. 1991; Ueki et al. 2004; Xu et al. 1994). Monocot and dicot introns have some structural differences (dicot introns tend to be more AU rich, for example), and the introns from one group are often incompletely spliced in the other (Goodall and Filipowicz 1991; Keith and Chua 1986). The retention of any intron in a transcript would interfere with expression and possibly destabilize the mRNA through NMD. Thus a direct comparison of IME in monocots and dicots by the same intron may not reveal true differences between the groups unless efficient splicing in both is demonstrated. Also, the species specificity of the IME machinery remains unknown, although some introns can clearly enhance expression in many widely divergent plants.

A potentially significant difference between the published descriptions of IME in monocots and dicots is that virtually all monocot studies have been done with transient expression assays, while many dicot studies have used stably transformed plants, which are easier to generate than in monocots. IME has been observed in transient assays in dicots, although in studies in which both methods could be compared the effect of the intron was more pronounced in the stable transformants (Jeong et al. 2006; Plesse et al. 2001). It will be interesting to see whether the magnitude of IME is greater in stably transformed monocots than in transient assays, as appears to be the case in rice (Tanaka et al. 1990).

Sequences Involved in IME

Further evidence that many introns stimulate mRNA levels without containing discrete enhancers or promoters came from attempts to define the intron sequences involved in IME through deletion analysis. Up to 75% of the sequences within the first intron from the maize *Adh1* gene can be deleted without affecting IME as long as the intron still can be spliced (Luehrsen and Walbot 1994). Similarly, deleting 883 nt from the middle of the 1028 nt first intron from the maize *Sh1* gene has little effect on the ability of this intron to stimulate expression (Clancy and Hannah 2002; Clancy et al. 1994). Deleting 655 of the 946 nt that comprise the rice *OstubA1* first intron causes the enhancement to decline, although the remaining 291-nt derivative still enhances roughly twentyfold (Jeon et al. 2000). In Arabidopsis, all sequences within the *TRP1* first intron can be individually deleted without significantly reducing IME (Rose and Beliakoff 2000), and more than 80% of *PRF2* intron 1 can be removed without eliminating its enhancing ability (Jeong et al. 2006). For these introns, most or all of the sequences responsible for IME must be redundant and distributed throughout the intron, or are limited to the ends of the intron.

One consequence of the inability of deletions to abolish IME is that no specific sequence has been shown to be absolutely required, and even the general nature of the sequences involved remains unclear. No motifs conserved between enhancing introns have been found. Deletions within a 145 nt truncated version of the maize *Sh1* first intron identified a 35 nt region whose removal reduces enhancement by half (from 24-fold to approximately 10-fold). Full enhancement can be restored by substituting this region with a similarly U-rich region from another part of the intron (Clancy and Hannah 2002). In addition, reducing the prevalence of short strings of U residues within *TRP1* intron 1 diminishes the ability of this intron to elevate mRNA accumulation in Arabidopsis (Rose 2002). In contrast to these U-rich candidates, a GC-rich octamer from the maize *GapA1* first intron was tentatively identified as an enhancing sequence (Donath et al. 1995). Our understanding of the mechanism of IME would be greatly advanced by a clear definition of the intron sequences responsible for stimulating expression.

The Role of Splicing in IME

The failure to identify sequences involved in IME could be explained if the increase in mRNA accumulation was a consequence of splicing. Early tests of the need for splicing in IME found that mutations that prevented splicing also eliminated the enhancing effect of an intron (Sinibaldi and Mettler 1992). However, the retained introns could interfere with expression independent of IME by inappropriately initiating (if in the 5'-UTR) or terminating (if in coding sequences) the reading frame. With a derivative of Arabidopsis *TRP1* intron 1 modified to preserve the reading frame of the adjacent exons, splicing was prevented by making the intron too short to be spliced, by mutations at the 5' splice site, or by eliminating all potential branchpoint sequences (Rose 2002; Rose and Beliakoff 2000). Each unspliceable "intron" is still able to enhance expression from within coding sequences relative to the intronless control, although the enhancement is reduced by about 50%. IME is destroyed completely only by the simultaneous elimination of branchpoints and the 5' splice site (Rose 2002), the structures involved in the first two steps of spliceosome assembly onto an intron. This suggests that the mechanism of IME requires that the splicing machinery be at least partly assembled onto an intron, even if it is unable to complete its task. In maize, derivatives of *Sh1* intron 1 with mutations in the 5' and 3' splice sites or that are too short to be spliced have less than 10% of the enhancing ability of the starting intron from the 5'-UTR (Clancy and Hannah 2002), indicating potential differences in the mechanism of IME between either these introns or these plants.

Introns might boost expression simply by providing an association between a transcript and the splicing machinery that increases the efficiency of other steps of pre-mRNA maturation, leading to greater mRNA production. The many reactions of gene expression are known to be interconnected and affect each other (Maniatis and Reed 2002). If interactions between pre-mRNA and the spliceosome are the sole basis of IME, then all efficiently spliced introns should enhance equally well.

Unfortunately, it has proven difficult to place an absolute or even a relative value on the stimulating ability of different introns. The magnitude of enhancement usually cannot be compared between publications because IME is influenced by many factors other than the intron, including the promoter and coding sequences of the gene used to monitor expression (Callis et al. 1987; Mascarenhas et al. 1990; Rethmeier et al. 1997; Sinibaldi and Mettler 1992), by the species and cell type in which expression is observed (Gallie and Young 1994; Xu et al. 1994), and because of the variability inherent in transient expression assays or differences between transgenic lines of uncertain copy number. Even within a single study, the intron may not be the only difference between constructs because the engineering used to insert an intron often creates additional changes in the encoded protein or mature mRNA. While usually minor, these sequence differences could affect mRNA stability, translation, or the activity of the reporter enzyme. Some genes essentially require introns in order to be expressed at all, precluding an accurate determination of the fold change in expression caused by an intron. In other cases, introns alter the tissue-specific expression pattern of a gene (Casas-Mollano et al. 2006; Chaubet-Gigot et al. 2001; Plesse et al. 2001), so that enhancement differs between tissues.

Despite these caveats, studies in which variables have been minimized clearly show that introns vary widely in their ability to stimulate expression. Six different introns were tested at the same location in a *TRP1:GUS* reporter gene in single-copy transgenic Arabidopsis (Rose 2002). Each of the six is efficiently spliced, and the resulting mature mRNA is identical in all cases. The degree to which mRNA accumulation is increased ranges from less than twofold for the first intron from the *TCH3* gene to more than 12-fold for introns from *UBQ10* or *atpk1*. In maize, the effects of seven introns on transient expression range from negligible to more than 20-fold, depending on the intron and reporter gene (Sinibaldi and Mettler 1992). The observation that some efficiently spliced introns fail to enhance means that simply providing an association between a pre-mRNA and the splicing machinery is not sufficient to cause IME, and suggests that the degree to which an intron stimulates expression is determined by elements other than the highly conserved sequences involved in the splicing reactions.

Intron Position

Investigations into the role of intron location have furnished a potentially significant clue about the mechanism of IME. The position of an intron can influence its ability to enhance expression in two ways. The first is the location of the intron within the gene from which it was isolated. Most of the introns shown to enhance expression in Arabidopsis and maize are first introns, many of which are found in the 5'-UTR of their native gene. First introns, particularly those in the 5'-UTR, tend to be longer than other introns (Chung et al. 2006; Hong et al. 2006), which may explain why enhancing introns also tend to be relatively long. Some later and shorter introns, such as *TRP1* intron 6 in Arabidopsis and *Adh1* introns 8 and 9 in maize (Callis et al. 1987; Rose 2002), can stimulate mRNA accumulation, although not as much as earlier introns. The other way in which position is important is that introns must be located near the start of a gene for maximal enhancement. Several introns that enhance expression from the 5'-UTR have no effect when inserted in the 3'-UTR (Clancy et al. 1994; Jeon et al. 2000; Snowden et al. 1996). When either *TRP1* intron 1 or *UBQ10* intron 1 is moved systematically down a *TRP1:GUS* fusion, its ability to increase mRNA accumulation declines with distance until it disappears entirely approximately one kilobase from the start of transcription (Rose

2004). Both of these position requirements may reflect the same mechanistic requirement for enhancing introns to be near the 5' end of a gene.

The Biochemistry of IME

Biochemical investigations have not solved the mystery of the mechanism of IME. While introns elevate mRNA accumulation, they apparently manage to do this without increasing either the rate of transcription or mRNA stability. Introns that significantly increase mRNA accumulation have at most a minor effect on the signal generated in nuclear run-on transcription assays (Dean et al. 1989; Rose and Last 1997). Similarly, the stability of mRNAs derived from intron-containing and intronless genes are indistinguishable (Nash and Walbot 1992; Rethmeier et al. 1997), although subtle changes may be difficult to detect if the mRNA already has a long half-life. Comparable results have been obtained in mammalian cells (Lu and Cullen 2003; Nott et al. 2003). How can these apparently incompatible results be reconciled? One possibility is that the run-on assays yield misleading results because delicate or transient intron-mediated changes in the transcription machinery could be lost during the isolation of the nuclei used in the assays. Another is that while run-on transcription assays measure mostly initiation, introns might primarily affect transcript elongation. While both could be true, the latter possibility is most consistent with the need for introns to be near the start of a gene for maximum effect, and forms the basis for the following model.

A Model for IME

All of the observations described above can be incorporated into a general model for IME. Enhancing introns may increase the processivity of the transcription machinery to make it more likely to reach the 3' end of a gene, where cleavage and polyadenylation create a stable mRNA. In the absence of introns, transcription would initiate at the same rate but the polymerase may be more likely to stall or fall off the template, leaving an abortive transcript that is rapidly degraded. Full-length transcripts from intronless genes could still be made and would be just as stable but would be less abundant. An increase in elongation would have the greatest effect on mRNA level if it happened near the start of a gene, and introns near the 5' end of a gene may have evolved to provide this function.

How might introns be able to affect the transcription machinery? Theoretically, they could operate at either the DNA or the RNA level. Intron sequences in the DNA might adopt a particular configuration or nucleosome association that eases passage of the transcription machinery or makes it more processive in some way. This may explain why the intron sequences involved in IME have been difficult to identify, as well as the need for introns to be in transcribed sequences to enhance. If the higher-order structure of the DNA makes introns easier to transcribe, they might be expected to function in either orientation because of the double-stranded nature of DNA, but this is not the case. Introns must be in the proper orientation for IME. However, the conserved splice site nucleotides and U-richness of most introns are not preserved in the other orientation, making it very unlikely that the backwards intron would be recognized as an intron and properly spliced. The retained "nortni" almost certainly would disrupt expression by terminating or altering the reading frame, negating any enhancement provided at the DNA level.

The need for splicing, or at least an association of the intron with the splicing machinery, is perhaps the strongest evidence that IME acts at the RNA level. In this case, some redundant and dispersed sequences in a newly transcribed intron might bind to factors that in turn communicate with the transcription machinery to render it more processive. An appealing target for such interactions is the carboxy-terminal domain (CTD) of the large subunit of RNA polymerase II (PolII). Transcription is regulated by changes in the phosphorylation of certain residues in the repeats that constitute the CTD, and splicing factors are known to interact with the CTD in a phosphorylation-dependent manner (Steinmetz 1997). This established link between introns and PolII provides a plausible opportunity for introns to affect the phosphorylation of the CTD and thus the activity of the transcription machinery. The crucial link-factors that bind to introns and modify PolII-currently remains missing. Identifying dispersed sequences that are involved in IME may provide a means to isolate such factors if they exist.

Regardless of whether introns act at the DNA or RNA level, an effect on transcript elongation would make biological sense if introns are one of the ways in which the transcription machinery can differentiate between genes and intergenic spaces. Transcripts that initiate spontaneously at sequences that are not genuine promoters or fail to terminate at the end of real genes are potentially dangerous because they can form antisense transcripts that could interfere with the expression of other genes. This risk would be reduced if the transcription machinery had the highest processivity only in actual genes, which could be recognized by the presence of introns.

Conclusions

It is not surprising that many different types of gene regulation involve introns. Introns are largely free of the sequence constraints inherent in coding sequences and therefore present many opportunities for the rapid evolution of new elements that control gene expression. Enough different kinds of regulation have been characterized to illustrate the danger in assuming that all introns affect expression by the same mechanism. The mysterious nature of IME suggests that introns still have a lot to teach us about how eukaryotic gene expression is regulated.

References

Behm-Ansmant I, Izaurralde E (2006) Quality control of gene expression: a stepwise assembly pathway for the surveillance complex that triggers nonsense-mediated mRNA decay. Genes Dev 20:391–398

Belostotsky DA, Rose AB (2005) Plant gene expression in the age of systems biology: integrating transcriptional and post-transcriptional events. Trends Plant Sci 10:347–353

Busch MA, Bomblies K, Weigel D (1999) Activation of a floral homeotic gene in *Arabidopsis*. Science 285:585–587

Callis J, Fromm M, Walbot V (1987) Introns increase gene expression in cultured maize cells. Genes Dev 1:1183–1200

Casas-Mollano JA, Lao NT, Kavanagh TA (2006) Intron-regulated expression of *SUVH3*, an *Arabidopsis Su(var)3-9* homologue. J Exp Bot 57:3301–3311

Chaubet-Gigot N, Kapros T, Flenet M, Kahn K, Gigot C, Waterborg JH (2001) Tissue-dependent enhancement of transgene expression by introns of replacement histone H3 genes of *Arabidopsis*. Plant Mol Biol 45:17–30

Chee PP, Klassy RC, Slightom JL (1986) Expression of a bean storage protein 'phaseolin minigene' in foreign plant tissues. Gene 41:47–57

Chung BY, Simons C, Firth AE, Brown CM, Hellens RP (2006) Effect of 5'UTR introns on gene expression in *Arabidopsis thaliana*. BMC Genomics 7:120

Clancy M, Hannah LC (2002) Splicing of the maize *Sh1* first intron is essential for enhancement of gene expression, and a T-rich motif increases expression without affecting splicing. Plant Physiol 130:918–929

Clancy M, Vasil V, Hannah LC, Vasil IK (1994) Maize *Shrunken-1* intron and exon regions increase gene expression in maize protoplasts. Plant Sci 98:151–161

Curi GC, Chan RL, Gonzalez DH (2005) The leader intron of *Arabidopsis thaliana* genes encoding cytochrome *c* oxidase subunit 5c promotes high-level expression by increasing transcript abundance and translation efficiency. J Exp Bot 56:2563–2571

Curie C, Axelos M, Bardet C, Atanassova R, Chaubet N, Lescure B (1993) Modular organization and developmental activity of an *Arabidopsis thaliana* EF-1α gene promoter. Mol Gen Genet 238:428–436

Dean C, Favreau M, Bond-Nutter D, Bedbrook J, Dunsmuir P (1989) Sequences downstream of translation start regulate quantitative expression of two petunia *rbcS* genes. Plant Cell 1:201–208

Deyholos MK, Sieburth LE (2000) Separable whorl-specific expression and negative regulation by enhancer elements within the *AGAMOUS* second intron. Plant Cell 12:1799–1810

Dimaano C, Ullman KS (2004) Nucleocytoplasmic transport: integrating mRNA production and turnover with export through the nuclear pore. Mol Cell Biol 24:3069–3076

Donath M, Mendel R, Cerff R, Martin W (1995) Intron-dependent transient expression of the maize *GapA1* gene. Plant Mol Biol 28:667–676

Fu D, Szucs P, Yan L, Helguera M, Skinner JS, von Zitzewitz J, Hayes PM, Dubcovsky J (2005) Large deletions within the first intron in *VRN-1* are associated with spring growth habit in barley and wheat. Mol Genet Genomics 273:54–65

Gallie DR, Young TE (1994) The regulation of gene expression in transformed maize aleurone and endosperm protoplasts. Plant Physiol 106:929–939

Garbarino JE, Oosumi T, Belknap WR (1995) Isolation of a polyubiquitin promoter and its expression in transgenic potato plants. Plant Physiol 109:1371–1378

Gidekel M, Jimenez B, Herrera-Estrella L (1996) The first intron of the *Arabidopsis thaliana* gene coding for elongation factor 1β contains an enhancer-like element. Gene 170:201–206

Goodall GJ, Filipowicz W (1991) Different effects of intron nucleotide composition and secondary structure on pre-mRNA splicing in monocot and dicot plants. EMBO J 10:2635–2644

Hong X, Scofield DG, Lynch M (2006) Intron size, abundance, and distribution within untranslated regions of genes. Mol Biol Evol 23:2392–2404

Isshiki M, Yamamoto Y, Satoh H, Shimamoto K (2001) Nonsense-mediated decay of mutant *waxy* mRNA in rice. Plant Physiol 125:1388–1395

Jeon JS, Lee S, Jung KH, Jun SH, Kim C, An G (2000) Tissue-preferential expression of a rice α-tubulin gene, *OsTubA1*, mediated by the first intron. Plant Physiol 123:1005–1014

Jeong YM, Mun JH, Lee I, Woo JC, Hong CB, Kim SG (2006) Distinct roles of the first introns on the expression of Arabidopsis profilin gene family members. Plant Physiol 140:196–209

Keith B, Chua N-H (1986) Monocot and dicot pre-mRNAs are processed with different efficiencies in transgenic tobacco. EMBO J 5:2419–2425

Kertesz S, Kerenyi Z, Merai Z, Bartos I, Palfy T, Barta E, Silhavy D (2006) Both introns and long 3′-UTRs operate as *cis*-acting elements to trigger nonsense-mediated decay in plants. Nucl Acids Res 34:6147–6157

Kim MJ, Kim H, Shin JS, Chung CH, Ohlrogge JB, Suh MC (2006) Seed-specific expression of sesame microsomal oleic acid desaturase is controlled by combinatorial properties between negative *cis*-regulatory elements in the *SeFAD2* promoter and enhancers in the 5′-UTR intron. Mol Genet Genomics 276:351–368

Kooiker M, Airoldi CA, Losa A, Manzotti PS, Finzi L, Kater MM, Colombo L (2005) BASIC PENTACYSTEINE1, a GA binding protein that induces conformational changes in the regulatory region of the homeotic Arabidopsis gene *SEEDSTICK*. Plant Cell 17:722–729

Kuhlemeier C, Fluhr R, Chua N-H (1988) Upstream sequences determine the difference in transcript abundance of pea *rbcS* genes. Mol Gen Genet 212:405–411

Last DI, Brettell, RIS, Chamberlain DA, Chaudhury A.M, Larkin PJ, Marsh E, Peacock WJ, Dennis ES (1991) pEmu: an improved promoter for gene expression in cereal cells. Theor Appl Genet 81:581–588

Le Hir H, Nott A, Moore MJ (2003) How introns influence and enhance eukaryotic gene expression. Trends Biochem Sci 28:215–220

Lu S, Cullen BR (2003) Analysis of the stimulatory effect of splicing on mRNA production and utilization in mammalian cells. RNA 9:618–630

Luehrsen KR, Walbot V (1994) Addition of A- and U-rich sequence increases the splicing efficiency of a deleted form of a maize intron. Plant Mol Biol 24:449–463

Lumbreras V, Stevens DR, and Purton S (1998) Efficient foreign gene expression in *Chlamydomonas reinhardtii* mediated by an endogenous intron. Plant J 14:441–447

Maniatis T, Reed R (2002) An extensive network of coupling among gene expression machines. Nature 416:499–506

Maquat LE (2004a) Nonsense-mediated mRNA decay: A comparative analysis of different species. Curr Genomics 5:175–190

Maquat LE (2004b) Nonsense-mediated mRNA decay: splicing, translation and mRNP dynamics. Nat Rev Mol Cell Biol 5:89–99

Mascarenhas D, Mettler IJ, Pierce DA, Lowe HW (1990) Intron-mediated enhancement of heterologous gene expression in maize. Plant Mol Biol 15:913–920

Morello L, Bardini M, Cricri M, Sala F, Breviario D (2006) Functional analysis of DNA sequences controlling the expression of the rice *OsCDPK2* gene. Planta 223:479–491

Morello L, Bardini M, Sala F, Breviario D (2002) A long leader intron of the *Ostub16* rice β-tubulin gene is required for high-level gene expression and can autonomously promote transcription both *in vivo* and *in vitro*. Plant J 29:33–44

Nash J, Walbot V (1992) *Bronze-2* gene expression and intron splicing patterns in cells and tissues of *Zea mays* L. Plant Physiol 100:464–471

Nott A, Le Hir H, Moore MJ (2004) Splicing enhances translation in mammalian cells: an additional function of the exon junction complex. Genes Dev 18:210–222

Nott A, Meislin SH, Moore MJ (2003) A quantitative analysis of intron effects on mammalian gene expression. RNA 9:607–617

Plesse B, Criqui MC, Durr A, Parmentier Y, Fleck J, Genschik P (2001) Effects of the polyubiquitin gene *Ubi.U4* leader intron and first ubiquitin monomer on reporter gene expression in *Nicotiana tabacum*. Plant Mol Biol 45:655–667

Reddy ASN (2001) Nuclear pre-mRNA splicing in plants. Crit Rev Plant Sci 20:523–571

Rethmeier N, Seurinck J, Van Montagu M, Cornelissen M (1997) Intron-mediated enhancement of transgene expression in maize is a nuclear, gene-dependent process. Plant J 12:895–899

Rose AB (2002) Requirements for intron-mediated enhancement of gene expression in *Arabidopsis*. RNA 8:1444–1453

Rose AB (2004) The effect of intron location on intron-mediated enhancement of gene expression in *Arabidopsis*. Plant J 40:744–751

Rose AB, Beliakoff JA (2000) Intron-mediated enhancement of gene expression independent of unique intron sequences and splicing. Plant Physiol 122:535–542

Rose A.B, Last RL (1997) Introns act post-transcriptionally to increase expression of the *Arabidopsis thaliana* tryptophan pathway gene *PAT1*. Plant J 11:455–464

Sheldon CC, Conn AB, Dennis ES, Peacock WJ (2002) Different regulatory regions are required for the vernalization-induced repression of *FLOWERING LOCUS C* and for the epigenetic maintenance of repression. Plant Cell 14:2527–2537

Sieburth LE, Meyerowitz EM (1997) Molecular dissection of the *AGAMOUS* control region shows that *cis* elements for spatial regulation are located intragenically. Plant Cell 9:355-365

Sinibaldi RM, Mettler IJ (1992) Intron splicing and intron-mediated enhanced expression in monocots. In Cohn, W.E. and Moldave, K. (eds.), Progress in Nucleic Acid Research and Molecular Biology. Academic Press, New York, Vol. 42, pp. 229–257

Snowden KC, Buchholz WG, Hall TC (1996) Intron position affects expression from the *tpi* promoter in rice. Plant Mol Biol 31:689–692

Sridhar VV, Surendrarao A, Liu Z (2006) *APETALA1* and *SEPALLATA3* interact with *SEUSS* to mediate transcription repression during flower development. Development 133:3159–3166

Stark K, Kirk DL, Schmitt R (2001) Two enhancers and one silencer located in the introns of *regA* control somatic cell differentiation in *Volvox carteri*. Genes Dev 15:1449–1460

Steinmetz EJ (1997) Pre-mRNA processing and the CTD of RNA polymerase II: the tail that wags the dog? Cell 89:491–494

Tanaka A, Mita S, Ohta, S, Kyozuka J, Shimamoto K, Nakamura K (1990) Enhancement of foreign gene expression by a dicot intron in rice but not in tobacco is correlated with an increased level of mRNA and an efficient splicing of the intron. Nucl Acids Res 18:6767–6770

Ueki J, Komari T, Imaseki H (2004) Enhancement of reporter-gene expression by insertions of two introns in maize and tobacco protoplasts. Plant Biotech 21:15–24

Vasil V, Clancy M, Ferl, RJ, Vasil IK, Hannah LC (1989) Increased gene expression by the first intron of maize *Shrunken-1* locus in grass species. Plant Physiol 91:1575–1579

Vitale A, Wu RJ, Cheng Z, Meagher RB (2003) Multiple conserved 5' elements are required for high-level pollen expression of the *Arabidopsis* reproductive actin *ACT1*. Plant Mol Biol 52:1135–1151

Weise A, Rodriguez-Franco M, Timm B, Hermann M, Link S, Jost W, Gorr G (2006) Use of *Physcomitrella patens* actin 5' regions for high transgene expression: importance of 5' introns. Appl Microbiol Biotechnol 70:337–345

Xu Y, Yu H, Hall TC (1994) Rice triosephosphate isomerase gene 5' sequence directs β-glucuronidase activity in transgenic tobacco but requires an intron for expression in rice. Plant Physiol 106:459–467

Zhang S-H, Lawton MA, Hunter T, Lamb CJ (1994) *atpk1*, a novel ribosomal protein kinase gene from *Arabidopsis*. I. Isolation, characterization, and expression. J Biol Chem 269:17586–17592

The Role of the Plant Nucleolus in Pre-mRNA Processing

J.W.S. Brown(✉), P.J. Shaw

Contents

Introduction	292
The Dynamic Nucleus	292
Nuclear Bodies in Animal Cells	293
Nuclear Bodies in Plant Cells	294
Traditional Functions of the Nucleolus	295
rRNA Transcription and Processing	295
Ribosome Assembly	296
Other Functions of the Nucleolus in RNA Metabolism	297
Signal Recognition Particle	297
tRNA and RNase P	298
Telomerase RNP	299
Small Nuclear RNPs	299
Heterochromatic Small Interfering RNAs	299
mRNA	300
The Nucleolus as a Major Centre for RNA Metabolism and Regulation of Gene Expression	305
References	306

Abstract The nucleolus is a multifunctional compartment of the eukaryotic nucleus. Besides its well-recognised role in transcription and processing of ribosomal RNA and the assembly of ribosomal subunits, the nucleolus has functions in the processing and assembly of a variety of RNPs and is involved in cell cycle control and senescence and as a sensor of stress. Historically, nucleoli have been tenuously linked to the biogenesis and, in particular, export of mRNAs in yeast and mammalian cells. Recently, data from plants have extended the functions in which the plant nucleolus is involved to include transcriptional gene silencing as well as mRNA surveillance and nonsense-mediated decay, and mRNA export. The nucleolus in plants may therefore have important roles in the biogenesis and quality control of mRNAs.

J.W.S. Brown
Plant Sciences Division, College of Life Sciences, University of Dundee at SCRI,
Invergowrie, Dundee DD2 5DA, Scotland, UK
e-mail: john.brown@scri.ac.uk

Introduction

The nucleolus is the largest and most visible sub-nuclear compartment. Its main function is the transcription and processing of ribosomal RNA (rRNA) and ribosomal subunit production. In recent years, molecular and proteomic approaches have begun to dissect the pathways of ribosomal subunit assembly and transport from the nucleolus and examine the composition of protein complexes and RNPs involved in these processes (Grandi et al. 2002; Dragon et al. 2002; Schafer et al. 2003; Gallagher et al. 2004; Grannemann and Baserga 2004, 2005). In addition, many lines of evidence point to the nucleolus having a wide range of other functions in both RNA metabolism and cell growth (for reviews see Pederson 1998; Olsen et al. 2000; Olsen 2004; Raška et al. 2006; Boisvert et al. 2007). The multifunctional nature of the nucleolus has also been highlighted by proteomic approaches which have identified numerous unexpected proteins in the nucleolus, indicative of new interactions or functions (Andersen et al. 2002, 2005; Scherl et al. 2002; Pendle et al. 2005). The presence of non-ribosomal RNAs such as the RNA moieties of the signal recognition particle (SRP) and telomerase and some tRNAs suggests roles in processing and assembly of other RNAs and RNPs (Pederson 1998; Olsen et al. 2000; Olsen 2004; Boisvert et al. 2007). In this aspect, plants are of growing importance through the recent linking of the nucleolus to production of heterochromatic siRNAs and transcriptional gene silencing (Li et al. 2006; Pontes et al. 2006) and to mRNA export, mRNA surveillance and nonsense-mediated decay (Kim et al., unpublished). Finally, the nucleolus has been shown to be crucial to cell division regulation and senescence and as a sensor of cell stress (Rubbi and Milner 2003; Olsen 2004; Boisvert et al. 2007). In this chapter, we examine the multifunctionality of the eukaryotic nucleolus and describe recent data on novel functions of the plant nucleolus and, in particular, its potential roles in mRNA processing and biogenesis.

The Dynamic Nucleus

The nucleus is the defining eukaryotic organelle. It is highly organised, with chromatin regions containing the genomic DNA and inter-chromatin domains containing the machinery for gene expression and regulation processes (Misteli 2005). The nucleus contains many sub-nuclear structures or bodies that vary in their components and functions (Lamond and Spector 2003; Cioce and Lamond 2005; Misteli 2005; Handwerger and Gall 2006). Such structures, rather than being fixed, unchanging entities, reflect the accumulation and steady-state flux of sub-populations of molecules and complexes, as has been clearly shown by fluorescence recovery after photobleaching (FRAP) studies. Their dynamic nature is also demonstrated by changes in size and distribution under different conditions and during differentiation and development reflecting the high mobility and trafficking of many components between the different bodies and nucleoplasm (Boudonck et al. 1998, 1999; Lamond and Spector 2003; Cioce and Lamond 2005; Misteli 2005; Handwerger and Gall 2006; Gorski et al. 2006; Raška et al. 2006). Although the functions of many nuclear bodies have yet to be fully elucidated, clearly some nuclear bodies have multiple functions and some functions may overlap. Thus the spatial and temporal organisation of the nucleus is highly dynamic and is integral to the regulation and integration of networks of gene expression processes which ultimately determine cellular function. Consistent with these considerations, the nucleolus, already recognised as multifunctional, continues to add to its list of functions and emerges as a dynamic and major centre for RNA metabolism in the cell.

Nuclear Bodies in Animal Cells

The nucleolus is not membrane-bound, has a high relative refractive index and thus appears under phase contrast microscopy as a dense region of the nucleus. It is formed around tens, hundreds or even thousands of tandemly repeated ribosomal RNA (rRNA) gene units and contains rRNA transcripts at different stages of transcription, processing and assembly into ribosomal subunits and the multitude of RNAs and proteins required for these processes. The observed nucleolar compartment therefore represents a snapshot of the steady-state import of components and export of the major products, the ribosomal subunits (Leung and Lamond 2003; Cheutin et al. 2004). The fundamental role of the nucleolus in production of ribosomes needed for translation is also reflected in the effects on rRNA transcription, ribosome production and nucleolar morphology of cell growth and environmental conditions (Leung and Lamond 2003; Leung et al. 2004; Mosgoeller 2004; Dove et al. 2006; Raška et al. 2006). This integral link between the nucleolus and normal functioning of the cell is further highlighted by its involvement in cell division, cell death and stress responses (Rubbi and Milner 2003; Olsen 2004; Raška et al. 2006; Boisvert et al. 2007). In addition to the factors required for ribosome biogenesis, the nucleolus also functions in aspects of the biogenesis of many other RNP complexes and contains proteins and protein complexes involved in a range of cellular functions as detailed below in this review.

Besides the nucleolus, other nuclear bodies such as Cajal bodies (CBs) and speckles have been identified in animal and plant nuclei (Cioce and Lamond 2005; Handwerger and Gall 2006). In particular, CBs, which are often closely associated with nucleoli, contain a number of different proteins including coilin, a protein essential for CB formation, and fibrillarin and dyskerin, nucleolar proteins required for pre-rRNA processing and modification (Cioce and Lamond 2005). CBs are involved in the maturation of spliceosomal snRNPs and snoRNPs, with newly assembled snRNPs and snoRNPs trafficking through CBs before accumulating in splicing speckles and the nucleolus, respectively (Narayanan et al. 1999; Sleeman and Lamond 1999). The modification of specific nucleotides in snRNAs, guided by small CB-specific RNAs (scaRNAs), occurs in CBs (Darzacq et al. 2002; Jady et al. 2003). CBs also contain U7snRNA, which functions in the 3' end processing of some histone mRNAs, and are enriched in the RNA of telomerase RNP (Frey and Matera 1995; Jady et al. 2004; Zhu et al. 2004; Pillai et al. 2005). CBs are dynamic structures whose presence and abundance are linked to the transcriptional activity of the cell and their association with the nucleolus is reflected in the trafficking of nucleolar proteins between the nucleolus and CBs. Thus nucleoli and CBs are both multifunctional and involved in RNP maturation.

Nuclear Bodies in Plant Cells

In plants, the best-characterised nuclear bodies are the nucleolus and CBs. The plant nucleolus differs to some extent in organisation and structure from the animal nucleolus, although its major function in rRNA and ribosomal subunit production remains the same (Brown and Shaw 1998; Shaw and Brown 2004). The plant nucleolus contains a large proportion of dense fibrillar component (DFC), much more than in typical animal nucleoli, which is surrounded by the granular component (GC) (Fig. 1). The DFC regions contain rDNA transcription units which are being actively transcribed and have been visualised as 'Christmas tree' structures around 300 nm in length (Shaw and Jordan 1995; Gonzalez-Melendi and Shaw 2001). The localisation by fluorescence microscopy and in situ hybridisation of nucleolar proteins, rRNA gene regions and snoRNA species to sub-domains within plant nucleoli have been correlated with early and late events in rRNA processing (Beven et al. 1996; Brown and Shaw 1998). A distinctive feature of most plant nucleoli is the nucleolar cavity, whose function is unknown but contains spliceosomal snRNAs and accumulates snoRNAs (Beven et al. 1995, 1996) and contains centres for the production of heterochromatic small interfering RNAs (siRNAs) (Pontes et al. 2006). The relationship between these centres and CBs is not yet clear (Li et al. 2006; Pontes et al. 2006) but highlights the interaction between CBs and the nucleolus.

Plant CBs vary in number and size in different cells and under different conditions (Boudonck et al. 1998; Collier et al. 2006). Recently, plant coilin was identified and, despite the sequence divergence from animal coilin, was shown to be required for CB formation. Mutants in plant coilin will be invaluable in addressing the functions

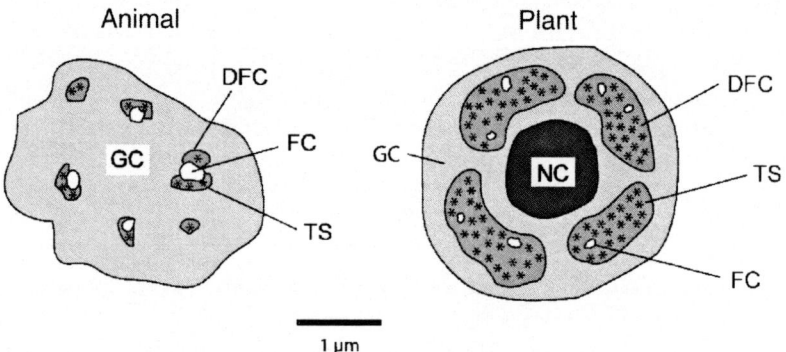

Fig. 1 Comparison of the structure of animal and plant nucleoli. *TS*, transcription sites; *FC*, fibrillar centres; *DFC*, dense fibrillar component; *GC*, granular component; *NC*, nucleolar cavity.

and properties of plant CBs (Collier et al. 2006). Plant CBs also contain snRNAs, snoRNAs and fibrillarin (Beven et al. 1995, 1996; Brown and Shaw 1998) and plant scaRNAs have been identified consistent with conserved roles in snRNP and snoRNP maturation (Marker et al. 2002). Moreover, the majority of plant snoRNAs are organised in gene clusters and expressed as polycistronic precursor snoRNAs (Brown et al. 2003). The detection of these polycistronic precursors in CBs and the nucleolus by in situ hybridisation suggests that processing occurs in both locations and/or that pre-snoRNAs traffic to the nucleolus via CBs (Shaw et al. 1998). The nucleolus is also involved in some virus infections (both animal and plant), with particular viral proteins localising to the nucleolus (Hiscox 2002, 2007). Although the function of the nucleolar localisation of many of these proteins remains to be established, the nucleolar accumulation of the ORF3 protein of the plant virus groundnut rosette virus (GRV) has been shown to involve interactions with CBs and fibrillarin and is essential for the formation of cytoplasmic RNPs and long-distance virus movement (Kim et al. 2007a, b). Finally, the Arabidopsis nucleolus contains exon junction complex proteins, involved in linking transcription and splicing to translation, export and mRNA surveillance (Pendle et al. 2005). Thus new functions in RNA metabolism have been demonstrated for both plant CBs and the nucleolus.

Other nuclear bodies in plants have been defined by the presence of particular proteins such as splicing factors (see the chapter by G. S. Ali and A. S. N. Reddy, this volume) and cyclophilins, HYL1, phytochrome and AKIP1. At present, the components and functions of these bodies are not well characterised but SR proteins and cyclophilins are found in distinct splicing speckles and cyclophilin-containing bodies can interact and regulate functions in splicing (Lorković 2004). HYL1, involved in microRNA biogenesis, localises to a small number of nuclear bodies while phytochrome- and AKIP-containing bodies are involved in signalling pathways interactions and may regulate mRNAs at the transcriptional or post-transcriptional levels (Yamaguchi et al. 1999; Han et al. 2002; Kircher et al. 2002; Chen et al. 2003; Ang et al. 1998; Li et al. 2002; Shaw and Brown 2004; Song et al. 2007). Thus nuclear bodies are sites where different factors accumulate, allowing efficient RNA-protein and protein-protein interactions or RNP assembly, to influence or regulate a particular process. The bodies can form or dissemble or increase and decrease in size and number depending on different conditions or activity of the cell, again reflecting the dynamic interplay between nuclear bodies and the surrounding nucleoplasm.

Traditional Functions of the Nucleolus

rRNA Transcription and Processing

The major recognised function of the nucleolus is transcription and processing of ribosomal RNA and assembly of ribosomal subunits. Eukaryotic ribosomal RNA genes are organised in large clusters, often containing hundreds or thousands of repeated genes, with each gene encoding one copy of the 18S, 5.8S and 25–28S rRNAs. The coding regions are separated by internal transcribed spacer (ITS) regions and flanked by external transcribed spacer (ETS) regions. Genes are transcribed by RNA polymerase I as polycistronic precursor rRNAs which are processed to the mature 5.8S, 18S and 25–28S rRNAs. Processing of pre-rRNAs involves (a) cleavage of the transcript into precursors of the mature rRNAs and trimming of these precursors to their final size and (b) modification of specific rRNA nucleotides. The best-characterised pathway of pre-rRNA processing is that of yeast, in which the step-wise cleavage and trimming reactions required for production of mature rRNAs have been demonstrated and in which many of the protein and snoRNP components involved have been identified (Venema and Tollervey 1999; Fatica and Tollervey 2002). Initial cleavage of the pre-rRNA involves the U3, U14, MRP, snR10 and snR30 snoRNAs and the resultant precursors are trimmed by the action of exonucleases: Rat1p, Xrn1p and the exosome (Venema and Tollervey 1999; Fatica and Tollervey 2002). The pathway of processing and modification is poorly understood in plants but is likely to contain many processing steps which are conserved with other eukaryotes. Nevertheless, significant advances in understanding rRNA processing and the function of nucleolar proteins have been made recently. In particular, a multiprotein complex containing nucleolin, fibrillarin, U3 and U14 has been shown to cleave the pre-rRNA in the 5′ external transcribed sequence (Saez-Vasquez et al. 2004). In addition, the absence of nucleolin expression causes disruption of the structural organisation of the nucleolus (Pontvianne et al. 2007).

Ribosomal RNAs contain numerous nucleotide modifications. The two major modifications in rRNA are 2′-O-ribose methylation and pseudouridylation. The sites of modification are determined by regions of complementarity to their cognate rRNA sequences. Box C/D and box H/ACA snoRNAs are responsible for generating 2′-O-ribose methylation and pseudouridylation of nucleotides, respectively (Kiss 2002). Thus snoRNAs act as guide sequences to define which nucleotides in rRNA are modified. In Arabidopsis and rice, snoRNA genes have been identified by both bioinformatics predictions and experimental approaches (Barneche et al. 2001; Brown et al. 2001, 2003; Qu et al. 2001; Marker et al. 2002). Many of the plant snoRNAs are conserved in yeast and/or human, in terms of the rRNA target sequences which are modified, but there are a number of plant-specific snoRNAs guiding modification of nucleotides in plant rRNAs which are unmodified in other organisms (Barneche et al. 2001; Brown et al. 2001; Qu et al. 2001; Marker et al. 2002). The fact that so many nucleotide modifications are conserved in eukaryotic rRNAs suggests important functions in rRNA processing or ribosome activity. The majority of nucleotide modifications are in the active sites of the ribosome and are thought to influence the efficiency of the ribosome and protein translation (Decatur and Fournier 2002).

Ribosome Assembly

The eukaryotic ribosome consists of a small (40S) and a large (60S) subunit containing the 5S, 5.8S, 18S and 26–28S rRNAs associated with around 80 ribosomal proteins (Tate and Poole 2004). As with other large RNP complexes such as the spliceosome, correct assembly occurs in a regulated and co-ordinated step-wise pathway involving non-ribosome factors and ribosomal proteins required for correct rRNA folding, protein association and ultimately export of the ribosomal subunits through the nuclear pore complex to the cytoplasm (Venema and Tollervey 1999; Fatica and Tollervey 2002; Grannemann and Baserga 2004). The pre-rRNA transcript is assembled into a 90S pre-ribosome containing factors required for formation of both ribosomal subunits. Cleavage of the primary transcript gives rise to pre-ribosomes which with further processing and assembly produce the 40S and 60S ribosomal subunits (Fatica and Tollervey 2002; Grannemann and Baserga 2004). One of these pre-ribosomal complexes includes the small subunit (SSU) processosome which contains the U3snoRNP along with ribosomal and non-ribosomal proteins including novel Utp proteins (Dragon et al. 2002; Grandi et al. 2002; Schafer et al. 2003). The SSU processosome complex is thought to be the RNP complex previously observed as knobs at the 5′ regions of pre-rRNA transcripts (Miller and Beatty 1969). In plants, very little is known about ribosomal subunit assembly apart from detailed analyses of the ribosomal protein complement (Barakat et al. 2001) and the identification of orthologues of some yeast and human associated factors by, for example, nucleolar proteomics (Pendle et al. 2005). Thus the biochemistry of rRNA processing and ribosome assembly lags significantly behind that of yeast and animal systems.

Other Functions of the Nucleolus in RNA Metabolism

Evidence for the nucleolus having functions other than rRNA transcription/processing and ribosome assembly continues to increase. These functions can be grouped into RNA processing activities, often associated with RNP assembly, mRNA metabolism, and regulatory cellular functions such as the control of the cell cycle and cell aging and as a sensor of cell stress (Fig. 2). In many cases, evidence currently comes from only one organism, making extrapolation to a general nucleolar function difficult. Nevertheless, the novel functions illustrate the potential complexity of the nucleolus. The various functions of the nucleolus have been reviewed previously (Pederson 1999; Olsen 2004; Boisvert et al. 2007) and here we briefly describe the roles of the nucleolus in RNA processing and RNP assembly.

Signal Recognition Particle

The signal recognition particle (SRP) is a small RNP involved in targeting the translation of specific proteins to the endoplasmic reticulum. SRP consists of the SRP RNA associated with specific proteins (Walter and Johnson 1994). At least part of the assembly of the SRP has been shown to occur in the nucleolus. Most, but not all, of the protein components of the mature cytoplasmic SRP (SRP19,

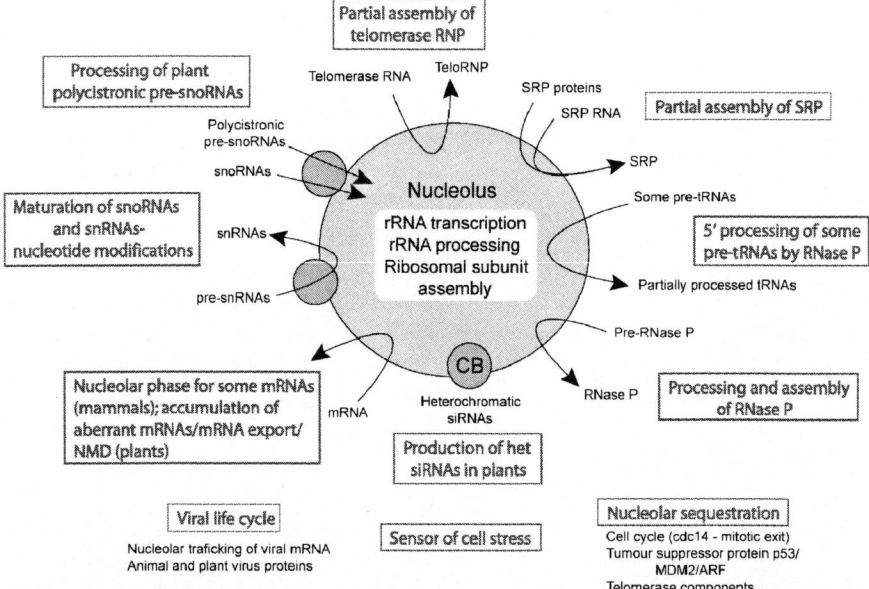

Fig. 2 The multifunctional nucleolus. Functions of the nucleolus and associated Cajal bodies in RNA metabolism and cell processes.

SRP68 and SRP72 but not SRP54) have been localised to the nucleolus in vertebrates (Politz et al. 2000) and in yeast (Grosshans et al. 2001). SRP RNA injected into mammalian cultured cells rapidly localised to the nucleolus and then gradually relocated to the cytoplasm, suggesting that SRP was being assembled in the nucleolus and then exported (Jacobsen and Pederson 1998). Thus the assembly of the SRP has a nucleolar phase, but must be completed in the cytoplasm.

tRNA and RNase P

Precursor tRNAs have been detected in the nucleolus by in situ hybridisation (Bertrand et al. 1998). Pre-tRNAs are trimmed at 5' and 3' ends and undergo base modifications such as pseudouridylation and 3' end processing (Hopper and Phizicky 2003). The 5' trimming of pre-tRNAs is catalysed by RNase P and both protein and RNA components of this RNP have been localised to the nucleolus (Jacobson et al. 1997; Jarrous et al. 1999), as has Cbf5p/Dyskerin which catalyses these and other pseudouridylations. The presence of components of the RNase P RNP also suggests that the assembly of this RNP could occur in the nucleolus. It has been shown that most tRNAs in *S. cerevisiae* are exported from the nucleus as aminoacyl tRNAs and that, in mutants defective in aminoacylation of tRNA, tRNAs accumulate in the nucleolus (Steiner-Mosonyi and Mangroo 2004), suggesting that the machinery for aminoacylation is also present in the nucleolus.

Telomerase RNP

Chromosome replication in eukaryotes requires the addition of multiple tandem repeats of a short sequence at the ends of each chromosome (TTAGGG in humans). This is catalysed by a reverse transcriptase (TERT), which copies the repeated sequence from an associated RNA (TERC) (Cech 2004). Most somatic cells do not have active telomerase, which is proposed to limit the number of rounds of cell division that they can support and has been suggested to be a factor in aging and carcinogenesis (Maser and DePinho 2002). The localisation of the RNA and the reverse transcriptase catalytic subunit of human telomerase, hTERT, suggests that the telomerase RNP may be assembled in the nucleolus (Mitchell and Collins 2000; Etheridge et al. 2002). Telomerase RNA contains a box H/ACA domain and a CB localisation signal allowing the telomerase RNP to localise to both CBs and the nucleolus in its maturation (Mitchell et al. 1999; Lukowiak et al. 2001; Jady et al. 2004). The localisation of telomerase in the nucleolus may reflect its assembly pathway and/or its sequestration as part of the regulation of its activity. Telomerase resides in the nucleolus but is specifically redistributed in late S-phase, when telomere replication takes place, thereby avoiding inappropriate expression throughout the nucleus which could cause new telomeric structures to be nucleated on replication intermediates or double strand breaks (Wong et al. 2002).

Small Nuclear RNPs

Small nuclear RNAs (snRNAs) are involved in splicing and form the cores of the spliceosomal complexes. There is good evidence that U4/U6.U5 and U2 snRNAs are also in the nucleolus and undergo part of their maturation or assembly processing there, and that the individual snRNAs are transported independently to the nucleolus (Lange and Gerbi 2000; Yu et al. 2001; Gerbi and Lange 2002). A nucleolar phase for snRNPs is strongly supported by the proteomic analyses of the nucleoli in both human and plant cells (Andersen et al. 2002, 2005; Scherl et al. 2002; Pendle et al. 2005). These studies showed the presence of many snRNA-associated proteins, particularly the Sm class of proteins, pointing to a role in the maturation of snRNAs and the assembly of snRNPs. U6snRNA differs from the other spliceosomal snRNAs in its cap structure and maturation pathway. Factors involved in U6 nucleotide modifications are found in the nucleolus, suggesting a nucleolar phase for the production of mature U6snRNA (Ganot et al. 1999).

Heterochromatic Small Interfering RNAs

Plants contain a novel RNA polymerase (RNA polymerase IV, Pol IV) involved in production of transcripts which give rise to double-stranded RNAs and small interfering

RNAs (siRNAs) which guide DNA methylation. The protein components required for the production of such heterochromatic siRNAs include the large subunit of Pol IV (NRPD1b), AGO3, DCL3 and RDR2. These proteins co-localised with siRNAs in foci in the nucleolus or in CBs (Li et al. 2006; Pontes et al. 2006). Thus the nucleolus/CBs are involved in siRNA production and assembly of silencing complexes. This role in the transcriptional gene silencing pathway in plants raises the question of whether other RNA silencing pathways or the cross talk among pathways also involves the nucleolus.

mRNA

The nucleolus has previously been implicated in mRNA export or surveillance (Schneiter et al. 1995; Pederson 1998). In yeast, in mRNA transport-defective mutants and mutants of some nucleolar proteins, the nucleolus was disrupted and fragmented and accumulated poly(A)+ RNA (Kadowaki et al. 1995; Schneiter et al. 1995). Similarly, in transport-defective mutants in *Schizosaccharomyces pombe*, an intron-containing transcript accumulated in the nucleolus, whereas transcripts from the intronless cDNA were exported (Ideue et al. 2004). Furthermore, heat shock of *S. pombe* cells resulted in accumulation of poly(A)+ RNA in the nucleolus (Tani et al. 1995) while ethanol stress caused the accumulation of specific transcripts in the nucleolus (Brodsky and Silver 2000). In *S. pombe* mutants which undergo aberrant cytokinesis generating nuclei with and without a nucleolus, poly(A)+ accumulated only in the latter, suggesting that poly A+ mRNA associates transiently with the nucleolus during export (Ideue et al. 2004). However, recent work on mechanisms of RNA quality control which involve polyadenylation of aberrant RNA species by the TRAMP complex before their degradation may explain the accumulation of polyadenylated RNA species in some of the yeast mutants (Carneiro et al. 2007). In mammalian cells, some spliced mRNAs (e.g. *c-myc*) were localised to the nucleolus while their unspliced versions were found in the nucleoplasm (Bond and Wold 1993). Finally, the nucleolar targeting of some animal viral proteins regulates viral mRNA export and is essential to the virus life cycle, being required for viral replication or diverting normal host cell functions (Hiscox 2002, 2007). Whether the nucleolus is directly involved in mRNA export is still an open question.

mRNA Biogenesis Proteins in the Nucleolus

Recent proteomic and RNomic analyses of nucleoli have again raised the question of the function of the nucleolus in mRNA biogenesis. Proteomics of human and plant nucleoli identified almost 700 and 217 proteins, respectively (Andersen et al. 2002, 2005; Scherl et al. 2002; Pendle et al. 2005; Leung et al. 2006). The plant proteome contained many expected nucleolar, ribosome biogenesis and ribosomal proteins as well as core snoRNP proteins and proteins involved in snoRNP biogenesis (Pendle et al. 2005). Unexpectedly, many components of mRNA splicing and translation were found in the plant nucleolus, an observation also made in the human nucleolar proteome (Andersen et al. 2002; Scherl et al. 2002). In particular, the plant nucleolar proteome contained exon junction complex (EJC) proteins which were also shown to be associated with nucleoli by localisation of GFP fusions (Pendle et al. 2005). In contrast, EJC proteins were detected in the human nucleolar proteome but have not been observed in the nucleolus (Custódio et al. 2004; Palacios et al. 2004), suggesting potential differences in EJC assembly, storage or function between animals and plants. The EJC is a multi-protein complex that is deposited upstream of splice junctions in mRNAs after intron splicing. It contains proteins that are recruited as part of the spliceosome, that interact with mRNP export adaptors, that enhance translation or that are involved in nonsense-mediated decay (Stutz and Izzuralde 2003; Maquat 2004; Lejeune and Maquat 2005; Tange et al. 2004; Andersen et al. 2006). As such, EJC proteins link gene transcription with splicing, export, translation and nonsense-mediated decay (Fig. 3). The discovery of EJC proteins, as well as splicing and translation factors, associated with the plant nucleolus suggested that aspects of mRNA biogenesis in plants involve the nucleolus.

mRNAs in the Plant Nucleolus

Whether mRNAs are present in the plant nucleolus was addressed by sequencing clones from cDNA libraries made from cell cultures, isolated nuclei and purified nucleoli. Not only were mRNAs detected in the nucleolar cDNA library but the abundance of different classes of mRNAs was significantly different between the three libraries (Fig. 4; Kim et al., unpublished). The mRNAs fell into three classes: fully spliced transcripts, transcripts from single exon genes (i.e. that had not undergone splicing) and aberrantly spliced transcripts. Aberrantly spliced transcripts include transcripts with potential splicing errors, incompletely spliced transcripts (where one or more introns is unspliced) and alternatively spliced transcripts. Although it is difficult to distinguish among these possibilities, the aberrant nature of this class of the transcripts in the nucleolus was demonstrated by the fact that over 90% contained premature stop codons. The proportion of fully spliced transcripts decreased from 82% in the whole cell library to 68% and 42% in the nuclear and nucleolar libraries, respectively. In contrast, the proportion of aberrantly spliced transcripts increased from 2% in the whole cell library to 13% and 38% in the nuclear and nucleolar libraries, respectively, while those of single exon genes remained around the same levels (16%–20%) (Fig. 4; Kim et al., unpublished). Thus the nucleolus contained enhanced levels of aberrant mRNAs. The unexpected discovery of mRNAs in the nucleolus suggests that mRNAs can move freely between the nucleoplasm and nucleolus and may point to a nucleolar mRNA export function. The majority of aberrant mRNAs contained premature termination codons, further

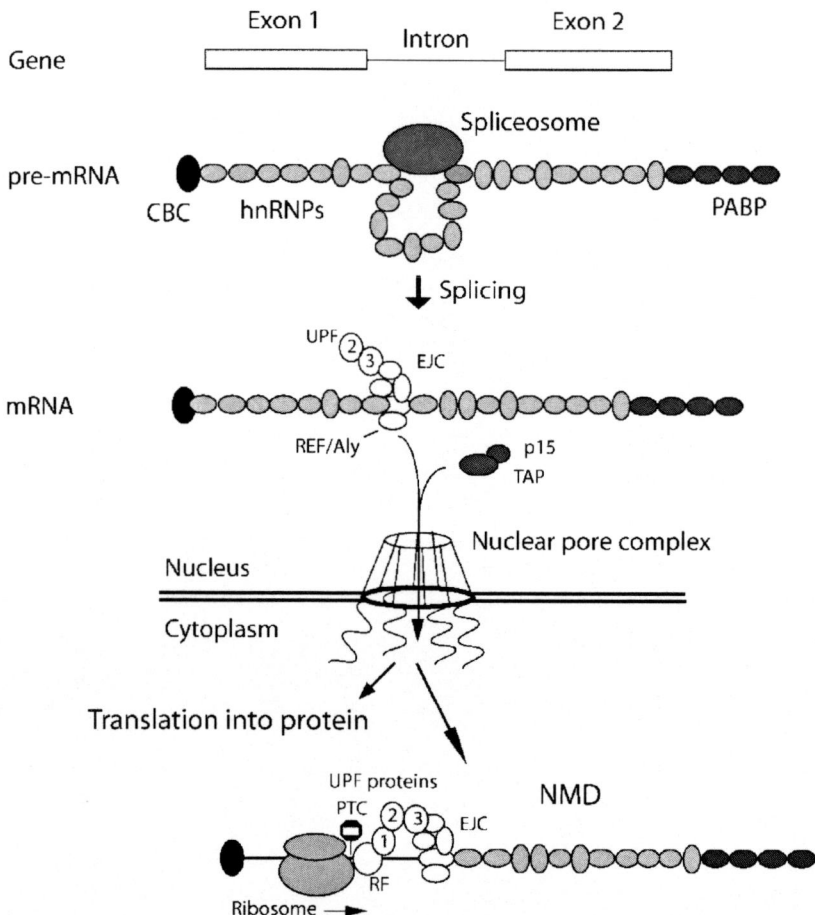

Fig. 3 mRNA biogenesis in mammalian cells—the exon junction complex links transcription, splicing, export, surveillance, translation and nonsense-mediated decay. As genes are transcribed, the pre-mRNAs are bound by hnRNP protein factors. After capping and polyadenylation, the cap and poly(A)+ tail are bound by the cap binding complex (CBC) and poly(A)-binding proteins (PABP). Introns and splice sites are recognized co-transcriptionally. Intron removal by the spliceosome deposits the exon junction complex (EJC) circa 24 nt upstream of the exon-exon junction. The EJC contains around 20 proteins including the UPF proteins UPF3 (3) and UPF2 (2) and the export factor Aly/REF. The latter interacts with the export adaptor dimmer TAP/p15 which, in turn, interacts with the nuclear pore complex, allowing the mRNP to be exported. Once exported, mRNAs are translated into proteins by ribosomes. If the mRNA contains a premature termination codon (PTC), pausing of the ribosome recruits a complex containing release factors (RF) and UPF1 (1). Phosphorylation of UPF1 is a key step in targeting the aberrant transcript for degradation.

suggesting a role in mRNA surveillance and nonsense-mediated decay. This is supported by the localisation of the NMD proteins UPF2 and UPF3 in the nucleolus (Kim et al., unpublished). Thus the plant nucleolus may function in mRNA export, surveillance and NMD.

Fig. 4 a-c mRNAs in the plant nucleolus. Distribution of the relative proportions of different classes of mRNAs from cDNA libraries of Arabidopsis whole cells, nuclei and nucleoli showing the decrease in fully spliced and increase in aberrantly spliced transcripts in the nucleolus. *SE*, single exon gene transcripts; *FS*, fully spliced transcripts; *AS*, aberrantly spliced transcripts.

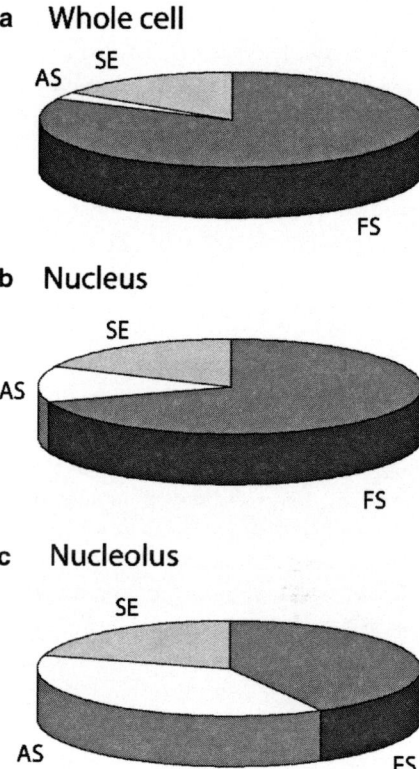

A Role in mRNA Export?

In eukaryotes, export of mRNPs from the nucleus to the cytoplasm is mediated by Aly/REF and UAP56 (Yra1p and Sub2p, respectively, in yeast) as part of the mRNP complex which interacts with the export receptor protein TAP (Mex67p in yeast) (see the chapter by V. Chinnusamy et al., this volume). This, in turn, interacts with components of the nuclear pore complex (Reed and Hurt 2002; Stutz and Izaurralde 2003; Aguilera 2005; Reed and Cheng 2005). In yeast, all three proteins (Mex67p, Yra1p and Sub2p) are required for export, but in *Drosophila* and *Caenorhabiditis elegans* Aly/REF is non-essential (Gatfield and Izaurralde 2002; Longmann et al. 2003; MacMorris et al. 2003). This may reflect the ability of arginine-serine-rich (SR) protein splicing factors to interact with TAP and act as alternative export adaptors (Reed and Cheng 2005). Little is known about mRNA export components and pathways in plants. In particular, it is not known whether Aly/REF or UAP56 is needed for mRNA export and only recently has a possible orthologue of TAP been identified (Hernandez-Pinzon et al. 2007). Nevertheless, three of the four Arabidopsis Aly/REF proteins and UAP56 localise strongly to the nucleolus (Pendle et al. 2005) and both Arabidopsis and tobacco Aly proteins shuttle between the nucleolus/nucleus and cytoplasm (Uhrig et al. 2004; Canto et al. 2006). In addition, besides EJC proteins, some splicing factors (including SR proteins), spliceosomal snRNP proteins and hnRNP proteins were identified in the Arabidopsis nucleolar proteome (Pendle et al. 2005) and the Arabidopsis SR protein RSZ22 has recently been shown to shuttle between the nucleoplasm, nuclear speckles and the nucleolus (Tillemans et al. 2006). These observations are consistent with the involvement of the nucleolus with mRNA export.

The potential export of mRNAs via the nucleolus parallels studies of viral RNA export in mammals. The export of herpesvirus saimiri (HVS) mRNA involves the expression of the viral ORF57 protein which causes Aly, UAP56 and other human export factors to be redistributed to the nucleolus (Boyne and Whitehouse 2006). ORF57 is a nucleocytoplasmic shuttling protein which binds viral intronless mRNAs and, by interacting with Aly and the transcription/export complex (TREX), accesses TAP-mediated export (Williams et al. 2005). Nucleolar trafficking of ORF57 is essential for efficient export of the viral mRNAs (Boyne and Whitehouse 2006). Similarly, the human immunodeficiency virus type 1 (HIV-1) exports unspliced and singly spliced viral mRNAs via the nucleolus for replication (Michienzi et al. 2000; Hiscox 2007). The virus-encoded protein Rev is a nucleocytoplasmic shuttling protein which accumulates in the nucleolus. It interacts with the export factor CRM1 and particular nucleoporins, causing their relocalisation to the nucleolus. By binding a specific sequence in viral unspliced mRNAs and interacting with CRM1 and the nucleoporins, it is thought to assemble an export-competent complex in the nucleolus (Hope 1999). Thus these viral mRNAs are targeted to the nucleolus, where they associate with export factors and are exported. These pathways may represent an inherent eukaryotic nucleolar mRNA export pathway, which is exploited by these viruses. The importance of mRNA export in abiotic stress responses is reviewed in the chapter by V. Chinnusamy et al. (this volume).

A Role in mRNA Surveillance and NMD?

Eukaryotic systems utilise NMD as a quality-control mechanism to recognise and destroy transcripts containing PTCs (PTC+ transcripts) (Maquat 2004; Lejeune and Maquat 2005) preventing the formation of truncated polypeptides which could be potentially harmful for the cell by acting as dominant-negative mutations (see the chapter by D. A. Belostotsky, this volume). The NMD pathway has been widely studied in yeast, *Caenorhabiditis elegans*, *Drosophila* and humans and there are differences in components and mechanisms by which PTCs are discriminated (Culbertson and Leeds 2003; Conti and Izzuralde 2005). In mammals, PTC-containing transcripts have classically been thought to be targeted for NMD by the interaction of a paused ribosome at the PTC with a downstream EJC (Maquat 2004; Lejeune and Maquat 2005). However, recent results suggest that the distance between the termination codon and the poly(A) tail is a major determinant of NMD such that premature translation termination in the wrong mRNP context, possibly through interaction with poly(A)-binding proteins, triggers NMD (Bühler et al. 2006).

An EJC downstream of a premature termination codon potentially acts as an enhancer of NMD in this model (Bühler et al. 2006). Thus the underlying mechanism of NMD in mammals may be more related to those of yeast and *Caenorhabiditis elegans*, which are EJC-independent and rely on the distance between the PTC and 3′ UTR, although, in yeast, this may also involve specific sequences being recognised by *trans*-acting factors.

More than 90% of the aberrant transcripts found in plant nucleoli are putative NMD substrates based on NMD determinants in plants (Kim et al., unpublished; Kertész et al. 2006). The majority of reported examples of NMD in plants derive from intron-containing genes and, for the most part, the positions of PTCs and exon junctions would be consistent with an EJC-dependent or -enhanced effect. However, it is also clear that EJC-independent NMD occurs in plants with reduced mRNA stability dependent on the position of the PTC in PTC-containing intronless transcripts (Van Hoof and Green 2006). Recently, long 3′ UTRs and introns in the 3′ UTR were shown to trigger NMD in plants (Kertész et al. 2006). In terms of the protein components of plant NMD, orthologues of UPF1, UPF2 and UPF3 have been identified (Pendle et al. 2005; van Hoof and Green 2006). Arabidopsis gene orthologues of *UPF1* and *UPF3* have also been shown to function in NMD. UPF3 suppresses the levels of aberrantly spliced, PTC-containing transcripts (Hori and Watanabe, 2005), and Arabidopsis *UPF1* is indispensable (Arciga-Reyes et al. 2006) and causes decreased levels of mRNAs when tethered in different positions in mRNAs (Kertész et al. 2006). Thus the rapid degradation of PTC-containing transcripts from both intron-containing and intronless genes, and the overall conservation of EJC and NMD proteins (Pendle et al. 2005; Kertész et al. 2006; van Hoof and Green 2006), suggests that both EJC-dependent and EJC-independent mechanisms are utilised in plants. The link between the nucleolus and NMD in plants, seen with the presence of aberrant mRNAs in the nucleolus, is further reflected by the localisation of UPF2 and UPF3 to the nucleolus (Kim et al., unpublished). Other aspects of mRNA decay are covered in the chapter by D. A. Belostotsky.

The Nucleolus as a Major Centre for RNA Metabolism and Regulation of Gene Expression

Over the last 40 years and particularly in the last 10–15 years, the nucleolus has grown in importance as a highly dynamic centre of RNA metabolism and RNP assembly. It has the major function in ribosomal subunit assembly needed to supply the cell with ribosomes for translation, and is involved in the processing and assembly of many other RNAs and RNPs. Not surprisingly, therefore, the nucleolus is involved in the regulation of key cellular processes and responds to growth conditions and external stresses. The recent links to transcriptional silencing, mRNA export and NMD in plants not only add to the multiple functions of the nucleolus but also enhance the concept of a role for the nucleolus in co-ordinating different RNA metabolism pathways.

References

Aguilera A (2005) Cotranscriptional mRNP assembly: from the DNA to the nuclear pore. Curr Opin Cell Biol 17:242–250

Ang LH, Chattopadhyay S, Wei N, Oyama T, Okada K, Batschauer A, Deng XW (1998) Molecular interaction between COP1 and HY5 defines a regulatory switch for light control of Arabidopsis development. Mol Cell 1:213–222

Andersen JS, Lyon CE, Fox AH, Leung AK, Lam YW, Steen H, Mann M, Lamond AI (2002) Directed proteomic analysis of the human nucleolus. Curr Biol 12:1–11

Andersen JS, Lam YW, Leung AK, Ogg SE, Lyon CE, Lamond AI, Mann M (2005) Nucleolar proteome dynamics. Nature 433:77–78

Andersen CBF, Ballut L, Johansen JS, Chamieh H, Nielsen KH, Oliveira CLP, Pedersen JS, Séraphin B, Le Hir H, Andersen GR (2006) Structure of the exon junction core complex with a trapped DEAD-box ATPase bound to RNA. Science 313:1968–1972

Arciga-Reyes L, Wootton L, Kieffer M, Davies B (2006) UPF1 is required for nonsense-mediated mRNA decay (NMD) and RNAi in Arabidopsis. Plant J 47:480–489

Barakat A, Szick-Miranda K, Chang L-F, Guyot R, Blanc G, Cooke R, Delseny M, Bailey-Serres J (2001) The organisation of cytoplasmic ribosomal protein genes in the Arabidopsis genome. Plant Physiol 127:398–415

Barneche F, Gaspin C, Guyot R, Echeverria M (2001) Identification of 66 box C/D snoRNAs in *Arabidopsis thaliana*: Extensive gene duplications generated multiple isoforms predicting new ribosomal RNA 2'-O-methylation sites. J Mol Biol 311:57–73

Bertrand E, Houser-Scott F, Kendall A, Singer RH, Engelke DR (1998) Nucleolar localization of early tRNA processing genes. Genes Dev 12:2463–2468

Beven AF, Lee R, Razaz M, Leader DJ, Brown JWS, Shaw PJ (1996) The organization of ribosomal RNA processing correlates with the distribution of nucleolar snRNAs. J Cell Sci 109:1241–1251

Beven AF, Simpson GG, Brown JWS, Shaw PJ (1995) The organization of spliceosomal components in the nuclei of higher plants. J Cell Sci 108:509–51

Bond VC, Wold B (1993) Nucleolar localisation of myc transcripts. Mol Cell Biol 13:3221–3230

Boyne JR, Whitehouse A (2006) Nucleolar trafficking is essential for nuclear export of intronless herpesvirus mRNA. Proc Natl Acad Sci USA 103:15190–15195

Boisvert F-M, van Koningsbruggen S, Navascués J, Lamond AI (2007) The multifunctional nucleolus. Nat Rev Mol Cell Biol 8:574–585

Boudonck K, Dolan L, Shaw PJ (1998) Coiled body numbers in the Arabidopsis root epidermis are regulated by cell type, developmental stage and cell cycle parameters. J Cell Sci 111:3687–3694

Boudonck K, Dolan L, Shaw PJ (1999) The movement of coiled bodies visualized in living plant cells by the green fluorescent protein. Mol Biol Cell 10:2297–2307

Brodsky AS, Silver PA (2000) Pre-mRNA processing factors are required for nuclear export. RNA 6:1737–1749

Brown JWS, Clark GP, Simpson, CG, Leader DJ, Lowe TM (2001) Multiple snoRNA gene clusters from *Arabidopsis*. RNA 7:5718–5732

Brown JWS, Echeverria M, Qu L-H (2003) Plant snoRNAs: functional evolution and new modes of gene expression. Trends Plant Sci 8:42–49

Brown JWS, Shaw PJ (1998) Small nucleolar RNAs and pre-rRNA processing in plants. Plant Cell 10:649–657

Bühler M, Steiner S, Mohn F, Paillusson A, Mühlemann O (2006) EJC-independent degradation of nonsense immunoglobulin-μ mRNA depends on 3' UTR length. Nat Struct Mol Biol 13:462–464

Canto T, Uhrig JF, Swanson M, Wright KM, MacFarlane SA (2006) Translocation of Tomato Bushy Stunt Virus p19 protein into the nucleus by ALY proteins compromises its silencing suppressor activity. J Virol 80:9064–9072

Carneiro T, Carvalho C, Braga J, Rino J, Milligan L, Tollervey D, Carmo-Fonseca M (2007) Depletion of the yeast nuclear exosome subunit Rrp6 results in accumulation of polyadenylated RNAs in a discrete domain within the nucleolus. Mol Cell Biol 27:4157–4165

Cech TR (2004) Beginning to understand the end of the chromosome. Cell 116:273–279

Chen M, Schwab R, Chory J (2003) Characterization of the requirements for localization of phytochrome B to nuclear bodies. Proc Natl Acad Sci USA 100:14493–14498

Cheutin T, Misteli T, Dundr M (2004) Dynamics of nucleolar components. In: The Nucleolus (ed. Olsen, M.O.J.) Landes, Georgetown, USA/Kluwer, New York, USA. pp 29–40

Cioce M, Lamond AI (2005) Cajal bodies: a long history of discovery. Annu Rev Cell Dev Biol 21:105–131

Collier S, Pendle A, Boudonck K, van Rij T, Dolan L, Shaw PJ (2006) A distant coilin homologue is required for the formation of Cajal bodies in Arabidopsis. Mol Biol Cell 17:2942–2951

Conti E, Izzuralde E (2005) Nonsense-mediated decay: molecular insights and mechanistic variations across species. Curr Opin Cell Biol 17:316–325

Culbertson MR, Leeds PF (2003) Looking at mRNA decay pathways through the window of molecular evolution. Curr Opin Genet Dev 13:207–214

Custódio N, Carvalho C, Condado I, Antoniou M, Blencowe BJ, Carmo-Fonseca M (2004) In vivo recruitment of exon junction complex proteins to transcription sites in mammalian cell nuclei. RNA 10:622–633

Darzacq X, Jády BE, Verheggen C, Kiss AM, Bertrand E, Kiss T (2002) Cajal body-specific small nuclear RNAs: a novel class of $2'$-O-methylation and pseudouridylation guide RNAs. EMBO J 21:2746–2756

Decatur WA, Fournier MJ (2002) rRNA modifications and ribosome function. Trends Biochem. Sci 7:344–351

Dove BK, You JH, Reed ML, Emmett SR, Brooks G, Hiscox JA (2006) Changes in nucleolar morphology and proteins during infection with the corona virus infectious bronchitis virus. Cell Microbiol 8:1147–1157

Dragon F, Gallagher JE, Compangnone-Post PA, Mitchell BM, Porwancher KA, Wehner KA, Wormsley S, Settlage RE, Shabanowitz J, Osheim Y, Beyer AL, Hunt DF, Baserga SJ (2002) A large nucleolar U3 ribonucleoprotein required for 18S ribosomal RNA biogenesis. Nature 417:967–970

Etheridge KT, Banik SS, Armbruster B N, Zhu Y, Terns RM, Terns MP, Counter CM (2002) The nucleolar localisation domain of of the catalytic subunit of human telomerase. J Biol Chem 277:24764–24770

Fatica A, Tollervey D (2002) Making ribosomes. Curr Opin Cell Biol 14:313–318

Frey MR, Matera AG (1995) Coiled bodies contain U7 small nuclear RNA and associate with specific DNA sequences in interphase human cells. Proc Natl Acad Sci USA 92:5915–5919

Gallagher JE, Dunbar DA, Grannemann S, Mitchell BM, Osheim Y, Beyer AL, Baserga SJ (2004) RNA polymerase I transcription and pre-rRNA processing are linked by specific SSU processosome components. Genes Dev 18:2506–2517

Ganot P, Jady BE, Bortolin ML, Darzacq X, Kiss T (1999) Nucleolar factors direct the $2'$-O-ribose methylation and pseudouridylation of U6 spliceosomal RNA. Mol Cell Biol 19:6909–6917

Gatfield D, Izurralde E (2002) REF1/Aly and the additional exon junction complex proteins are dispensible for nuclear mRNA export. J Cell Biol 159:579–588

Gerbi SA, Lange TS (2002) All small nuclear RNAs (snRNAs) of the [U4/U6.U5] tri-snRNP localize to nucleoli; Identification of the nucleolar localization element of U6 snRNA. Mol Biol Cell 13:3123–3137

Gonzalez-Melendi P, Wellis B, Beven AF, Shaw PJ (2001) Single ribosomal transcription units are linear, compacted Christmas trees in plant nucleoli. Plant J 27:223–233

Gorski, SA, Dundr M, Misteli T (2006) The road much travelled: trafficking in the cell nucleus. Curr Opin Cell Biol 18:284–290

Grandi P, Rybin V, Bassker J, Petfalski E, Strauss D, Marzioch M, Schafer T, Kuster B, Tschochner H, Tollervey D, Gavin A-C, Hurt E (2002) 90S pre-ribosomes include the 35S pre-rRNA, the U3 snoRNP, and 40S subunit processing factors but predominantly lack 60S synthesis factors. Mol Cell 10:105–115

Grannemann S, Baserga SJ (2004) Ribosome biogenesis: of knobs and RNA processing. Exp Cell Res 296:43–50

Grannemann S, Baserga SJ (2005) Crosstalk in gene expression: coupling and co-regulation of rDNA transcription, pre-ribosome assembly and pre-rRNA processing. Curr Opin Cell Biol 17:281–286

Grosshans H, Deinart K, Hurt E, Simos G (2001) Biogenesis of the signal recognition particle (SRP) involves import of SRP proteins into the nucleolus, assembly with SRP-RNA, and Xpo1p-mediated export. J Cell Biol 153:745–762

Han MH, Goud S, Song L, Fedoroff N (2004) The Arabidopsis double-stranded RNA-binding protein HYL1 plays a role in microRNA-mediated gene regulation. Proc Natl Acad Sci USA 101:1093–1098

Handwerger KE, Gall JG (2006) Subnuclear organelles: new insights into form and function. Trends Cell Biol 16:19–26

Hernandez-Pinzon I, Yelina NE, Schwach F, Studholme DJ, Baulcombe D, Dalmay T (2007) SDE5, the putative homologue of a human mRNA export factor, is required for transgene silencing and accumulation of *trans*-acting endogenous siRNA. Plant J 50:140–148

Hiscox JA (2002) The nucleolus—a gateway to viral infection? Arch Virol 147:1077–1089

Hiscox JA (2007) RNA viruses: hijacking the dynamic nucleolus. Nat Rev Microbiol 5:119–127

Hope TJ (1999) The ins and outs of HIV. Rev Arch Biochem Biophys 365:186–191

Hopper AK, Phizicky EM (2003) tRNA transfers to the limelight. Genes Dev 17:162–180

Hori K, Watanabe Y (2005) UPF3 suppresses aberrant spliced mRNA in *Arabidopsis*. Plant J 43:530–540

Ideue T, Azad AK, Yoshida J, Matsusaka T, Yanagida M, Ohshima Y, Tani T (2004) The nucleolus is involved in mRNA export from the nucleus in fission yeast. J Cell Sci 117:2887–2895

Jacobson MR, Cao LG, Taneja K, Singer RH, Wang YL, Pederson T (1997) Nuclear domains of RNA subunit of RNase P. J Cell Sci 110:829–837

Jacobsen MR, Pederson T (1998) Localization of signal recognition particle RNA in the nucleolus of mammalian cells. Proc Natl Acad Sci USA 95:7981–7986

Jady BE, Darzacq X, Tucker KE, Matera AG, Bertrand E, Kiss T (2003) Modification of Sm small nuclear RNAs occurs in the nucleoplasmic Cajal body following import from the cytoplasm. EMBO J 22:1878–1888

Jady BE, Bertrand E, Kiss T (2004) Human telomerase RNA and box H/ACA scaRNAs share a common Cajal body-specific localization signal. J Cell Biol 164:647–652

Jarrous N, Wolenski JS, Wesolowski D, Lee C, Altmann S (1999) Localisation in the nucleolus and coiled bodies of protein subunits of the ribonucleoprotein ribonuclease P. J Cell Biol 146:559–572

Kadowaki T, Schneiter R, Hitomi M, Tartakoff AM (1995) Mutations in nucleolar proteins lead to nucleolar accumulation of poly(A)+RNA in Saccharomyces cerevisiae. Mol Biol Cell 6:1103–1110

Kertész S, Keréyi Z, Mérai Z, Bartos I, Palfy T, Barta E, Silhavy D (2006) Both introns and long 3′-UTRs operate as *cis*-acting elements to trigger nonsense-mediated decay in plants. Nucl Acids Res 34:6147–6157

Kim SH, Ryabov EV, Kalinina NO, Rakitina DV, Gillespie T, Haupt S, MacFarlane S, Brown JWS, Taliansky M (2007a) Cajal bodies, the nucleolus and fibrillarin are required for a plant virus systemic infection. EMBO J 26:2169–2179

Kim SH, MacFarlane S, Kalinina NO, Rakitina DV, Ryabov EV, Gillespie T, Haupt S, Brown JWS, Taliansky M (2007b) Interaction of a plant virus-encoded protein with the major nucleolar protein fibrillarin is required for systemic virus infection. Proc Natl Acad Sci USA 104:11115–11120

Kircher S, Gil P, Kozma-Bogna L, Fejes E, Speth V, Husselstein-Muller T, Bauer D, Adam E, Schafer E, Nagy F (2002) Nucleocytoplasmic partitioning of the plant photoreceptors phytochrome A, B, C, D, and E is regulated differentially by light and exhibits a diurnal rhythm. Plant Cell 14:1541–1555

Kiss T (2002) Small nucleolar RNAs: An abundant group of non-coding RNAs with diverse cellular functions. Cell 109:145–148

Lange TS, Gerbi SA (2000) Transient nucleolar localization of U6 small nuclear RNA in *Xenopus laevis* oocytes. Mol Biol Cell 11:2419–2428

Lamond AI, Spector DL (2003) Nuclear speckles: a model for nuclear organelles. Nat Rev Mol Cell Biol 4:605–612

Lejeune F, Maquat LE (2005) Mechanistic links between nonsense-mediated mRNA decay and pre-mRNA splicing in mammalian cells. Curr Opin Cell Biol 17:309–315

Leung AK, Lamond AI (2003) The dynamics of the nucleolus. Crit Rev Eukaryot Gene Expr 13:39–54

Leung AKL, Gerlich D, Miller G, Lyon CE, Lam YW, Lleres D, Daigle N, Zomerdijk J, Ellenberg J, Lamond AI (2004) Quantitative kinetic analysis of nucleolar breakdown and reassembly during mitosis in live human cells. J Cell Biol 166:787–800

Leung AK, Trinkle-Mulcahy L, Lam YW, Andersen JS, Mann M Lamond AI (2006) NoPdb: Nucleolar Proteome Database Nucleic Acids Res. 34:218–20

Li CF, Pontes O, El-Shami M, Henderson IR, Bernatavichute YV, Chan SW-L, Lagrange T, Pikaard CS, Jacobsen SE (2006) An ARGONAUTE4-containing nuclear processing center colocalized with Cajal bodies in *Arabidopsis thaliana*. Cell 126:93–106

Li J, Kinoshita T, Pandey S, Ng CK, Gygi SP, Shimazaki K, Assmann SM (2002) Modulation of an RNA-binding protein by abscisic-acid-activated protein kinase. Nature 418:793–797

Longman D, Johnstone JL, Cáceres JF (2003) The Ref/Aly proteins are dispensable for mRNA export and development in *Caenorhabditis elegans*. RNA 9:881–891

Lorković ZJ, Hilscher J, Barta A (2004) Use of fluorescent protein tags to study nuclear organisation of the spliceosomal machinery in transiently transformed living plant cells. Mol Biol Cell 15:3233–3243

Lukowiak AA, Narayanan A, Li ZH, Terns RM, Terns MP (2001) The snoRNA domain of vertebrate telomerase RNA functions to localize the RNA within the nucleus. RNA 7:1833–1844

MacMorris M, Brocker C, Blumenthal T (2003) UAP56 levels affect viability and mRNA export in *Caenorhabditis elegans*. RNA 9:847–857

Maquat LE (2004) Nonsense-mediated mRNA decay: splicing, translation and mRNA dynamics. Nat Rev Mol Cell Biol 5:89–99

Marker C, Zemann A, Terhorst T, Kiefmann M, Kastenmayer JP, Green P, Bachellerie JP, Brosius J, Huttenhofer A (2002) Experimental RNomics: identification of 140 candidates for small non-messenger RNAs in the plant Arabidopsis thaliana. Curr Biol 12:2002–2013

Maser RS, DePinho RA (2002) Connecting chromosomes, crisis and cancer. Science 297:565–569

Matera AG (1999) Nuclear bodies: multifaceted subdomains of the interchromatin space. Trends Cell Biol 19:302–309

Michienzi A, Cagnon L, Bahner I, Rossi JJ (2000) Ribozyme-mediated inhibition of HIV 1 suggests nucleolar trafficking of HIV-1 RNA. Proc Natl Acad Sci USA 97:8955–8960

Miller OLJ, Beatty RR (1969) Visualisation of nucleolar genes. Science 164:955–957

Misteli T (2005) Concepts in nuclear architecture. BioEssays 27:477–487

Mitchell JR, Cheng J, Collins K (1999a) A box H/ACA small nucleolar RNA-like domain at the human telomerase RNA 3' end. Mol Cell Biol 19:567–576

Mitchell JR, Collins K (2000) Human telomerase activation requires two independent interactions between telomerase RNA and telomerase reverse transcriptase. Mol Cell 6:361–371

Mosgoeller W (2004) Nucleolar ultrastructure in vertebrates. In: The Nucleolus (ed. Olsen, M.O.J.) Landes, Georgetown, USA/Kluwer, New York, USA. pp 10–19

Narayanan A, Speckmann W, Terns R, Terns MP (1999) Role of the box C/D motif in localization of small nucleolar RNAs to coiled bodies and nucleoli. Mol Biol Cell 10:2131–2147

Olsen MOJ, Dundr M, Szebeni A (2000) The nucleolus: an old factory with unexpected capabilities. Trends Cell Biol 10:189–196

Olsen MOJ (2004) Non-traditional roles of the nucleolus. In: The Nucleolus (ed. Olsen, M.O.J.) Landes, Georgetown, USA/Kluwer, New York, USA. pp 329–342

Palacios I M, Gatfield G, St. Johnston D, Izaurralde E (2004) An eiF4AIII-containiung complex required for mRNA localization and nonsense-mediated decay. Nature 427:753–757

Pederson T (1998) The plurifunctional nucleolus. Nucl Acids Res 26:3871–3876

Pendle AF, Clark GP, Boon R, Lewandowska D, Lam YW, Andersen J, Mann M, Lamond AI, Brown JWS, Shaw PJ (2005) Proteomic analysis of the Arabidopsis nucleolus suggests novel nucleolar functions. Mol Biol Cell 16:260–269

Pillai RS, Will CL, Luhrmann R, Schumerli D, Muller B (2001) Purified U7 snRNPs lack the Sm proteins D1 andD2 but contain Lsm10, a new 14 kDa Sm D1-like protein. EMBO J 20:5470–5479

Politz JC, Polena I, Trask I, Bazett-Jones DP, Pederson T (2005) A nonribosomal landscape in the nucleolus revealed by the stem cell protein nucleostemin. Mol Biol Cell 16:3401–3410

Pontes O, Li CF, Nunes PC, Haag J, Ream T, Vitins A, Jacobsen SE, Pikaard CS (2006) The *Arabidopsis* chromatin-modifying nuclear siRNA pathway involves a nucleolar RNA processing center. Cell 126:79–92

Pontvianne F, Matia I, Douet J, Tourmente S, Medina FJ, Echeverria M, Saez-Vasquez J (2007) Characterization of AtNUC-L1 reveals a central role of nucleolin in nucleolus organization and silencing of AtNUC-L2 gene in Arabidopsis. Mol Biol Cell 18:369–379

Qu LH, Meng Q, Zhou L, Chen Y-Q (2001) Identification of 10 novel snoRNA gene clusters from *Arabidopsis thaliana*. Nucl Acids Res 29:1623–1630

Reed R, Hurt E (2002) A conserved mRNA export machinery coupled to pre-mRNA splicing. Cell 108:523–531

Reed R, Cheng H (2005) TREX, SR proteins and export of mRNA. Curr Opin Cell Biol 17:269–273

Raška I, Shaw PJ, Cmarko D (2006) Structure and function of the nucleolus in the spotlight. Curr Opin Cell Biol 18:325–334

Rubbi CP, Milner J (2003) Disruption of the nucleolus mediates stabilisation of p53 in response to DNA damage and other stresses. EMBO J 22:6068–6077

Saez-Vasquez J, Caparros-Ruiz D, Barneche F, Echeverria M (2004) A plant snoRNP complex containing snoRNAs, fibrillarin and nucleolin-like proteins is competent for both rRNA gene binding and pre-rRNA processing in vitro. Mol Cell Biol 24:7284–7297

Scherl A, Couté Y, Déon C, Callé A, Kindbeiter K, Sanchez J-C, Greco A, Hochstrasse D, Diaz J-J (2002) Functional proteomics analysis of human nucleolus. Mol Biol Cell 13:4100–4109

Schafer T, Strauss D, Petfalski E, Tollervey D, Hurt E (2003) The path from nucleolar 90S to cytoplasmic 40S pre-ribosomes. EMBO J 22:1370–1380

Schneiter R, Kadowaki T, Tartakoff AM (1995) mRNA transport in yeast: time to reinvestigate the functions of the nucleolus. Mol Biol Cell 6:357–370

Shaw PJ, Brown JWS (2004) Plant nuclear bodies. Curr Opin Plant Biol 7:614–620

Shaw PJ, Jordan EG (1995) The nucleolus. Annu Rev Cell Dev Biol 11:93

Shaw PJ, Beven AF, Leader DJ, Brown JWS (1998) Localization and processing from a polycistronic precursor of novel snoRNAs in maize. J Cell Sci 111:2121–2128

Sleeman JE, Lamond AI (1999) Newly assembled snRNPs associate with coiled bodies before speckles, suggesting a nuclear snRNP maturation pathway. Curr Biol 9:1065–1074

Song L, Han MH, Lesicka J, Fedoroff NV (2007) Arabidopsis primary microRNA processing proteins HYL1 and DCL1 define a nuclear body distinct from the Cajal body Proc Natl Acad Sci USA 104:5437–5442

Steiner-Mosonyi M, Mangroo D (2004) The nuclear tRNA aminoacylation-dependent pathway may be the principal route used to export tRNA from the nucleus in *Saccharomyces cerevisiae*. Biochem J 378:809–816

Stutz F, Izaurralde E (2003) The interplay of nuclear mRNP assembly, mRNA surveillance and export. Trends Cell Biol 13:319–327

Tange TO, Nott A Moore MJ (2004) The ever-increasing complexities of the exon junction complex. Curr Opin Cell Biol 16:279–284

Tani T, Derby RJ, Hiraoka Y, Spector DL (1995) Nucleolar accumulation of poly(A)+RNA in heat-shocked yeast cells: implication of nucleolar involvement in mRNA transport. Mol Biol Cell 6:1515–1534

Tate WP, Poole ES (2004) My favorite organelle: The ribosome: lifting the veil from a fascinating organelle. Bioessays 26:582–588

Tillemans V, Leponce I, Rausin G, Dispa L, Motte P (2006) Insights into nuclear organisation in plants as revealed by the dynamic distribution of Arabidopsis SR splicing factors. Plant Cell 18:3218–3234

Uhrig JF, Canto T, Marshall DF, MacFarlane SA (2004) Relocalisation of nuclear ALY proteins to the cytoplasm by the Tomato Bushy Stunt Virus p19 pathogenicity protein. Plant Physiol 135:2411–2423

van Hoof A, Green PJ (2006) NMD in plants. NMD in plants. In Nonsense-mediated mRNA decay, L. Maquat, ed. (Landes Bioscience, USA)

Venema J, Tollervey D (1999) Ribosome synthesis in *Saccharomyces cerevisiae*. Annu Rev Genet 33:261–311

Wong JM, Kusdra L, Collins K (2002) Subnuclear shuttling of human telomerase induced by transformation and DNA damage. Nat Cell Biol 4:731–736

Walter P, Johnson AE (1994) Signal sequence recognition and protein targeting to the endoplasmic reticulum membrane. Annu Rev Cell Biol 10:87–119

Williams BJL, Boyne JR, Goodwin DJ, Roaden L, Hautbergue GM, Wilson SA, Whitehouse A (2005) The prototype gamma-2 herpesvirus nucleocytoplasmic shuttling protein, ORF 57, transports viral RNA through the cellular mRNA export pathway. Biochem J 387:295–308

Yamaguchi R, Nakamura M, Mochizuki N, Kay SA, Nagatani A (1999) Light-dependent translocation of a phytochrome B-GFP fusion protein to the nucleus in transgenic Arabidopsis. J Cell Biol 145:437–445

Yu YT, Shu MD, Narayanan A, Terns RM, Terns MP, Steitz JA (2001) Internal modification of U2 small nuclear (sn)RNA occurs in nucleoli of *Xenopus* oocytes. J Cell Biol 152:1279–1288

Zhu Y, Tomlinson, RL, Lukowiak AA, Terns RM, Terns MP (2004) Telomerase RNA accumulates in Cajal bodies in human cancer cells. Mol Biol Cell 15:81–90

Index

A

AAT, 21
ABA, 248–250
    hypersensitivity, 140–142, 145–147
    signaling, 140–142, 147
ABA hypersensitive 1 (ABH1), 140–144, 146, 147
Aberrantly spliced transcripts, 301, 303
Abiotic stress(es), 236, 237, 245, 248, 251, 258, 260–262, 269–271
Abscisic acid signal transduction, 140–147
Adaptor protein, 226
Adenylate uridylate-rich element (ARE), 239, 246
Agamous, (AG) 281
AHG2/PARN, 140
AKIP1, 140, 141
Alternate acceptor splice site, 19, 28, 29, 32, 33
Alternate donor, 26, 28, 29
Alternate donor splice site, 19, 28, 29, 32, 33
Alternative 5′ or 3′ splice site selection, 224
Alternate splice site, 32
Alternate 3′ terminal exons, 20
Alternate 5′ terminal exons, 20
Alternate terminal exons, 19, 26, 27
Alternative exons, 224
Alternative polyadenylation, 155, 158, 159, 164, 165, 169
Alternative splicing, 18–33, 73, 74, 220, 225, 227–229
    cold regulated, 262, 263
    drought regulate, 265, 266
    heat regulated, 262
    osmotic regulated, 265
    radiation and light, 267, 268
    salt regulated, 265, 266
    signalling affecting, 269, 270
    SR splicing factors, 262, 263
    temperature regulated, 261, 262, 264, 265
    wound, 267
Arabidopsis, 18, 21, 24, 29–33, 86, 88, 89, 91, 92, 94–96, 202–204, 206, 209–214
Arabidopsis CBP80, 139
Arabidopsis *R* gene complement, 221
Arabidopsis *RPS4*, 224, 225
Argonaute (AGO) family proteins, 244
ASAP II, 33
ASTI, 27, 28, 31, 32, 34
ASTRA, 34
AtCBP20, 142, 146
AtCPSF100, 160, 162, 163, 165–168
AtCPSF160, 162, 163, 166, 168
AtCPSF30, 162–164, 166, 168–170
AtCPSF73(I), 153, 162, 163, 166
AtCPSF73(II), 153, 163
AtCstF77, 165, 166
AtPAP, 163, 165
AtPP2C, 142
Auxin, 247, 248
Avirulence, 220, 222, 225, 226

B

Barley *Mla* genes, 222
BiFC, 105–108, 116
Biotic stress, 258, 260
BLAST, 19, 21
BLAT, 19, 21, 26
Bulk mRNA export, 240, 242, 246, 251

C

Cajal bodies, 104, 108, 109, 112, 293–295, 298–300
Carboxy-terminal domain (CTD), 240, 287
CBP20, 140, 142, 145–147
Chlamydomonas reinhardtii, 88, 96

Chromatin immunoprecipitation, 240
Chromosome region maintenance 1 (CRM1), 242
cis-natural antisense transcripts (*cis*-NATs), 145
ClusterMerge, 25
Co-chaperones, 222
Coiled-coil, 220
Cold, 248–250
Conserved gene structure, 221, 224
CONSTANS, 145
CPSF, 152, 154, 160–162, 169, 170
CPSF30, 153, 159, 161, 163, 164, 169
CPSF73, 160, 163
CstF, 152, 154, 165
CstF50, 153, 165, 170
CstF64, 153, 154, 165–168
CstF77, 153, 165, 166
CTD. *See* Carboxy-terminal domain
Cyclophilins, 91

D
DAG, 24, 25
DEAD-box protein 5 (Dbp5), 241–243, 245, 248, 250, 251
DEAD-box RNA helicase, 248
Deadenylation, 181–185, 192, 194
Decapping, 181–185, 194
Degradation, 141
Deletions, 281, 283
Dicer, 211
Dicer-like 1 (DCL1), 244
Dicot, 279, 282, 283
Disease resistance, 18
Domain, 220–224, 226–228
Drought tolerance, 145
DScam, 18
Dynamic programming, 25

E
ECgene, 33
Effector, 220, 226, 229
    virulence function, 229
Elongation, 280, 286, 287
Enhancer, 280–283
Ensembl, 25
EPGA, 19
EST_GENOME, 21
Exon, definition, 71–73, 227
Exon junction complex (EJC), 128–130, 241, 279, 295, 301, 302
Exon skipping, 19, 26, 27
Exon-splicing enhancers, 71

Exonuclease, 182, 183, 189
Exosome, 181, 182, 187, 190–192
Export, 141, 142, 145
Export-competent mRNA ribonucleoprotein (mRNP), 235, 236, 239–243, 248, 251
Expressed sequence tags (ESTs), 19
Expressed transcript sequences, 19, 25, 27

F
FCA, 203–209, 214
FG repeats, 238, 243
Flax, 223, 225
FLC, 202–214
FLIP, 106, 114
FLM, 145
Flowering, 93, 94
    repressors, 202, 210, 212
    time, 144, 145, 201–214
FPA, 203, 205, 213, 215
FRAP, 105–107, 112–114
Full-length cDNAs (FL-cDNAs), 19
FY, 153, 159, 162–166, 168, 169, 203–208

G
Gene expression, 62, 73, 76
Genes, 220–222, 224, 227–229
GeneSeqer, 21
Genetic analysis, 243, 249
Global climate, 236
GMAP, 21, 26
Guard cells, 141, 142

H
Heat shock protein 70 (HSP70), 245
Heat shock protein, 222
Heat stress, 245, 249
Heterochromatic Small Interfering RNAs, 294, 299
Heterogenous nuclear RNPs (hnRNPs), 240
Heaviest bundling, 24
Hollywood, 33
Host-plant recognition, 247, 248
Human, 18, 30–33
HYL1, 140, 141

I
Initiate or terminate within an intron, 19
Intron, 84, 92
    alternatively spliced, 95, 96
    conserved, 95, 96

definition, 71–73
location, 279, 285
mediated enhancement, 280, 282
plant, 85, 92, 95
recognition, 84, 85
retention, 224, 227, 228
IP6-, 242
Isoforms, 18, 22–27, 32, 33

K
Karyopherin (importin β), 236, 238
Kinases, 84, 91

L
L6, 223–225
LAMMER kinase, 124, 126, 130–134
Leptomycin B, 130
Leucine-rich repeat, 220
LOS4, 140, 141
Low-temperature, 248

M
Maize, 95, 96
Microarray(s), 142–144, 180, 187, 188, 191, 194
microRNA(s), 211, 212, 236–238, 243, 244, 251
MIR genes, 244
miRNAs/siRNAs, 237, 238, 251
Monocot, 282–283
3′ mRNA processing, 142
Mouse, 31–33
mRNA
  cap, 212, 213
  cap binding proteins, 139, 141, 143, 145, 149
  export, 210, 211, 291, 292, 300–305
  metabolism, 140, 144
  polyadenylation, 202–204, 206–208, 213
  processing, 139–142, 145, 147
  stability, 180, 181, 193, 194, 285, 286
  surveillance, 291, 292, 295, 302, 304
mRNP, 239–243, 248, 251
  assembly, 295, 297, 305

N
NAGNAG, 32
Nat-siRNAs, 244
Nonsense mediated decay (NMD), 33, 77, 85, 92, 96, 130, 181–183, 186, 187, 205, 228, 291, 292, 301, 302, 304, 305

Nonsense-mediated mRNA decay, 278
Nonsense-mediated RNA decay (NMD), 141
Nuclear bodies, 293–295
Nuclear-cytoplasmic shuttling protein, 242
Nuclear export factors (NEFs), 237–239, 243, 245
Nuclear import, 242, 249
Nuclear mRNA, 142, 145
Nuclear mRNA export, 237, 246, 249, 251
Nuclear pore complex (NPC), 237
Nucleolar proteome, 301, 304
Nucleolus, 292–305
Nucleoporins (NUPs), 210, 237, 243, 245, 248
Nucleotide
  binding site, 220
  modifications, 296
NUP107–160 complex, 246, 248, 251

O
Organism complexity, 18, 30, 32
Origin, 77, 78
Oryza sativa, 88

P
(p14), 69
PAP. *See* Poly(A) polymerase
Pap1p, 167
PARN. *See* Poly(A) ribonuclease)
P-body, 183
Pfs2p, 206, 208
Physcomitrella patens, 88, 95, 96
Pinus taeda, 88
Plant development, 18
Plant polyadenylation signals, 155–157
Pokeweed antiviral protein, 187
Poly(A) polymerase, 152–154, 160, 166, 167, 170
Poly(A) tail formation, 241
Polyadenylation
  alternative, 204
  signal, 152, 154–157, 159, 165, 167, 168, 170
Poly(A) ribonuclease), 184, 185, 190, 192, 194
Position, 285, 286
PP2CA, 142
Premature termination codon, 92, 95, 96
Pre-mRNA, 141, 142, 145, 284, 285
Presplicing complexes, 70
Profilin, 279
Program to assemble spliced alignments (PASA), 24–27
Promoter, 278–284, 287

Proteins, 220–222, 224–229
Proteomic
  analysis, 242
  diversity, 23, 30
Proteomics, 300
PTC. *See* Premature termination codon

R
Ran-GTPase-, 236
Ratio, 223, 225, 227
Receptor-cargo complexes, 243
Resistance, 220–229
Retained intron, 19, 26, 27, 32
Ribosomal
  RNA, 291–293, 295, 296
  subunits, 291, 293, 295, 297
Rice, 18, 21, 25, 29, 31–33, 86, 88, 92, 94
RNA
  export factors, 238, 240
  helicase, 209, 211, 240, 241, 248, 251
RNA-dependent RNA polymerases (RDRs), 244
RNAi-induced transcriptional silencing (RITS), 244
RNA-induced silencing complex (RISC), 244
RNA polymerase, 119, 121, 124
RNA polymerase (Pol) II, 240, 287
RNase P, 298
RUST, 33

S
SAD1, 140, 141
SAD2, 140, 141
scaRNAs, 293, 295
Schizosaccharomyces pombe, 91
Self-inhibition, 226
Sequence alignment, 19, 21
Signal recognition particle, 292, 297
Silencing, 204, 211
sim4, 21
siRNAs, 237, 238, 243, 244, 251
Size, 65, 68, 70
Skipped exons, 28, 32
Small nuclear RNPs, 299
Small RNAs, 243, 244, 251
Sm core protein, 3
snoRNAs, 294–296
snoRNPs, 293
snRNAs, 2, 293–295, 299
SnRNPs, 3, 84, 90, 91, 208, 209
  localization, 109
Speckles, 104, 108–112, 114, 115, 125, 127–129, 132

Splice factors, 227
Splice site(s), 19, 21, 32
  alternative, 92, 96
  selection, 92
Splice site consensus sequences, 40
Spliceosome, 84, 241, 278, 284
  assembly, 66, 69–73, 79, 84, 90
  major (U2) type, 2
  major and minor, 91
  minor (U12) type, 2
Splicing, 84–97, 141, 142, 145, 147, 240–242, 278–280, 283–285, 287
  alternative, 92–95, 202, 203, 208–210
  conserved, 95–97
  enhancer, 11
  factor, 5
  graph, 21–25
  machinery, 284, 285, 287
  plant, 85
  regulator, 10
  regulatory elements, 85, 89, 90, 92
  signals, 62, 63, 65, 68, 70, 74, 77, 78
SR proteins, 6–9, 71–73, 84–97, 119, 123, 124, 126–134, 210
  alternative splicing of, 93–97
  domain organization, 86
  evolution, 89, 95
  expression, 89, 93–96
  function, 84, 86, 89–91, 97
  genes of, 88, 89, 94–97
  phenotype, 94
  phosphorylation, 84, 91
  plant-specific, 86, 90, 94–97
  protein-protein interactions, 90
  redundancy, 84, 89, 97
SR splicing factors, 262
  ATP-dependent dynamics, 112, 113
  cell cycle-dependent dynamics, 104, 110
  cold-induced localization, 110, 111
  diffusion co-efficients, 106, 112
  dynamics of, 104, 105, 107, 109–111
  heat-induced localization, 111
  IGC, 109
  immunocytochemistry, 105, 109, 110
  localization of, 107, 111, 112, 116
  phosphorylation-dependent localization, 127
  in speckles, 109, 111, 112
  subcellular distribution, 270
  transcription-dependent localization, 111
SRPK, 124, 129–134
Splice site mutants, 48
STABILIZED1, 140, 141

Index 317

Subcellular localization, 221, 227
Symplekin, 153, 154, 160, 162, 163, 166, 167

T
ta-siRNAs, 244
Telomerase RNP, 293, 299
Temperature, 202, 209, 210
Thermotolerance, 246
Tissue-specificity, 280, 281
Tobacco *N* gene, 223
Toll/interleukin-1 receptor, 221
Toll interleukin1 receptor-nucleotide-binding-Leu-rich repeat-type resistance gene, 246
Transcript assembly program (TAP), 23
Transcription, 237, 239–241, 243, 245, 246, 248, 251, 280, 285–287
Transcriptome profiling, 188–190
Transcripts, 220, 229
Translational efficiency, 279
Transcription-export complex (TREX), 240
Triticum aestivum, 88, 91
tRNA, 292, 298
Truncated, 222–229

U
U11 snRNA, 67, 70
U11/U12 di-snRNP, 69, 70

U12-dependent introns, 62, 73, 74
U12 snRNAs, 67, 68, 70
U12 spliceosome, 66, 68–70, 73, 76–79
U12-type introns, 40–42, 44, 48, 55, 56
U2AF, 205, 209
U2-type introns, 40–42, 44–46, 48, 53–56
U4atac, 66–71, 77
U4atac/U6atac·U5 tri-snRNP, 69, 70
U6atac snRNAs, 67, 68, 71
UA-richness, 65, 66, 71
Ubiquitin ligase complex, 222
Upstream ORFs, 223
U-rich sequences, 85, 90
5′-UTR, 223
UTR, 29

V
Viral RNA export, 304

W
Wheat, 88, 91

X
XRN4, 181, 186, 187, 189, 190, 192, 194

Z
Zea mays, 88

Current Topics in Microbiology and Immunology

Volumes published since 2002

Vol. 271: **Koehler, Theresa M. (Ed.):** Anthrax. 2002. 14 figs. X, 169 pp. ISBN 3-540-43497-6

Vol. 272: **Doerfler, Walter; Böhm, Petra (Eds.):** Adenoviruses: Model and Vectors in Virus-Host Interactions. Virion and Structure, Viral Replication, Host Cell Interactions. 2003. 63 figs., approx. 280 pp. ISBN 3-540-00154-9

Vol. 273: **Doerfler, Walter; Böhm, Petra (Eds.):** Adenoviruses: Model and Vectors in VirusHost Interactions. Immune System, Oncogenesis, Gene Therapy. 2004. 35 figs., approx. 280 pp. ISBN 3-540-06851-1

Vol. 274: **Workman, Jerry L. (Ed.):** Protein Complexes that Modify Chromatin. 2003. 38 figs., XII, 296 pp. ISBN 3-540-44208-1

Vol. 275: **Fan, Hung (Ed.):** Jaagsiekte Sheep Retrovirus and Lung Cancer. 2003. 63 figs., XII, 252 pp. ISBN 3-540-44096-3

Vol. 276: **Steinkasserer, Alexander (Ed.):** Dendritic Cells and Virus Infection. 2003. 24 figs., X, 296 pp. ISBN 3-540-44290-1

Vol. 277: **Rethwilm, Axel (Ed.):** Foamy Viruses. 2003. 40 figs., X, 214 pp. ISBN 3-540-44388-6

Vol. 278: **Salomon, Daniel R.; Wilson, Carolyn (Eds.):** Xenotransplantation. 2003. 22 figs., IX, 254 pp. ISBN 3-540-00210-3

Vol. 279: **Thomas, George; Sabatini, David; Hall, Michael N. (Eds.):** TOR. 2004. 49 figs., X, 364 pp. ISBN 3-540-00534X

Vol. 280: **Heber-Katz, Ellen (Ed.):** Regeneration: Stem Cells and Beyond. 2004. 42 figs., XII, 194 pp. ISBN 3-540-02238-4

Vol. 281: **Young, John A. T. (Ed.):** Cellular Factors Involved in Early Steps of Retroviral Replication. 2003. 21 figs., IX, 240 pp. ISBN 3-540-00844-6

Vol. 282: **Stenmark, Harald (Ed.):** Phosphoinositides in Subcellular Targeting and Enzyme Activation. 2003. 20 figs., X, 210 pp. ISBN 3-540-00950-7

Vol. 283: **Kawaoka, Yoshihiro (Ed.):** Biology of Negative Strand RNA Viruses: The Power of Reverse Genetics. 2004. 24 figs., IX, 350 pp. ISBN 3-540-40661-1

Vol. 284: **Harris, David (Ed.):** Mad Cow Disease and Related Spongiform Encephalopathies. 2004. 34 figs., IX, 219 pp. ISBN 3-540-20107-6

Vol. 285: **Marsh, Mark (Ed.):** Membrane Trafficking in Viral Replication. 2004. 19 figs., IX, 259 pp. ISBN 3-540-21430-5

Vol. 286: **Madshus, Inger H. (Ed.):** Signalling from Internalized Growth Factor Receptors. 2004. 19 figs., IX, 187 pp. ISBN 3-540-21038-5

Vol. 287: **Enjuanes, Luis (Ed.):** Coronavirus Replication and Reverse Genetics. 2005. 49 figs., XI, 257 pp. ISBN 3-540- 21494-1

Vol. 288: **Mahy, Brain W. J. (Ed.):** Foot-and-Mouth-Disease Virus. 2005. 16 figs., IX, 178 pp. ISBN 3-540-22419X

Vol. 289: **Griffin, Diane E. (Ed.):** Role of Apoptosis in Infection. 2005. 40 figs., IX, 294 pp. ISBN 3-540-23006-8

Vol. 290: **Singh, Harinder; Grosschedl, Rudolf (Eds.):** Molecular Analysis of B Lymphocyte Development and Activation. 2005. 28 figs., XI, 255 pp. ISBN 3-540-23090-4

Vol. 291: **Boquet, Patrice; Lemichez Emmanuel (Eds.):** Bacterial Virulence Factors and Rho GTPases. 2005. 28 figs., IX, 196 pp. ISBN 3-540-23865-4

Vol. 292: **Fu, Zhen F. (Ed.):** The World of Rhabdoviruses. 2005. 27 figs., X, 210 pp. ISBN 3-540-24011-X

Vol. 293: **Kyewski, Bruno; Suri-Payer, Elisabeth (Eds.):** CD4+CD25+ Regulatory T Cells: Origin, Function and Therapeutic Potential. 2005. 22 figs., XII, 332 pp. ISBN 3-540-24444-1

Vol. 294: **Caligaris-Cappio, Federico, Dalla Favera, Ricardo (Eds.):** Chronic Lymphocytic Leukemia. 2005. 25 figs., VIII, 187 pp. ISBN 3-540-25279-7

Vol. 295: **Sullivan, David J.; Krishna Sanjeew (Eds.)**: Malaria: Drugs, Disease and Post-genomic Biology. 2005. 40 figs., XI, 446 pp. ISBN 3-540-25363-7

Vol. 296: **Oldstone, Michael B. A. (Ed.)**: Molecular Mimicry: Infection Induced Autoimmune Disease. 2005. 28 figs., VIII, 167 pp. ISBN 3-540-25597-4

Vol. 297: **Langhorne, Jean (Ed.)**: Immunology and Immunopathogenesis of Malaria. 2005. 8 figs., XII, 236 pp. ISBN 3-540-25718-7

Vol. 298: **Vivier, Eric; Colonna, Marco (Eds.)**: Immunobiology of Natural Killer Cell Receptors. 2005. 27 figs., VIII, 286 pp. ISBN 3-540-26083-8

Vol. 299: **Domingo, Esteban (Ed.)**: Quasispecies: Concept and Implications. 2006. 44 figs., XII, 401 pp. ISBN 3-540-26395-0

Vol. 300: **Wiertz, Emmanuel J.H.J.; Kikkert, Marjolein (Eds.)**: Dislocation and Degradation of Proteins from the Endoplasmic Reticulum. 2006. 19 figs., VIII, 168 pp. ISBN 3-540-28006-5

Vol. 301: **Doerfler, Walter; Böhm, Petra (Eds.)**: DNA Methylation: Basic Mechanisms. 2006. 24 figs., VIII, 324 pp. ISBN 3-540-29114-8

Vol. 302: **Robert N. Eisenman (Ed.)**: The Myc/Max/Mad Transcription Factor Network. 2006. 28 figs., XII, 278 pp. ISBN 3-540-23968-5

Vol. 303: **Thomas E. Lane (Ed.)**: Chemokines and Viral Infection. 2006. 14 figs. XII, 154 pp. ISBN 3-540-29207-1

Vol. 304: **Stanley A. Plotkin (Ed.)**: Mass Vaccination: Global Aspects – Progress and Obstacles. 2006. 40 figs. X, 270 pp. ISBN 3-540-29382-5

Vol. 305: **Radbruch, Andreas; Lipsky, Peter E. (Eds.)**: Current Concepts in Autoimmunity. 2006. 29 figs. IIX, 276 pp. ISBN 3-540-29713-8

Vol. 306: **William M. Shafer (Ed.)**: Antimicrobial Peptides and Human Disease. 2006. 12 figs. XII, 262 pp. ISBN 3-540-29915-7

Vol. 307: **John L. Casey (Ed.)**: Hepatitis Delta Virus. 2006. 22 figs. XII, 228 pp. ISBN 3-540-29801-0

Vol. 308: **Honjo, Tasuku; Melchers, Fritz (Eds.)**: Gut-Associated Lymphoid Tissues. 2006. 24 figs. XII, 204 pp. ISBN 3-540-30656-0

Vol. 309: **Polly Roy (Ed.)**: Reoviruses: Entry, Assembly and Morphogenesis. 2006. 43 figs. XX, 261 pp. ISBN 3-540-30772-9

Vol. 310: **Doerfler, Walter; Böhm, Petra (Eds.)**: DNA Methylation: Development, Genetic Disease and Cancer. 2006. 25 figs. X, 284 pp. ISBN 3-540-31180-7

Vol. 311: **Pulendran, Bali; Ahmed, Rafi (Eds.)**: From Innate Immunity to Immunological Memory. 2006. 13 figs. X, 177 pp. ISBN 3-540-32635-9

Vol. 312: **Boshoff, Chris; Weiss, Robin A. (Eds.)**: Kaposi Sarcoma Herpesvirus: New Perspectives. 2006. 29 figs. XVI, 330 pp. ISBN 3-540-34343-1

Vol. 313: **Pandolfi, Pier P.; Vogt, Peter K. (Eds.)**: Acute Promyelocytic Leukemia. 2007. 16 figs. VIII, 273 pp. ISBN 3-540-34592-2

Vol. 314: **Moody, Branch D. (Ed.)**: T Cell Activation by CD1 and Lipid Antigens, 2007, 25 figs. VIII, 348 pp. ISBN 978-3-540-69510-3

Vol. 315: **Childs, James, E.; Mackenzie, John S.; Richt, Jürgen A. (Eds.)**: Wildlife and Emerging Zoonotic Diseases: The Biology, Circumstances and Consequences of Cross-Species Transmission. 2007. 49 figs. VII, 524 pp. ISBN 978-3-540-70961-9

Vol. 316: **Pitha, Paula M. (Ed.)**: Interferon: The 50th Anniversary. 2007. VII, 391 pp. ISBN 978-3-540-71328-9

Vol. 317: **Dessain, Scott K. (Ed.)**: Human Antibody Therapeutics for Viral Disease. 2007. XI, 202 pp. ISBN 978-3-540-72144-4

Vol. 318: **Rodriguez, Moses (Ed.)**: Advances in Multiple Sclerosis and Experimental Demyelinating Diseases. 2008. XIV, 376. ISBN 978-3-540-73679-9

Vol. 319: **Manser, Tim (Ed.)**: Specialization and Complementation of Humoral Immune Responses to Infection. 2008. XII, 174. ISBN 978-3-540-73899-2

Vol. 320: **Paddison, Patrick J.; Vogt, Peter K. (Eds.)**: RNA Interference. 2008. VIII, 273. ISBN 978-3-540-75156-4

Vol. 321: **B. Beutler (Ed.):** Immunology, Phenotype First: How Mutations Have Established New Principles and Pathways in Immunology. 2008. ISBN 978-3-540-75202-250

Vol. 322: **Romeo, Tony (Ed.):** Bacterial Biofilms. 2008. XII, 299. ISBN 978-3-540-75417-6

Vol. 323: **Tracy, Steven; Oberste, M. Steven; Drescher, Kristen M. (Eds.):** Group B Coxsackieviruses. 2008. ISBN 978-3-540-75545-6

Vol. 324: **Nomura, Tatsuji; Watanabe, Takeshi; Habu, Sonoko (Eds.):** Humanized Mice. 2008. ISBN 978-3-540-75646-0

Vol. 325: **Shenk, Thomas E.; Stinski, Mark F.; (Eds.):** Human Cytomegalovirus. 2008. ISBN 978-3-540-77348-1

Vol. 326: **Reddy, Anireddy S.N; Golovkin, Maxim (Eds.):** Nuclear pre-mRNA processing in plants. 2008. ISBN 978-3-540-76775-6